Studies in Theoretical and Applied Statistics
Selected Papers of the Statistical Societies

For further volumes:
http://www.springer.com/series/10104

António Pacheco • Rui Santos •
Maria do Rosário Oliveira •
Carlos Daniel Paulino
Editors

New Advances in Statistical Modeling and Applications

Springer

Editors

António Pacheco
CEMAT and Departamento de Matemática
Instituto Superior Técnico
Universidade de Lisboa
Lisboa, Portugal

Maria do Rosário Oliveira
CEMAT and Departamento de Matemática
Instituto Superior Técnico
Universidade de Lisboa
Lisboa, Portugal

Rui Santos
CEAUL and School of Technology
    and Management
Polytechnic Institute of Leiria
Leiria, Portugal

Carlos Daniel Paulino
CEAUL and Departamento de Matemática
Instituto Superior Técnico
Universidade de Lisboa
Lisboa, Portugal

ISSN 2194-7767   ISSN 2194-7775 (electronic)
ISBN 978-3-319-05322-6   ISBN 978-3-319-05323-3 (eBook)
DOI 10.1007/978-3-319-05323-3
Springer Cham Heidelberg New York Dordrecht London

Library of Congress Control Number: 2014938629

Printed on acid-free paper

Springer is part of Springer Science+Business Media (www.springer.com)

Foreword

Dear reader, On behalf of the four Scientific Statistical Societies – the SEIO, Sociedad de Estadística e Investigación Operativa (Spanish Statistical Society and Operation Research); SFdS, Société Française de Statistique (French Statistical Society); SIS, Società Italiana di Statistica (Italian Statistical Society); and the SPE, Sociedade Portuguesa de Estatística (Portuguese Statistical Society) – we would like to inform you that this is a new book series of Springer entitled "Studies in Theoretical and Applied Statistics," with two lines of books published in the series: "Advanced Studies" and "Selected Papers of the Statistical Societies."

The first line of books offers constant up-to-date information on the most recent developments and methods in the fields of theoretical statistics, applied statistics, and demography. Books in this series are solicited in constant cooperation between the statistical societies and need to show a high-level authorship formed by a team preferably from different groups so as to integrate different research perspectives.

The second line of books presents a fully peer-reviewed selection of papers on specific relevant topics organized by the editors, also on the occasion of conferences, to show their research directions and developments in important topics, quickly and informally, but with a high level of quality. The explicit aim is to summarize and communicate current knowledge in an accessible way. This line of books will not include conference proceedings and will strive to become a premier communication medium in the scientific statistical community by receiving an Impact Factor, as have other book series such as "Lecture Notes in Mathematics."

The volumes of selected papers from the statistical societies will cover a broad range of theoretical, methodological as well as application-oriented articles, surveys and discussions. A major goal is to show the intensive interplay between various, seemingly unrelated domains and to foster the cooperation between scientists in different fields by offering well-founded and innovative solutions to urgent practice-related problems.

On behalf of the founding statistical societies I wish to thank Springer, Heidelberg and in particular Dr. Martina Bihn for the help and constant cooperation in the organization of this new and innovative book series.

Rome, Italy   Maurizio Vichi

Preface

The material of this volume was inspired by selected papers presented at SPE 2011, the XIX Annual Congress of the Portuguese Statistical Society. The annual congress of SPE is the most important statistics meeting taking place every year in Portugal, constituting a primary forum for dissemination of Statistics in Portugal and a privileged channel for scientific exchange between members of SPE and other statistical societies.

SPE 2011 was organized by Instituto Superior Técnico—Technical University of Lisbon and the School of Technology and Management—Polytechnic Institute of Leiria, by invitation from the Directive Board of the Portuguese Statistical Society (SPE). It took place from September 28 to October 1, at Hotel Miramar Sul, in the beautiful picturesque Portuguese sea town of Nazaré.

SPE 2011 continued paving the success of previous SPE congresses having an attendance in excess of 200 participants and included 140 communications from authors from 11 countries (Argentina, Austria, Brazil, England, Germany, the Netherlands, Portugal, Scotland, Spain, Switzerland, and the USA), aside from a 1-day mini-course on Longitudinal Data Analysis, given by M. Salomé Cabral (University of Lisbon) and M. Helena Gonçalves (University of Algarve).

For the pleasant and stimulating scientific and social environment enjoyed by participants in the event, we must thank in a very special way the members of the Organising Committee (Alexandra Seco, António Pacheco, Helena Ribeiro, M. Rosário de Oliveira, Miguel Felgueiras, and Rui Santos) and the Scientific Committee (António Pacheco, António St. Aubyn, Carlos A. Braumann, Carlos Tenreiro, and M. Ivette Gomes).

Last but not least, we must also thank the following four distinguished invited plenary speakers who have honoured us by contributing the first four papers of the volume: Fernando Rosado (University of Lisbon), Graciela Boente (Universidad de Buenos Aires), João A. Branco (Technical University of Lisbon), and M. Ivette Gomes (University of Lisbon).

The publication of this volume, which is the last stone in the SPE 2011 building, aims to disseminate some of the most important contributions presented at SPE 2011 to the international scientific community. The papers included in the volume mix in a nice way new developments in the theory and applications of Probability and Statistics. There is a total of 27 papers which, for the convenience of readers, were arranged into the following four parts:

- Statistical Science
- Probability and Stochastic Processes
- Extremes
- Statistical Applications

The editors would like to thank all authors who submitted papers to the volume, the anonymous referees for their insightful criticism and excellent reviewing work that contributed to improve the scientific quality and presentation of the accepted papers, and the current Directive Board of SPE, especially Vice President Pedro Oliveira, for assistance during the preparation of this volume. The included papers were accepted for publication after a careful international review process that involved a minimum of 2 referees per paper and a total of more than 70 referees from 10 countries (Argentina, Belgium, France, Germany, Italy, Portugal, Russia, Spain, Switzerland, and the USA).

The editors are very pleased that their work comes to an end at the International Year of Statistics (Statistics 2013), which is a moment of worldwide celebration and recognition of the contributions of statistical science to the humanity. In addition, for them as well as for all participants in SPE 2011, it is a fond remembrance the fact that SPE 2011 paid tribute to the following former presidents of SPE who had retired in the previous year:

- M. Ivette Gomes (1990–1994)
- João A. Branco (1994–2000)
- Fernando Rosado (2000–2006)

SPE wanted, with the tribute, to thank these former popular presidents of SPE for their invaluable work for the progress of the Portuguese Statistical Society and its national and international recognition, as well as for the development of statistics in Portugal, and pay homage also to their strong personal qualities. In this respect, we and SPE would like to provide our most sincere thanks to Isabel Fraga Alves, Manuela Souto de Miranda, and M. Manuela Neves for having promptly and very kindly accepted to be first instance spokespersons in the tribute sessions of M. Ivette Gomes, João A. Branco, and Fernando Rosado, respectively. It was also very moving to the editors and the Organising Committee of SPE 2011 the fact that this event took place close to the end of the mandate as president of SPE of

- Carlos A. Braumann (2007–2011)

whose support, as well as that of the Directive Board of SPE, was invaluable and decisive for the great success of SPE 2011.

Lisbon, Portugal   António Pacheco, M. Rosário de Oliveira,
                                                     Carlos Daniel Paulino
Leiria, Portugal   Rui Santos
July 2013

Contents

Contributors

Anabela Afonso Department of Mathematics and Research Center of Mathematics and Applications (CIMA-UE), University of Évora, Évora, Portugal

Airlane P. Alencar IME, University of São Paulo, São Paulo, Brazil

Valeska Andreozzi Centro de Estatística e Aplicações da Universidade de Lisboa, FCUL, Lisboa, Portugal

Nelson Antunes FCT of University of Algarve and CEMAT, Faro, Portugal

Paulo Araújo Santos Departamento de Informática e Métodos Quantitativos, Escola Superior de Gestão e Tecnologia, Instituto Politécnico de Santarém, Santarém, Portugal

Juan Lucas Bali Facultad de Ciencias Exactas y Naturales, Universidad de Buenos Aires and CONICET, Buenos Aires, Argentina

Graciela Boente Facultad de Ciencias Exactas y Naturales, Universidad de Buenos Aires and CONICET, Buenos Aires, Argentina

João A. Branco Department of Mathematics and CEMAT, Instituto Superior Técnico, TULisbon, Portugal

Carlos A. Braumann Department of Mathematics, Centro de Investigação em Matemática e Aplicações, Universidade de Évora, Évora, Portugal

Maria de Fátima Brilhante CEAUL and Universidade dos Açores, DM, Ponta Delgada, Portugal

Frederico Caeiro Faculdade de Ciências e Tecnologia da Universidade Nova de Lisboa and CMA, Caparica, Portugal

Clara Carlos Escola Superior de Tecnologia do Barreiro, Instituto Politécnico de Setúbal, Lavradio, Portugal

Luísa Carvalho University of Évora, Évora, Portugal

Clara Cordeiro University of Algarve and CEAUL, Faro, Portugal

José Carlos Dias BRU-UNIDE and ISCTE-IUL Business School, Lisboa, Portugal

Sara Simões Dias Departamento Universitário de Saúde Pública, FCM-UNL, Lisboa, Portugal

Luís Ferreira dos Santos Serviço de Cardiologia, Tondela-Viseu Hospital Center, Viseu, Portugal

Marta Ferreira University of Minho, DMA/CMAT, Braga, Portugal

Patrícia A. Filipe Centro de Investigação em Matemática e Aplicações, Colégio Luís Verney, Universidade de Évora, Évora, Portugal

Maria Isabel Fraga Alves Faculdade de Ciências, Departamento de Estatística e Investigação Operacional, Universidade de Lisboa, Lisboa, Portugal

M. Ivette Gomes Universidade de Lisboa, CEAUL and DEIO, FCUL, Lisboa, Portugal

Délia Gouveia CEAUL, CIMO/IPB and University of Madeira, Funchal, Portugal

Carla Henriques CMUC and Escola Sup. Tecnologia e Gestão, Inst. Polit. de Viseu, Viseu , Portugal

Lígia Henriques-Rodrigues CEAUL and Instituto Politécnico de Tomar, Tomar, Portugal

Paulo Infante Department of Mathematics and Research Center of Mathematics and Applications (CIMA-UE), University of Évora, Évora, Portugal

Gonçalo Jacinto CIMA-UE and ECT/DMAT of University of Évora, Évora, Portugal

Sandra Lagarto Colégio Luís Verney, CIMA-University of Évora, Évora, Portugal

Manuela Larguinho Department of Mathematics, ISCAC, Bencanta, Coimbra, Portugal

Luiz Guerreiro Lopes CIMO/IPB, ICAAM/UE and University of Madeira, Funchal, Portugal

Susete Marques Instituto Superior de Agronomia (UTL) and CEF, Tapada da Ajuda, Lisboa, Portugal

João Paulo Martins School of Technology and Management, Polytechnic Institute of Leiria, CEAUL-Center of Statistics and Applications of University of Lisbon, Lisbon, Portugal

Maria Oliveira Martins Unidade de Parasitologia e Microbiologia Médicas, IHMT-UNL, Lisboa, Portugal

Ana Cristina Matos Escola Sup. Tecnologia e Gestão, Inst. Polit. de Viseu, Viseu, Portugal

Sandra Mendonça CEAUL and University of Madeira, Funchal, Portugal

Manuel Cabral Morais CEMAT and Mathematics Department, Instituto Superior Técnico, Technical University of Lisbon, Lisbon, Portugal

Isabel Natário Faculdade de Ciências e Tecnologia (UNL) and CEAUL, Quinta da Torre, Caparica, Portugal

M. Manuela Neves ISA, Technical University of Lisbon and CEAUL, Tapada da Ajuda, Lisboa, Portugal

Rui Nunes Research Centre for Spatial and Organizational Dynamics (CIEO), University of Algarve, Faro, Portugal

M. Manuela Oliveira Universidade de Évora and CIMA, Évora, Portugal

Patrícia Oom do Valle Research Centre for Spatial and Organizational Dynamics (CIEO), University of Algarve, Faro, Portugal

António Pacheco CEMAT and Departamento de Matemática, Instituto Superior Técnico, Universidade de Lisboa, Lisboa, Portugal

Claudia Pereira MMEAD/ECT of University of Évora, Évora, Portugal

Dinis Pestana Faculdade de Ciências (DEIO), Universidade de Lisboa and CEAUL, Lisboa, Portugal

Alexandra Pinto Faculty of Medicine of Lisbon, Laboratory of Biomathematics, Lisboa, Portugal

Efigénio Rebelo Research Centre for Spatial and Organizational Dynamics (CIEO), University of Algarve, Faro, Portugal

M. Luísa Rocha Universidade dos Açores (DEG) and CEEAplA, Ponta Delgada, Portugal

Carlos J. Roquete Instituto de Ciências Agrárias e Ambientais Mediterrânicas, Universidade de Évora, Évora, Portugal

Fernando Rosado CEAUL-Center of Statistics and Applications, University of Lisbon, Lisbon, Portugal

Thelma Sáfadi DEX, Federal University of Lavras, Lavras, Brazil

Tiago Salvador Instituto Superior Técnico, Technical University of Lisbon, Lisboa, Portugal

Rui Santos CEAUL and School of Technology and Management, Polytechnic Institute of Leiria, Leiria, Portugal

Ricardo Sousa Higher School of Health Technology of Lisbon, Polytechnic Institute of Lisbon, CEAUL-Center of Statistics and Applications of University of Lisbon, Lisbon, Portugal

Part I

Statistical Science

The Non-mathematical Side of Statistics

João A. Branco

Abstract

It is well recognized and accepted that mathematics is vital to the development and the application of statistical ideas. However, statistical reasoning and proper statistical work are grounded on types of knowledge other than mathematics. To help to understand the nature of statistics and what its goals are some major aspects that make statistics different from mathematics are recalled. Then non-mathematical features are considered and it is observed how these are diverse and really indispensable to the functioning of statistics. Illustrations of various non-mathematical facets are brought about after digging into statistical analyses attempting to end the Mendel–Fisher controversy on Mendel's data from breeding experiments with garden peas. Any serious statistical study has to take into account the mathematical and the non-mathematical sources of knowledge, the two sides that form the pillars of statistics. A biased attention to one side or the other not only impoverishes the study but also brings negative consequences to other aspects of the statistical activity, such as the teaching of statistics.

1 Statistics and Mathematics

Although there is a general consensus among statisticians that mathematics is essential to the development and practice of statistics there is also disagreement and confusion about the amount and the level of sophistication of mathematics used in connection to statistical work. The role of mathematics has been viewed differently throughout the times within the statistical community.

When statistics was at its beginnings, the need for some mathematics was felt, surely because a theoretical basis for statistics was missing. The precise nature of

J.A. Branco (✉)
Department of Mathematics and CEMAT, Instituto Superior Técnico, TULisbon, Portugal
e-mail: jbranco@math.ist.utl.pt

A. Pacheco et al. (eds.), *New Advances in Statistical Modeling and Applications*,
Studies in Theoretical and Applied Statistics, DOI 10.1007/978-3-319-05323-3_1,
© Springer International Publishing Switzerland 2014

the mathematical statistics that arose in the early twentieth century was naturally associated with little data available and the lack of computing power. In William Newmarch Presidential address to the Statistical Society of London on "Progress and Present Conditions of Statistical Inquiry" [22] we can appreciate this concern. Newmarch examines, together with seventeen other fields of statistical interest, the topic "Investigations of the mathematics and logic of Statistical Evidence" saying that it

> ...relates to the mathematics and logic of Statistics, and therefore, as many will think, to the most fundamental enquire with which we can be occupied...This abstract portion of the enquiries we cultivate is still, however, in the first stages of growth. (p. 373)

Ronald Fisher's celebrated book *Statistical Methods for Research Workers* [5] begins with a lapidary first sentence that had tremendous impact on the future development of statistics:

> The science of statistics is essentially a branch of Applied Mathematics, and may be regarded as mathematics applied to observational data. (p. 1)

This view may be quite natural knowing that Fisher was involved in deep mathematical thinking to establish the foundations of statistics [4]. This potential definition of statistics could have had the same importance as any other, but coming from such an outstanding scientist it had decisive influence in valuing, possibly too highly, the role of mathematics and of mathematicians in the progress of statistics. Too many mathematical abstractions invade the realm of statistics. Mathematical Statistics was born and grew so strongly that it was identified, in some quarters, with Statistics itself. Even today statistical courses and statistical research continue to take place under the umbrella organization of departments of mathematics and the teaching of statistics at school is conducted mainly by teachers of mathematics.

John Tukey was one of the first statisticians to perceive that this line of thought was leaving aside crucial aspects of the subject matter of statistics. He opens his revolutionary paper on "The Future of Data Analysis" [27], by showing his dissatisfaction with the inferential methodology as well as the historical development of mathematical statistics and announcing a new era for statistics:

> For a long time I have thought I was a statistician, interested in inferences from the particular to the general. But as I have watched mathematical statistics evolve, I have had cause to wonder and to doubt.... All in all, I have come to feel that my central interest is in data analysis,... (p. 1)

Tukey's insight of the nature of statistics appears even more profound if we take into consideration the fact that he was a former pure mathematician and that his paper was published in the *Annals of Mathematical Statistics*, a true sanctuary of mathematical statistics. His ideas took time to be assimilated by the community but they raise immediate discussions about the purpose of statistics and the role of mathematics in the development of statistics. One wonders if Tukey's paper had any influence in the decision by the Institute of Mathematical Statistics to split the *Annals of Mathematical Statistics*, just a few years later in 1972, into two different journals, erasing the words "Mathematical Statistics" from the title list of its journals. Many distinguished statisticians have made contributions to these discussions (see, for example, [2, 3, 15, 17, 18, 29]). In 1998 the statistical journal *The Statistician* published four papers giving a critical appraisal on statistics and mathematics [1, 11, 25, 26] commented by 21 qualified discussants. The debate is very illuminating and reveals a general consensus among the four authors and the discussants, possibly extensible to a great majority of statisticians, that mathematics is a necessary tool to be used as much as statistics needs it, but no more than that in what concerns statistical practice. More recently another author [20] gives a retrospective of the influential role of Tukey's 1962 paper, connected with the issue of statistics and mathematics.

The idea of freeing statistics from the rigidity and limitations of classical inference and going in the direction of data analysis, as advanced by Tukey, was followed by others, as in the area of robust statistics. In robust statistics [10, 14] a unique true model is not accepted. Instead robust statistics considers a family of approximate models, a supermodel, and it tries to find estimates that are as good as possible for all models of the family. The search for procedures that are optimal under a given model gives way to the search for procedures that are stable under a supermodel.

The advent of new technologies has opened the door to the production of a multitude of huge data sets of all kinds and in all fields of activity. It is evident that classical statistical methods often relying on assumptions of independence, normality, homogeneity and linearity, are not ready for a direct analysis of unstructured and non-homogenous data sets with a very large number of observations and variables and where the number of dimensions sometimes exceeds the number of observations. In these circumstances what can be done to analyse this kind of data? According to the spirit emanating from Tukey's concept of data analysis one should do everything sensible, using all means, arguments and tools, including all standard statistical methods and computing facilities, to get to the point, to answer as better as possible the question that we think only the data can help to answer properly. There was a time when statisticians were very interested in the study of methods to analyse small data sets and the discovery of asymptotic behaviours was a temptation that many were willing to try. Today we look around and see floods of data coming from every field such as astronomy, meteorology, industry, economy, finance, genomics and many others. Much of the immense work needed to analyse this sea of data is being done by professionals other than statisticians, working in areas traditionally not considered within the statistical arena (data mining, neural networks, machine learning, data visualization, pattern recognition and image analysis) but having some overlap with statistics. Since most of the tools of data analysis are of statistical nature it comes as a surprise when we see that statisticians are somehow reluctant to get involved with the analysis of large data sets [16]. That position can have negative consequences for the future of statistics. At a time when new statistical methods are needed, to face the complexity of modern data, statisticians should be committed to develop necessary theoretical studies to guarantee the progress of statistics. But to search for the convenient methods that statistics is needing, statisticians should first understand what are the problems and difficulties one encounters when analysing large data sets, and that can only be achieved with a steady involvement in the analysis of this type of data.

In this quick journey on a long road we have seen statistics and mathematics always together but apart, showing different phases of engagement because they are actually distinct undertakings. Next, in Sect. 2, differences between statistics and mathematics are highlighted. In Sect. 3, we discuss an example to illustrate that statistics appreciates its mathematics companionship but needs other companions to arrive at its purposes. Final remarks are presented in Sect. 4.

2   Statistics Is Not Mathematics

No one is interested in discussing whether economics or physics is not mathematics, but statisticians are usually attracted and concerned with the subject of the title of this section. The reason may be found in the recurrent historical misunderstanding between statistics and mathematics, as perceived by the contents of the previous section and of the references found there. To distinguish statistics from mathematics one could start from their definitions but we would find that there is not a unique definition for any of the subjects. To avoid a long discussion and some philosophical considerations we consider only a few characteristics typical of statistics that will serve to highlight the differences of the two subjects. These characteristics have been referred to by many statisticians at large, in particular by those interested in the teaching of statistics, and are:

1. *Origin*: Mathematics and Statistics are both very old. One might say that mathematics was born when primitive men first started to count objects and living things. The origin of statistics is associated with the moment when men first felt the need to register the results of counting, with the interest to remember the past and try to foresee the future. However, statistics, as an academic discipline, is much younger than mathematics, only a little over a century old, while one can speak of hundreds and hundreds of years for the various branches that form the present undergraduate students' mathematical curriculum. Statistics grew outside mathematics prompted primarily by questions and problems arising in all aspects of human activity and all branches of science. The first statisticians were truly experimental scientists [7]. Experimental scientists needing to analyse complex data had—as they have now and will always have—an important catalyst role in broadening the field of statistics and the development of new statistical methods.

2. *Variability*: In a world without variability, or variation, there would be no statistics. But variability abounds in our world and the uncertainty it generates is everywhere. The role of statistics is to understand variability by caring about identifying sources, causes, and types of variation, and by measuring, reducing and modelling variation with the purpose of control, prediction or simple explanation. Statistical variation does not matter much to mathematics, a relevant fact that has to be used to distinguish the two disciplines.

3. *No unique solutions*: Statistics results depend on many factors: data under analysis, model choice and model assumptions, approach to statistical inference, method employed and the personal views of the statistician who does the analysis. Instead of well-identified solutions as is usual in mathematics, various

solutions of a nondeterministic nature, leading to different interpretations and decisions is a common scenario in statistics.

4. *Inductive reasoning*: Two types of reasoning that work in opposite directions can be found in Mathematics and Statistics. Deductive reasoning is used in mathematics: from a general principle known to be true we deduce that a special case is true. Statistical inference uses inductive reasoning: the evidence we find in a representative sample from a population is generalized to the whole population.

5. *Scientific method*: In [19], Mackay and Oldford view the statistical method as a series of five stages: Problem, Plan, Data, Analysis and Conclusion (PPDAC). By comparison with the scientific method for the empirical sciences they conclude that the statistical method is not the same as the scientific method. But, although statistics is a unique way of thinking, it follows the general principles of scientific method. It is embedded in almost every kind of scientific investigations adding rigor to the scientific process and scientific results. To the contrary, mathematics, as generally accepted, does not follow the scientific method.

6. *Context*: Data needs statistics to be analysed and statistics needs data to work. With no data one does not need any statistics. But data are numbers in a context, and that is why context is essential in statistical work. Context is crucial even before we have data because knowing context one can decide how data may be collected to better conduct the statistical analysis. The conclusions of a statistical study have to recall the context to answer properly the questions formulated in the beginning of the study. The case of mathematics is different. The work of mathematics is mainly abstract, it deals with numbers without a context. While context is the soil that makes statistics grow well it may be a drawback that disrupts the natural development of mathematics. That is why context may be sometimes undesirable for mathematicians.

Other aspects typical of statistics but not of mathematics, and certainly not the only ones, are the terminology and the language, the measurement and design issues associated with the collection of proper statistical data, the interpretation of statistical results and the communication of statistical ideas and statistical results to a large and diverse audience.

The idea to set apart statistics from mathematics is not intended to say that mathematics is not important to statistics but to justify that statistical knowledge and statistical reasoning, specific as they are, must be envisioned and cared about as a unique scientific process that must be let to develop freely without any constraints from other fields, in particular from mathematics with which it has a strong connection.

Any inattention to this is likely to distort the natural progress of statistics. One area where this may happen is the teaching of statistics. If teachers and scholars fail to explain clearly the true nature of statistics and the specificity of statistical thinking, their students, detached from the reality of statistics, will tend to propagate a wrong message. And this state of affairs is not uncommon if we think that, on the one hand, the statistics taught at school level is often part of the mathematics curriculum and the teachers who are trained to teach mathematics have, in general, little contact with statistics and no experience whatsoever with the practice of statistics. On the other hand, at the university level, introductory courses of statistics face the limitations of time allocated to these courses and with little time the syllabus concentrates on formal methods putting sometimes more weight on mathematical aspects than should be the case. Besides, statistics is a difficult subject to teach: students don't feel comfortable with uncertainty and probability, and how do we teach, in the first instance, the ideas of variability or data analysis? How can lecturers, in a limited amount of time, make their students understand that: (1) to have a good knowledge of the context is important, (2) good interpretation of statistical results requires ability, (3) making final decisions about a problem has to be based upon conclusions of statistical analyses, generally not unique and (4) they must exert good communication skills to dialogue with those who have posed the problem, know well the context and expect to follow the statistical arguments and results? Some aspects can only be learned by getting involved with problems of the real world, that is, by doing statistics.

Interest in the teaching of statistics is not new [13, 28] but it grew tremendously when it was felt that citizens living in a modern society should be statistically literate and statistics was then introduced in the school mathematics curriculum. Many obvious questions, that are not easy to answer, were then put forward: Who is going to teach statistics? Who can? What to teach? How to do it? and so on. The International Statistical Institute realizing the scale of the problem and its interest to the community created IASE (International Association of Statistical Education) to promote statistical education. IASE organizes conferences and other actions concerning the teaching and the learning of statistics. Today statistical education is a topic of research that attracts a large number of people who publish the results of their investigations in specialized journals. Ideas of changing curricula, styles and methods of teaching and learning are in the air [9, 12]. Although school elementary courses and university introductory courses are very distinct and run in completely different scenarios there are reasons to believe that the difficulties encountered in passing the statistical message in both cases share some form of influence of two general conditions: mathematical and non-mathematical aspects of statistics and the relative importance that is given to each one of them.

Next, we look at a statistical article [23] trying to identify and discuss various non-mathematical aspects of the analysis. Any other non-theoretical work could be used to illustrate the role of the non-mathematical aspects but this particular one has the advantage that the author of the present work is a joint author of that paper and then he can review and quote from it more freely.

3 The Non-mathematical Side of Statistics

The title of the paper mentioned at the end of the previous section, "A statistical model to explain the Mendel–Fisher controversy", is self-explanatory in what the authors want to do. The question is how they arrive at that model and what they are doing with it. Let us review the various phases of this work.

3.1 The Problem

Gregory Mendel, known as the founder of scientific genetics, published, as early as 1866, his two laws of heredity (the principle of segregation and the principle of independence) [21]. This amazing discovery was arrived at after continuous meticulous work, during more than 7 years, on controlled experiments by cross-breeding garden pea plants. Mendel cultivated and tested around 29,000 plants. Inspired by good judgement and based on empirical calculations (proper statistical methods did not exist at the time) on the registered data of the artificial fertilization Mendel worked out the laws of hereditary. But despite being an extraordinarily revolutionary achievement it was forgotten until 1900, during 35 years, when it was rediscovered by independent researchers.

Ronald Fisher, known as the founder of modern statistics [24], and a great geneticist, soon got interested in Mendel's work. In 1911 he made a first analysis of Mendel's results, and having found that they were exceptionally good questioned the authenticity of the data presented by Mendel. Twenty five years later, in 1936, Fisher came back to review the problem and performed a thorough and rigorous analysis of the same data and of all the Mendel experiments supposed to generate that data. He reinforced his previous opinion concluding that the data are simply "too good to be true" [6], what became a truly demolishing assessment for Mendel's image. Apparently Fisher's veiled accusation of forgery was ignored until the centennial celebration of Mendel's 1866 publication when it suddenly came to light and a stream of controversial opinions, about the relevant question, started to flood the publication spaces with tens of papers, including the recent book "Ending the Mendel–Fisher Controversy" [8] which really does not manage to accomplish what its title promises. Pires and Branco [23] present a short chronological account of the major facts of this controversy almost century-old controversy. They refer to the vast bibliography that has been produced, some of which is very illuminating for the sake of understanding the problem and the discussion of the analyses proposed by the various contributors.

The relevant question is: is Fisher right? That is, has Mendel's data been faked? Since Mendel's laws are right, we must start by asking if Fisher's analysis is correct, because if it is not then the reason for the accusation would be lost. A second question is: if Fisher is right can we think of other possible reasons why Mendel's data conforms so well to his model, instead of immediately accusing him of scientific misconduct?

3.2 Data in Context

To answer the first question Fisher's analysis must be reviewed. As mentioned in [23] only the part of Fisher' paper related to a chi-square analysis is considered here. It is in fact the extremely high p-value obtained by Fisher in that analysis that served mostly to support his attack on Mendel and that has also been the bone of contention among the scientists involved in the debate.

To understand the data as prepared by Fisher in order to apply the chi-square goodness-of-fit test it is necessary to get into the genetic background, to follow the details of a large number of complex and delicate experiments, to be aware of the subtle problems of measurement of the experimental results and finally to understand Mendel's theory. Mendel's paper is simple, clear and certainly the best to help the reader in these matters, but there are other useful sources. Pires and Branco [23] give an organized summary of relevant aspects of the experimentation and give comments on the data that help to understand why and how the chi-square can be used. Knowing the context and understanding the data is essential also to follow the arguments advanced by many researchers to defend their proposals to solve the controversy.

Mendel concentrated on the transmission of seven binary traits or characteristics of garden pea plants (two traits observed in the seeds and five observed in the plants). One trait has two forms (phenotypes), A (named dominant) and a (named recessive), just like seed shape (round, A, or wrinkled, a) and flower colour (purple, A or white, a). He tried various types of cross fertilization and observing the traits of the offsprings and comparing the results with his expectations he consolidated his theory. Following [23] and a classification used by Fisher, the experiments can be classified into single trait experiments, bifactorial experiments and trifactorial experiments according to the number of traits considered in each crossing, one, two or three. Fisher included in his analysis more complex experiments classified into two new categories: gametic ratios and illustrations of plant variation experiments. In accordance with Mendel's theory, crossing a number of plants pure lines (those whose offsprings are always similar to their parents) and then crossing the resulting offsprings (no pure lines any more, called hybrids) then the offsprings of this last crossing will be of the two original phenotypes A and a in the proportion of 3:1. That is, in a population of n offsprings the number of phenotypes A (success), n_A, will be distributed as a binomial distribution, $n_A \sim Bin(n, p)$, where p is the probability of success ($p = 3/4$ in this case of the ratio 3:1), under the standard hypotheses: each observation is considered a Bernoulli and trials are independent. A more thorough and complete description of this interpretation, extended to all cases of cross breading included in the study, is in [23]. To test Mendel's theory we consider the number of successes, $n_A \sim Bin(n, p)$, and the hypothesis H_0: $p = p_0$, where p_0 is the true probability of success under Mendel's theory. The observed value of the test statistic to test H_0 against H_1: $p \neq p0$ is given by $\chi = (n_1 - np_0)/\sqrt{np_0(1 - p_0)}$. Assuming n is large the p-value of the test is $P\left(\chi_1^2 > \chi^2\right)$.

3.3   Fisher's Chi-Square Analysis

Having assumed the binomial model (in some cases a multinomial model was assumed) and independence of experiments Fisher tested H_0 applying a chi-square goodness of fit test and then he summed up all the chi-square statistics and degrees

Table 1 Fisher's chi-square analysis ("Deviations expected and observed in all experiments")

Experiments	Expectation	χ^2	Probability of exceeding deviations observed
3:1 ratios	7	2.1389	0.95
2:1 ratios	8	5.1733	0.74
Bifactorial	8	2.8110	0.94
Gametic ratios	15	3.6730	0.9987
Trifactorial	26	15.3224	0.95
Total	64	29.1186	0.99987
Illustrations of plant variation	20	12.4870	0.90
Total	84	41.6056	0.99993

of freedom in each class of experiments, having arrived at the results of Table 1 (Table 5 of [6]).

The first two lines of Table 1 referred to the single trait experiments. Gametic ratios are 1:1 and the rest of the experiments have ratios of 2:1 or 3:1, which means that in H_0: $p = p_0$, the possible values of p_0 are: 3/4, 2/3 and 1/2. The final value of the chi-square, after aggregation, leads to the amazing p-value of 0.99993. This means that, with data collected and reported correctly and under the assumed distributional assumptions, the probability of getting an overall better result, in a new sample of experiments, is 0.007 %. If 100,000 scientists, including Mendel, were to repeat these experiments, only seven lucky ones would observe data that agree with the theory better than Mendel's original data. Fisher decided that Mendel was not one of the lucky scientists and preferred to suggest that the data may have been massaged, not necessarily by Mendel but by some of his dedicated assistants. And what about Fisher, has he done a correct analysis? Pires and Branco [23] repeated Fisher' work and did a simulation study based on 1,000,000 random replications of the experiments. They concluded that Fisher is correct if the assumed assumptions are met. Pires and Branco [23] also did a study of a model based on 84 separated binomials and the results are similar to the results found in the two previous alternatives but as it attains a larger overall p-value, 0.99998, the evidence against Mendel is even greater than that worked out by Fisher. This is the model selected by Pires and Branco [23] for the rest of their investigations, the less favourable to Mendel. Fisher's accusation started a large debate involving many scientists. According to Pires and Branco [23] most of the arguments that have been put forward can be classified into three categories: (1) those who do not believe in Fisher's analysis, (2) those who accept that Fisher is correct but look for other ways to analyse Mendel's data and (3) those who believe in Fisher's analysis but, not accepting that Mendel has deliberately faked his data, look for other methods or arguments that could explain the observed high p-value. Despite so much interest and effort and so many, biological, statistical or methodological explanations, attempting to solve the controversy no one has succeeded in refuting Fisher's analysis so that the hostile remark "too good to be true" could be dismissed.

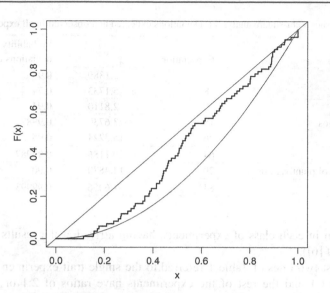

Fig. 1 Empirical cumulative distribution function of the *p*-values (*stair steps line*); cumulative distribution function of the uniform (0,1) random variable (*straight line*); cumulative distribution function of the maximum of two (0,1) uniform random variables (*curve*)

3.4   A Cute Little Theorem

Having concluded, based on the study of two models and a Monte Carlo simulation, that Fisher's analysis is correct, the authors turn to the distribution of the *p*-values from the 84 binomial model. They realize that, accepting all model assumptions are true and that Mendel's theory is right, the application of the chi-square test is correct and the *p*-values follow a uniform distribution. And in that case its empirical cumulative distribution function (e.c.d.f.) is graphed as the diagonal of the square $(0, 1) \times (0, 1)$, as represented in Fig. 1, reproduced from [23]. The plot of the e.c.d.f. of the *p*-values (stair steps line) is far from that diagonal, and this discrepancy is also supported by the Kolmogorov–Smirnov goodness of fit test. Then the question arises: what is the c.d.f. of the *p*-values and what is the reason that justifies that distribution?

The authors, inspired by the visual of Fig. 1, notice that the plot of the e.c.d.f. of the observed *p*-values lies between the diagonal (plot of the c.d.f. of the uniform) and the plot of the c.d.f. of the maximum of two uniforms (0,1), $y = x^2$. This triggers the following hypothesis: the c.d.f. of the *p*-values that best fits the data (e.c.d.f. of the observed *p*-values) probably lies between the two lines, $y = x$ and $y = x^2$. One possible justification of this proposal is: Mendel repeated those experiments that produced results that deviate markedly from his theory, and choose only the best of the two. To model this proposal one assumes that an experiment is repeated whenever its *p*-value is smaller than a value of a parameter, α, selected by the experimenter. Then the experiment with the largest *p*-value is chosen.

Assuming this selection procedure the authors proved that (a "cute little theorem", as the editor of Statistical Science called it during the revision process)

$$F_\alpha(x) = \begin{cases} x^2, & \text{if } 0 \le x \le \alpha \\ (1+\alpha)x - \alpha, & \text{if } \alpha < x \le 1. \end{cases}$$

Repeating Fisher's analysis under the proposed model (α is estimated from the sample of 84 binomial p-values) the p-value, root of the controversy, drops down to a reasonable value, not too close to one. This means that the conditions under which Fisher did his analysis may not be right because the selection of the data may have been done according to a procedure distinct from the one he was assuming. Fisher overlooked other possible procedures as the one suggested in [23] to support of their proposed model. In [23] it is shown that the proposed model is better than any model within the family of models whose c.d.f. is a linear combination of the extreme models, x and x^2.

The authors of the paper we are discussing present a list of quotations from Mendel's paper that clearly support their proposed model. In the end, regarding Mendel's position in the controversy, the authors try to dismiss the accusation of fraud and attribute his motivation to pure unconscious bias.

One of the latest attempts to solve the Mendel–Fisher controversy is described in [23]. From this quick review of that paper one can see that there are some pieces of mathematical statistics: the theorem, essential to the formulation of the model on which the authors based their analysis, and eventually some formal aspects behind the probability distributions used in the text, including those associated with the chi-square and Kolmogorov–Smirnov tests. The rest of the paper can be considered of non-mathematical nature, including the application of the statistical tests and all the statistical reasoning used in the analysis. The computation needs are not great: we should remember that Fisher did his analysis as rigorous as it can be without any of the computational means we have today, but without a computer the simulation study would not be possible. The context involving the controversy is briefly described in Sects. 3.1 and 3.2 but it is constantly behind most of the decisions concerning the global statistical analysis. Judgements of non-statistical nature permeate this research and when correctly formulated they are illuminating for the progress of the work. Bad judgements can cause irreparable damage to the research. Fisher's "too good to be true" is probably an imprudent judgement because he could not prove his accusation and did not look for other possible reasons that could explain the high observed p-value. That is what many people are doing now, by trying to solve the controversy he started.

4   Final Remarks

The discipline of statistics is the result of many endeavours and ideas from many people working in a great variety of fields, some interested in immediately solving their real problems, some eager to devise rigorous and efficient methods of analysis and others devoted to the teaching and the spread of statistics. That is probably why talking about statistics is a matter of experience and sometimes a controversial subject. This text is a personal opinion about central features of statistics and should be viewed only as such. The main point is to highlight aspects of statistics that are not of a mathematical nature. It is unfortunate that these aspects are not given the importance they actually have in the statistical work. Of course, mathematics is an essential tool: formal statistical methods are a product of mathematics and they would be of little value without the contribution of probability theory. But a good knowledge of the context of the statistical problem, the efficient manipulation of the computational tools and the ability to use other pieces of knowledge are crucial for the selection of a convenient model and for the global analysis. Doing statistics is like making a cake: various ingredients are needed and if they are combined in the right proportions and mixed with artistic hands the expected product will have a good chance of being delicious. In the discussion of the example presented in Sect. 3 we can recognize statistical ingredients and observe how they are used to arrive at a final product, just as in this allegory. Along the path of statistics we have spotted important moments of change: the need for mathematics, the appearance of mathematical statistics, the arrival of data analysis and the current wave of massive data sets that is presenting enormous methodological challenges to statistics.

Statistics is a dynamic discipline that is continuously evolving and adapting to different realities. It will find a new paradigm to succeed in the present demand and in what will come next.

References

1. Bailey, R.A.: Statistics and mathematics: the appropriate use of mathematics within statistics. Statistician **47**, 261–271 (1998)
2. Benzécri, J.P.: Analyse des Données, 2 vols.. Dunod, Paris (1994)
3. Box, G.E.P.: Some problems of statistics and everyday life. J. Am. Stat. Assoc. **74**, 1–4 (1979)
4. Fisher, R.A.: On the mathematical foundations of theoretical statistics. Philos. Trans. R. Soc. Lond. A **222**, 309–368 (1922)
5. Fisher, R.A.: Statistical Methods for Research Workers. Oliver and Boyd, Edinburgh (1925)
6. Fisher, R.A.: Has Mendel's work been rediscovered? Ann. Sci. **1**, 115–137 (1936)
7. FitzPatrick, P.J.: Leading British statisticians of the nineteenth century, J. Am. Stat. Assoc. **55**, 38–70 (1960)
8. Franklin, A., Edwards, A.W.F., Fairbanks, D.J., Hartl, D.L., Seidenfeld, T.: Ending the Mendel-Fisher Controversy. University of Pittsburgh Press, Pittsburgh (2008)
9. Garfield, J.B. (ed.): Innovations in Teaching Statistics. The Mathematical Association of America, Washington, DC (2005)
10. Hampel, F.R.: Robust estimation: a condensed partial survey. Z. Wahrscheinlichkeitstheorie verw. Geb. **27**, 87–104 (1973)
11. Hand, D.J.: Breaking misconceptions—statistics and its relationship to mathematics. Statistician **47**, 245–250 (1998)
12. Higgins, J.J.: Nonmathematical statistics: a new direction for the undergraduate discipline. Am. Stat. **53**, 1–6 (1999)
13. Hotelling, H.: The teaching of statistics. Ann. Math. Stat. **11**, 457–470 (1940)
14. Huber, P.J.: Robust statistics: a review. Ann. Math. Stat. **43**, 1041–1067 (1972)

15. Huber, P.J.: Applications vs. abstraction: the selling out of mathematical statistics. Adv. Appl. Prob. **7**, 84–89 (1975)
16. Huber, P.J.: Data Analysis, What Can Be Learned from the Past 50 Years. Wiley, Hoboken (2011)
17. Kendall, M.G.: The history and the future of statistics. In: Bancroft, T.A. (ed.) Statistical Papers in Honour of George Snedecor, pp. 193–210. Iowa State University Press, Ames (1972)
18. Kiefer, J.: Review of M.G. Kendall and A. Stuart: The advanced theory of statistics, vol. 2. Ann. Math. Stat. **35**, 137–1380 (1964)
19. MacKay, R.J., Oldford, R.W.: Scientific method. Statistical method and the speed of light. Stat. Sci. **15**, 254–278 (2000)
20. Mallows, C.: Tukey's paper after 40 years. Technometrics **48**, 319–336 (2006)
21. Mendel, G.: Versuche über Plflanzenhybriden Verhandlungen des naturforschenden Vereines in Brünn, Bd. IV für das Jahr 1865. English translation (1909): Experiments in plant hybridization. In: Bateson, W. (ed.) Mendel's Principles of Heredity, pp. 317–361. Cambridge University Press, Cambridge (1866)
22. Newmarch, W.: Inaugural address on the progress and present condition of statistical inquiry, delivered at the Statistical Society of London. J. Stat. Soc. Lond. **32**, 359–390 (1869)
23. Pires, A.M., Branco, J.A.: A statistical model to explain the Mendel-Fisher controversy. Stat. Sci. **25**, 545–565 (2010)
24. Rao, C.R.: R. A. Fisher: The founder of modern statistics. Stat. Sci. **7**, 34–48 (1992)
25. Senn, S.: Mathematics: governess or handmaid? Statistician **47**, 251–259 (1998)
26. Sprent, P.: Statistics and mathematics - trouble at the interface? Statistician **47**, 239–244 (1998)
27. Tukey, J.: The future of data analysis. Ann. Math. Stat. **33**, 1–67 (1962)
28. Vere-Jones, D.: The coming of age of statistical education. Int. Stat. Rev. **63**, 3–23 (1995)
29. Wolfowitz, J.: Reflections on the future of mathematical statistics. In: R.C. Bose et al. (eds.) Essays in Probability and Statistics, pp. 739–750. University of North Carolina Press, Chapel Hill (1969)

Outliers: The Strength of Minors

Fernando Rosado

Abstract
Let us think, particularly, about Statistics. Statistics is simply the science of data. It is usually also "applied" because research, most of the times, also implies an application. Statistics is interesting and useful because it provides strategies and instruments to work the data in a way that we can better understand real problems. Data are numbers (or the lack of numbers) inserted in a certain context or experience. To determine the average of 50 numbers is pure algebra, it's not Statistics. To decide over that value of 50 and to choose whether we have a small or a large sample is, in each case, to assume the difference of a certain value (even if we use the average determined before!)—is, in fact, Statistics.

Moreover, let us think, what are the main topics of Statistics for the twenty-first century? What is now "important" comes from the previous century? In one recent ISI congress—International Statistical Institute—(now called ISI WSC—World Statistics Congress) the topic "Water" was elected for a whole day of scientific lectures. Why?

Thinking about investigation, let us bring up the expression "quos fama obscura recondit". [Thinking about investigation. This great expression by Virgil, The Aeneid (Eneid, V, 302) is used, among many other, by Saint Augustine, De civitate dei, (*The City of God*, volume I, Book VII, Chapter III, p. 611 and so on. Education Service. Calouste Gulbenkian Foundation, 1991).] On the one hand, in the dichotomy between the "minor reason" and "a higher reason", should the statistician have, as a goal, (only) the knowledge which allows him to cover all the basic scientific requirements? On the other hand, that knowledge should be the beginning and statistics still assumes the great importance of "scientific

F. Rosado (✉)
CEAUL-Center of Statistics and Applications, University of Lisbon, Lisbon, Portugal
e-mail: fernando.rosado@fc.ul.pt

A. Pacheco et al. (eds.), *New Advances in Statistical Modeling and Applications*,
Studies in Theoretical and Applied Statistics, DOI 10.1007/978-3-319-05323-3_2,
© Springer International Publishing Switzerland 2014

details" of those who are hidden by an obscure fame—let's call them *outliers*.
They are stimulators of research and they can be originated by different values
of the same sample. A minority!

Are these "minors" who make science go forward?! The strength is in them!

1   Statistics as a Science

Science in general and Statistics in particular is a noble occupation, necessary to
the body and mind, compulsory to the well-being and to happiness. But the fact is
that science is expensive. This way, only the rich can make use of it... and the poor
become poorer if they employ it. Though great effort and dedication are demanded,
the solution (in spite of all this) should probably be in making science to walk out of
that dilemma. And this is what it is required to Portuguese statisticians brought
together by a leading project—SPE! The Portuguese Statistical Society (SPE in
Portuguese) joins both researchers and makers of Science. In the *Memorial* [6],
several authors wrote about research. The fact that the chapters are so up-to-date
makes that edition an important document in the search and creation of a story on
investigation. This topic is crucial for Science and for Statistics.

Statistics affects all and touches life in many situations. As citizens we help
to provide statistical information—our own birth and death are recorded to create
indexes and rates. Moreover, advertising leads us to believe in something or fools
us using statistical facts and figures to support their product.

That's the importance of the individual/of an observation.

Leading a community, through its governmental institutions and trading, depends
a lot on the statistical information and that dependence increases even more as the
trade influences the economic and social life planning. The advertisers, managers
and administrative leaders who use (and sometimes abuse) statistics are a strong
number of people. But there are others, considering social science students and
politicians. All these people apply statistical facts and methods to build the starting
point of politics. Such facts and methods also have a very important place in the
development of sociology and economics as sciences. They are also relevant for
scientific researchers, considering biology, for instance, and for those who work
with the most exact sciences like physics, chemistry or engineering these facts and
methods become fundamental.

That's the importance of Statistics.

The statistical ideas are the centre of many theories and, in fact, a "statistical
approach" is maybe one of the most distinguishing features of modern science.
Finally, statistics as a subject is naturally very interesting for the relatively small
group of professional statisticians. As a result of the various ways we found to look
at the topic, the word "statistics" and the ones connected to it ("statistical" as an
adjective and "statistician" as a noun) have several meanings. First of all, we have
the dictionary definitions in which statistics is referred to the topic as a whole and,
in a broader meaning, numeric data.

Common sense says that statistics are just numbers. The ordinary user has the tendency to think that a statistician is mainly someone who counts the number of things.

For an economist, used to the economical theory ideas, "statistical" is almost a synonym of "amount". For a physicist, "statistical" is the opposite of the exact, as for him statistics is a subject that above all considers groups and possibilities, more than certainties. For the scientist and investigator who is used to get knowledge from controlled experiments, statistical methods are those which he applies when a rigorous control of an experiment is impossible or very difficult to maintain. The field of application of statistics is mostly economical, but not totally economical— that's why the statistician is often considered an economist. On the other hand, as statistical methods are basically mathematical, many people still think—even today—that the statistician is a sort of mathematician. We could almost assert that the mathematician accepts the statistician as an economist and that the economist considers him a mathematician. Some (few?!) think that statistical methods are so poorly rigorous that anyone can "prove" no matter what; others acknowledge that, because they are such harsh methods, they prove nothing. The third group unites those who state that, as a way to increase knowledge, the power of statistics is unlimited and almost magical.

It is normal to start a book about Statistics, for instance, by defining and illustrating the topic we are referring to. A book in which Statistics is the main topic is obviously not an exception. A (random) reading of the first pages of a book suggests two perspectives for the introductory definitions. These definitions are brief and shallow most of the times. Others interleave identifiable subjects which restrict the text. Let us reflect on the issue overall. Facing the topic of Statistics we can consider many thoughts. The first one is considering that Statistics is at the same time a science and an art. It is a science because its methods are basically systematic and have a general application; it is an art because the success of its application can (also) depend on the experience and the skill of the statistician and his knowledge of the field of application he works in. However, it is not necessary to be a statistician to appreciate the main beliefs that are its basis. As a science, Statistics and particularly the statistical methods are a part of the scientific method in general and is based in the same background and processes.

Thus Statistics, as many other subjects, is always evolving. It is sustained by a theory . . . hence, it is also and above all, progressive! A theory is a set of primary guidelines of a science or of an art with a certain doctrine about them.

Statistics is a science because, basically, it develops a rigorous and rational understanding of a wide range of knowledge in a large sundry of purposes. So, it must be an organised set of information rooted on objective verifiable interactions and with a universal value.

It is obvious, commonly accepted and in a good dictionary we may find a definition for Statistics: it is a science that studies the methodical grouping of social facts which are usable for a numeric evaluation (of the population, of the birth rate and mortality, of the income rate and taxes, of the agricultural crops, of the criminal rate, of the religious beliefs, etc.). In a slender perspective, sometimes one also elects the word Statistic to refer to a part of applied mathematics which uses probabilities to establish hypothesis based upon real events, in order to predict occurrences. Progress has proved the first is stronger than the latter.

Statistics is the science of data, also applied because research often wants an application. Statistics is interesting and useful because it provides strategies and instruments to work the data in order to best "deal with" real issues. Data are numbers (or the lack of them) inserted in a given context or experience. But to determine the average of 50 numbers is pure arithmetic, not Statistics. To reason over that value of 50, decide whether we have a small or a big sample and in each case to conclude about the disparity of a specific value (even if one uses the average calculated before!) is really Statistics. Though Statistics may be considered a mathematical science, it is not a branch of maths and it shouldn't be taught like one. We can reason about statistical thought which stands and supports the decision theory.

2   Statistical Science: Inference and Decision

Statistics, practically demands judgements. It is easy to list the mathematical hypothesis that justify the use of a certain methodology but it is not so easy to decide about that method, when it can "surely" be used in an empirical point of view. Here, experience becomes crucial. Even in the simplest scientific analysis— and less disputable?—assuming Statistics as a branch of applied mathematics, the final goal is mostly related to predictions. Thus is the pragmatic point of view of Statistics that we are talking about. However, prediction is directly connected to Inference and Decision. Every theory illuminates an empirical point of view and this informs the theory, in a dialectic correlation. Every time we question the users (mainly those who are pragmatically dependent) there are "suggestions" that emerge with "a lot of case studies" and "job offers" with "less theory and more practice". Yet, theoretical support is always acknowledged and it should always be around and available!

We have therefore reached the Theory of Statistical Decision. It is the one where it is founded and where the genesis of the "statistician's job" remains. About this topic—theory—we chose some key ideas by Murteira [4].

"Although man is called to make decisions on a daily basis, only recently have problems with these decisions appeared and were dealt with under a scientific close look." (ib, p. 97). Historically, "the theory of statistical decision is essentially due to A. Wald who followed Neyman–Pearson tradition and enhanced the horizons, using the development of the game theory, by von Neumann and Morgenstern. The great worth of Wald (...) is in a contribution for a debate where (...) generally speaking, the standard procedures are peculiar cases of statistical decision" (ib, pp. 108–109). Nevertheless, in order to avoid confusions, we should make clear that "(...) the theory that is about to be thought about is connected to the individual decision, not to a group decision. (...) the theory that is about to be analysed isn't trying to substitute the decision-maker—but to bestow a set of rules which help the decision-maker (...). Generally one might say that this is a problem of decision when it becomes imperative to choose or opt at least between two courses of action" (*ib*, p. 97). But are there others who believe Statistics is not a theory? Could one accept that it is "an instrument or a tool in which its most relevant applications are naturally on the scientific sphere of influence"?[1] "In the scientific assertion problems—or the ones related to statistical decision—one works most of the times with a probabilistic model or, at least with a strong probabilistic component" (*ib*, p. 23).

3 The Need of *Outliers*

All models[2] are very important (they are fundamental!) in scientific research. Besides, in modelling there is (also) the sample and each of its parts—(particularly) its observations and its dimension. Searching for *outliers* in a sample is a research issue that can be fixed... And it will generate (or create the need for support in at least) a Theory! Some scientists assume Theory as a synonym for hypothesis. But theory is different from hypothesis and from science as a global system. As, in the scientific method, the hypothesis is a previous stage of the Theory. And it is an integrant part of science, either theoretical or applied. Theory opposes *praxis* or action, and yet they are complementary. As we know, the scientific method goes through several stages: observation and experiences, hypothesis and drawing conclusions, a general law or a Theory. This is a hypothesis which has been confirmed by experience and is a part of Science. But there are several kinds of Theory and Sciences. Nevertheless, we can only find possible two kinds of Theory: the deductive kind and the inductive kind. As far as the deductive are concerned, there is a series of valid statements or true premises (theorem) which is built upon a group of primitive premises (axioms) by the application of certain rules of inference. In the inductive we find a set of real or probable premises (theorems, axioms and definitions) which is developed according to several particular cases under a process of immediate and generalising inference. To many, the conclusion of induction is only a probability. And the "probability of a law" grows with the number of cases that confirm it. If after careful tests we "confirm a *discordant value*", we have the set or the core of rules that may form—one or the—theory of *outliers*.

One of the most general ways to define Statistics is to consider it a method to choose between alternatives of action facing uncertainty, by collecting and interpreting data about what it is being studied. Thus, in general, the Theory of *Outliers* in Statistical Data becomes a capital gain for Statistical Science. Some worries may follow its construction; yet, obstacles don't seem insurmountable and it is likely that a great future is ahead. The methods of scientific interpretation,

---

[1]Murteira in [4] quoting Gustavo de Castro, 1952, *Mathematical Statistics as a Scientific Tool*, pp. 52–64.

[2]Brief summary on this topic. For further information read [7].

theoretically speaking, are obviously meticulous and lead us to valid conclusions from a scientific point of view. Besides, the quality of the data used—which affects conclusions—cannot be the support of the accusation of those methods. Bad quality can question methodology. Good data quality is a wishful statistical benefit. And surely quality improves with the Theory of *Outliers*.

In conclusion, (one or the) theory of *outliers* should not only provide a set of rules which helps the decision-maker but also build instruments that can evaluate the quality of the decision. And with that hope . . . we must proceed, although theory should be transmitted in order to grow and to develop itself.

4   Fortune/Chance Decide!?

Considering[3] an analysis of *outliers* in statistical data, let us divide—or split up—the data we are studying into two (two, and only two groups alone?): "the selected", which supposedly has the larger number of observations, the majority, and "the suspicious". The latter (always present?!) has less data points because we ordinarily[4] believe in only one or two *discordant values*. There is no evident reason for that choice. Yet, it is (nearly always) done. The confirmation—of a *discordant/suspected value* as an *outlier*—is (stronger as we proceed) in the use of *outliers significance tests* in most scientific spheres and according to experts on the applications, including statistical packages. Everybody wishes to improve the quality of their work and conclusions through a "purification of the data". Still, can the suspicious hold more, better information than the selected? What is the reason why we choose the selected and not the others? Why aren't the suspicious—which are hidden by an obscure fame—those which are selected in the most eloquent statistical analysis? After all, this is a very important issue: showing[5] which are "the true ones", although we don't get all (the best!) clues out of them, and they allow us to see all the frailties of conclusions. Once we divided the statistical data between the selected and the suspicious, we should question "which is" or "why it is" selected. Who provides that "statistical circumstance"?

Chance[6] is the only thing that doesn't happen randomly. Statistics[7] is very old but it has a very short history. It was considered a subject in schools[8] only in the second

---

[3]The Roman goddess who, such as the Greek analogous Tyche, operated as she pleased, both happiness and sorrow, according to her wit.

[4]Also for scientific reasons!

[5]It is always the search for Truth that is the issue!

[6]Talking about *outliers*, let us remind this topic. The sentence is from Almada Negreiros (p. 125, *Mathematics and Culture*. Furtado Coelho et al., 1992. Edições Cosmos). This was a topic of discussion in a conference held by Tiago de Oliveira (*ib*, pp. 125–149). Statistics goes well with chance and they both create need. It is a recurring topic which entitled an edition of SPE—Statistics with Chance and Need; Proceedings of the 11th Annual Congress.

[7]About this topic, read the "small expedition" presented by Tiago de Oliveira (*ib*, pp. 125–128).

[8]The articles by Efron and Rao on this matter are important, in [5].

quarter of the twentieth century and its main architect was Fisher—also called the founder of modern statistical science. Certainly Fortune is the leader; she assures us all "the selection" or, on the other hand, "the obscurity", more according to her wit, rather than justice.

But how many are selected and how many are suspicious? If it is Fortune which decides which are the obscure—those which hold "strength"—why doesn't she get that credit for herself? Is it because she suffered a difficult fate herself? In that case, she honours others when she can't honour herself—she is her own opponent! Should the selected always be the majority? And do they deserve more attention? They are often chosen because they are important for the study. Among the selected, which are "the weakest"? Are they all just as good? What is the reason (or the cause) for some selected be considered of minor importance (because they surely exist!)? Moreover, how do we compare "minor selected" and "suspicious"? The latter, already called *outliers*, can detain much more value. It is an *outlier* which allows a deeper statistical analysis. This might be the origin of a work of excellence. It is the choice of a study of *outliers* that can make the difference between a statistician and a user of statistics. This also applies to research. *Outliers*—which are hidden by an obscure fame—bestow data with life. Indeed chance is the one which assures each statistician's fame or obscurity. Let not the one worthy of honour be judged, for he is among the selected. Stronger are those which chance—Mother Nature—has provided with much more (statistical) information. To select them, let us create (at least) a theory! Does Fortune/Chance also decide in research!?

5   *Outliers*: A Path in Research

Experience can turn research into passion. Basic scientific investigation, as well as applied scientific investigation, represents a significant illustration of man over Nature. On the one hand, investigation is *leitmotiv* on the search for Truth. This investigation, on the other hand, stimulates the great(est) thought (ever) that wishes for a reconciliation between reason and faith, which are often questionable. They are like two wings on which the human spirit rises to the contemplation of truth. If scientific investigation follows strict methods and keeps faithful to its own object there is no room for discrepancy. And if the research is based on smaller support—the background ones—it becomes easier to understand the two. They don't actually reach (apparently) contradictory goals.

All great theories start with a small step and, frequently based upon (statistical) data. Because it involves so many subjects, Statistics is shared in Science and for that reason, Research in Statistics is of great importance.

In Statistical Research, such as in many other subjects, one searches for truth. Truth? Search is a permanent course patterned by small steps—unsure and fragile at first—yet firm when experience and wisdom allows it. Notwithstanding, "there are no paths, we have to walk!" (says the poet) because "a path only results in traces . . ." and thus with Statistics.

Let us walk!

As we walk—more this way, less that way—questions about "goals", "interest" or "value" of the "produced science" occur. In the end, there's the "search for truth" and its value . . .

Certainly small contributions are solid because they are easily "controlled", "assimilated" and "ordered". So they occupy their place! Larger contributions, on the other hand, may be more fragile because of the vulnerability of their small support. And the latter, as a whole, create a theory.

Just the same, there are *outliers* in Statistics and many other subjects—from Physics to Metaphysics. There are profound doubts in all of them. Looking at *outliers* wherever they are, we create Science–particularly Statistics. Consequently, we open a (another) path on the search for Truth.

It is the strength of the weakest—the minors!

6   In Perspective

Statistics is also associated with the collecting and use of data in order to support the organisation of a certain state. The justice system is, in fact, one of the fundamental pillars of a modern state and it is basics in the politics of most of the countries. The most recent advance in the theory of *outliers* has emerged from the statistical inference to interpret data in a legal point of view. The legal courts are introducing new challenges for statisticians who are thus asked to speak upon non-traditional lines of work—for instance, the correct application of laws involving authors' rights or, much stronger, the biostatistics or genetic evidences in certain proofs. It is the rise of forensic statistics; probably the most recent topic of *outlier* study.

As we know, the definition of *outlier* depends on the area of statistics that we are working on. According to this, it is not possible to find a "general definition". Time series, for instance, demand a difference between *additive outliers* and *innovation outliers*. Spatial data, on the other hand, ask for a generalisation of the few existing results for circular data, where the influence of its dimension leads us to multivariate data. In multivariate data we are confronted with the additional difficulty of ordering data, which had been crucial for the research of univariate data.

The *outlier* issue can also relate to the general challenge of teaching statistics—from the conceptual point of view and from the pragmatic point of view. This subject is firstly posed when the most part of students' practical education is based on academic exercises. We are much aware that the statistician is also educated by professional practice. Yet, it is important to alert for the actual problems from the experiment's point of view, mainly in the "final education" subjects which involve statistical modelling, for example.

Though with different difficulty levels, many areas of investigation are opened to the study of *outliers* in statistical data. The choice[9] we have made—according

---

[9]For details: Chapters 2–4, in [7].

to a general approach—has the obvious advantage of turning the field of possible applications vaster. On the other hand, it has limited the study of some primary topics. Among these—has it has been said before—we underline the studies in time series and in surveys or census, where the first developments are very recent. Sure we can say that the general methodology we talked about before is applied here, despite the specificities that come from there.

The existence of an *outlier* is always related to a certain model and an observation can be discordant to a model and not to others.

The great goal in any *outlier* study will always be: *What is an outlier and how to deal with that statement.*

Once we define a theory, as we have said before, it is very important to evaluate the performance of the several tests of discordancy. This is also a sphere where there is much work to do.

The Beckman and Cook [1] study—although it is over 30 years old—made an excellent summary of the statistical approach of *outliers*, either from the historical point of view or from the application of standard models of statistics. Maybe now it is the moment to make a new up-to-date statement. In that study mentioned before, Beckman and Cook ironically conclude that "Although much has been written, the notion of outlier seems as vague today as it was 200 years ago".

What would we say today?

Of course, from then on we have registered an advance, but there is yet much to do.

The development of modern statistical theory has been a three—sided tug war between the Bayesian, frequentist and Fisherian viewpoints. In 1975, Lindley [3] foretold that the twenty-first century would be Bayesian—because 2020 was a crucial year. The Bayesian methods are complicated mainly for the theory of *outliers* where, as we have seen before, there is (always) much subjectivity involved a priori. Is there a great topic of investigation here as well?

Symbolically, as a counteraction, in 1998, Efron [2] predicts that "the old Fisher will have a very good 21st century".

The theory and practice of Statistics span a range of diverse activities, which are motivated and characterised by varying degrees of formal intent. Activity in the context of initial data exploration is typically rather informal; activity relating to concepts and theories of evidence and uncertainty is somewhat more formally structured and activity directed at the mathematical abstraction and rigorous analysis of these structures is intentionally highly formal. The world of applied statistics demands an arrangement between the Bayesian and frequentist ways of thinking and, for now, there is no substitute for the Fisher concept. It is interesting to register ideas about the modified likelihood functions or the pseudo-likelihoods, that is to say, functions of some or of all the data and a part of or all the parameters which can be widely treated as genuine likelihoods.

How do all these questions relate to the statistical study of *outliers*? What is the nature and scope of Bayesian Statistics within the "outlier problem"?

In the study mentioned above, where is this topic in the "statistical triangle"?[10]

This is a scientific challenge for the future. Possibly, this challenge has extra difficulty because we don't know the number of *outliers* in a sample—"outlier problem" or "outliers problem"?

Several topics are in need of more enhancement: the causes (deterministic and statistic) of the presence of *outliers* and the question of their existence in structured models (univariate and multivariate); the differences between simple *outliers* and multiple *outliers*.

On the other hand, different goals we intend to reach when we study *outliers* in a sample influence the conclusions. The outcome of the work done will be varied if we only wish to approach the detection of *outliers* in a set of data, or, if we want to put it together with more complex statistical models, involving for instance the presence of influent observations. Here we will be addressing issues of strength that intersect with the study of *outliers* but which are not the same. According to this, we are not far from the theory of extreme values.

The general theory of *outliers in statistical data*, in several directions, has much advanced in the last 40 years, and in it a great part of the first challenges found the contributions that made it an area of knowledge which already existed as a field of study. Once we reach that phase, we should proceed with wider developments in the (already) explored areas—and the multivariate area will be one of them— as other topics begin to show; and, among them, the most important seems to be performance appraisal. In fact, statistical analysis of multivariate data requires our work in a double way—the tests and models of discordancy. In this topic it is important to produce new ideas because the complex structure of these data is an enemy of the scientific simplicity we need to obtain the greatest success, especially in applications.

In the future, *outliers* will increasingly continue to occupy a place in the centre of statistical science and in statistical methods, because a discordant observation will always be a challenge for the analyst and it can widely influence their final report for the most important decision making. We are talking about excellence!

However, when everything is said and done, the main issue in the study of (supposedly) suspicious observations continues to be the one which defied the first investigators—*What is an outlier and how should one work with that observation?*

In the end of the second millennium, Time magazine organised a list of significant figures of the last thousand years. The names were ordered according to a vote. The first place of "the millennium person" was given to Saint Francis of Assisi, followed by Gutenberg, Christopher Columbus, Michelangelo, Martin Luther, Galileo, Shakespeare, Thomas Jefferson, Mozart and, in tenth place, Einstein.

A winner gathers values that give him distinction. Well, with the goal of electing the person of the millennium, the voters would have ordered their own criteria. The latter, coming from a set of rules, allowed the definition of a first place.

---

[10][2] for details.

Thinking about those variables—set of rules or "reasons for the election"—let us bring up the expression "quos fama obscura recondit". On the one hand, in the dichotomy between the "minor reason" and "a higher reason", should the statistician have, as a goal, (only) the knowledge which allows him to cover all the basic scientific requirements? On the other hand, that knowledge should be the beginning and statistics still assumes the great importance of "scientific details" of those who are hidden by an obscure fame—let's call them *outliers*. They are stimulators of research and they can be originated by different values of the same sample—one or more. A minority!

Are these "minors" who make science go forward?! The strength is in them!

Saint Francis is always seen as a reference and a simple life role model. "Francis poverty" is many times mentioned. His name is also connected to "ecology"—and to "peace". *Which would have been, and how can we find out, the most important variables that made Saint Francis the elected one?*

The knowledge of the statistical components that allow to find (and define) a discordant value in a sample is also a topic for the theory of *outliers*. In every model, whatever the criteria of discordancy, to be in first place is to be an *outlier*! Facing the demanding topics on *outliers in statistical data* described above—we quote this "last outlier"—"at least let's start working, because up to now we have done very little".

Acknowledgements Research partially funded by FCT, Portugal, through the project Pest-OE/MAT/UI0006/2011.

References

1. Beckman, R.J., Cook, R.D.: "Outlier. . .s" (with discussion). Technometrics **25**, 119–163 (1983)
2. Efron, B.: R. A. Fisher in the 21st century. Stat. Sci. **13**, 95–122 (1998)
3. Lindley, D.V.: The future of statistics – a Bayesian 21st century. Supp. Adv. Appl. Prob. **7**, 106–115 (1975)
4. Murteira, B.J.: Estatística: Inferência e Decisão. Imprensa Nacional Casa da Moeda, Lisboa (1988)
5. Rao, C.R., Székely, G.J.: Statistics for the 21st Century. Methodologies for Applications of the Future. Marcel Dekker, New York (2000)
6. Rosado, F. (ed.): Memorial da Sociedade Portuguesa de Estatística. Edições SPE, Lisboa (2005)
7. Rosado, F.: Outliers em Dados Estatísticos. Edições SPE, Lisboa (2006)

Resampling Methodologies in the Field of Statistics of Univariate Extremes

M. Ivette Gomes

Abstract

In the field of statistics of univariate extremes, we deal with the importance of resampling methodologies, such as the generalised jackknife and the bootstrap in the derivation of a reliable semi-parametric estimate of a parameter of extreme or even rare events. Among those parameters, we can refer high quantiles, expected shortfalls, return periods of high levels or the primary parameter of extreme events, the extreme value index (EVI), the parameter considered in this article. In order to illustrate such topics, we consider minimum-variance reduced-bias estimators of a positive EVI.

1 Extreme Value Theory: A Brief Introduction

We use the notation γ for the extreme value index (EVI), the shape parameter in the extreme value distribution function (d.f.),

$$EV_\gamma(x) = \begin{cases} \exp\{-(1+\gamma x)^{-1/\gamma}\}, \ 1+\gamma x > 0 \text{ if } \gamma \neq 0 \\ \exp\{-\exp(-x)\}, \ x \in \mathbb{R} \qquad \text{if } \gamma = 0, \end{cases} \tag{1}$$

and we consider models with a heavy right-tail. Note that in the area of statistics of extremes, and with the notation RV_a standing for the class of regularly varying functions at infinity with an index of regular variation equal to $a \in \mathbb{R}$, i.e. positive measurable functions $g(\cdot)$ such that for any $x > 0$, $g(tx)/g(t) \to x^a$, as $t \to \infty$ (see [3], for details on regular variation), we usually say that a model F has a heavy right-tail $\overline{F} := 1 - F$ whenever $\overline{F} \in RV_{-1/\gamma}$, for some $\gamma > 0$. Then, as first proved in [14], F is in the domain of attraction for maxima of a Fréchet-type d.f., the EV_γ d.f.

M.I. Gomes (✉)

CEAUL and DEIO, FCUL, Universidade de Lisboa, Campo Grande, 1749-016 Lisboa, Portugal

e-mail: ivette.gomes@fc.ul.pt

A. Pacheco et al. (eds.), *New Advances in Statistical Modeling and Applications*, Studies in Theoretical and Applied Statistics, DOI 10.1007/978-3-319-05323-3_3, © Springer International Publishing Switzerland 2014

in (1), but with $\gamma > 0$, and we use the notation $F \in \mathscr{D}_M(EV_{\gamma>0}) =: \mathscr{D}_M^+$. This means that given a sequence $\{X_n\}_{n\geq 1}$ of independent and identically distributed random variables (r.v.'s), it is possible to normalise the sequence of maximum values, $\{X_{n:n} := \max(X_1, \ldots, X_n)\}_{n\geq 1}$ so that it converges weakly to an r.v. with the d.f. EV_γ, with $\gamma > 0$.

In this same context of heavy right-tails, and with the notation $U(t) := F^{\leftarrow}(1 - 1/t), t \geq 1$, being $F^{\leftarrow}(y) := \inf\{x : F(x) \geq y\}$ the generalised inverse function of F, we can further say that

$$F \in \mathscr{D}_M^+ \iff \overline{F} \in RV_{-1/\gamma} \iff U \in RV_\gamma, \tag{2}$$

the so-called first-order condition. The second equivalence in (2), $F \in \mathscr{D}_M^+$ if and only if $U \in RV_\gamma$, was first derived in [7].

For a consistent semi-parametric EVI-estimation, in the whole $\mathscr{D}_M^+$, we merely need to assume the validity of the first-order condition, in (2), and to work with adequate functionals, dependent on an intermediate tuning parameter k, related to the number of top order statistics involved in the estimation. To say that k is intermediate is equivalent to say that

$$k = k_n \to \infty \quad \text{and} \quad k_n = o(n), \text{ i.e. } k/n \to 0, \text{ as } n \to \infty. \tag{3}$$

To obtain information on the non-degenerate asymptotic behaviour of semi-parametric EVI-estimators, we further need to work in $\mathscr{D}_{M|2}^+$, assuming a second-order condition, ruling the rate of convergence in the first-order condition, in (2). The second-order parameter $\rho(\leq 0)$ rules such a rate of convergence, and it is the parameter appearing in the limiting result,

$$\lim_{t \to \infty} \frac{\ln U(tx) - \ln U(t) - \gamma \ln x}{A(t)} = \begin{cases} \frac{x^\rho - 1}{\rho} & \text{if } \rho < 0 \\ \ln x & \text{if } \rho = 0, \end{cases} \tag{4}$$

which we often assume to hold for all $x > 0$, and where $|A|$ must be in RV_ρ [13]. For technical simplicity, we usually further assume that $\rho < 0$, and use the parameterisation

$$A(t) =: \gamma \beta t^\rho. \tag{5}$$

We are then working with a class of Pareto-type models, with a right-tail function

$$\overline{F}(x) = Cx^{-1/\gamma}\left(1 + D_1 x^{\rho/\gamma} + o\left(x^{\rho/\gamma}\right)\right), \tag{6}$$

as $x \to \infty$, with $C > 0$, $D_1 \neq 0$ and $\rho < 0$.

In order to obtain full information on the asymptotic bias of corrected-bias EVI-estimators, it is further necessary to work in $\mathscr{D}_{M|3}^+$, assuming a general third-order condition, which guarantees that, for all $x > 0$,

$$\lim_{t\to\infty} \frac{\frac{\ln U(tx) - \ln U(t) - \gamma \ln x}{A(t)} - \frac{x^\rho - 1}{\rho}}{B(t)} = \frac{x^{\rho+\rho'} - 1}{\rho + \rho'}, \tag{7}$$

where $|B|$ must then be in $RV_{\rho'}$. More restrictively, and equivalently to the aforementioned third-order condition, in (7), but with $\rho = \rho' < 0$, we often consider a Pareto third-order condition, i.e. a class of Pareto-type models, with a right-tail function

$$\overline{F}(x) = C x^{-1/\gamma}\left(1 + D_1 x^{\rho/\gamma} + D_2 x^{2\rho/\gamma} + o\left(x^{2\rho/\gamma}\right)\right),$$

as $x \to \infty$, with $C > 0, D_1, D_2 \neq 0$ and $\rho < 0$, a large sub-class of the classes of models in [26,27]. Then we can choose in the general third-order condition, in (7),

$$B(t) = \beta' t^\rho = \frac{\beta' A(t)}{\beta \gamma} =: \frac{\xi A(t)}{\gamma}, \quad \beta, \beta' \neq 0, \quad \xi = \frac{\beta'}{\beta}, \tag{8}$$

with β and β' "scale" second and third-order parameters, respectively.

2 EVI-Estimators Under Consideration

For models in $\mathcal{D}_M^+$, the classical EVI-estimators are the Hill estimators [28], averages of the scaled log-spacings or of the log-excesses, given by

$$U_i := i\left\{\ln \frac{X_{n-i+1:n}}{X_{n-i:n}}\right\} \quad \text{and} \quad V_{ik} := \ln \frac{X_{n-i+1:n}}{X_{n-k:n}}, \quad 1 \leq i \leq k < n,$$

respectively. We thus have

$$H(k) \equiv H_n(k) := \frac{1}{k}\sum_{i=1}^{k} U_i = \frac{1}{k}\sum_{i=1}^{k} V_{ik}, \quad 1 \leq k < n. \tag{9}$$

But these EVI-estimators have often a strong asymptotic bias for moderate up to large values of k, of the order of $A(n/k)$, with $A(\cdot)$ the function in (4). More precisely, for intermediate k, i.e. if (3) holds, and under the validity of the general second-order condition in (4), $\sqrt{k}\,(H(k) - \gamma)$ is asymptotically normal with variance γ^2 and a non-null mean value, equal to $\lambda/(1 - \rho)$, whenever $\sqrt{k}\,A(n/k) \to \lambda \neq 0$, finite, the type of k-values which lead to minimal mean square error (MSE). Indeed, it follows from the results in [8] that under the second-order condition in (4), and with the notation $\mathcal{N}(\mu, \sigma^2)$ standing for a normal r.v. with mean μ and variance σ^2,

$$\sqrt{k}\,(H(k) - \gamma) \stackrel{d}{=} \mathcal{N}(0, \sigma_H^2) + b_H \sqrt{k}\,A(n/k) + o_p\left(\sqrt{k}\,A(n/k)\right),$$

where $\sigma_H^2 = \gamma^2$, and the bias $b_H \sqrt{k} A(n/k)$, equal to $\gamma \beta \sqrt{k} (n/k)^\rho/(1-\rho)$, whenever (5) holds, can be very large, moderate or small (i.e. go to infinity, constant or zero) as $n \to \infty$. This non-null asymptotic bias, together with a rate of convergence of the order of $1/\sqrt{k}$, leads to sample paths with a high variance for small k, a high bias for large k, and a very sharp MSE pattern, as a function of k. The optimal k-value for the EVI-estimation through the Hill estimator, i.e. $k_{0|H} :=$ $\arg\min_k \text{MSE}(H(k))$, is well approximated by $k_{A|H} := \arg\min_k \text{AMSE}(H(k))$, with AMSE standing for asymptotic MSE, defined by

$$\text{AMSE}(H(k)) = \frac{\gamma^2}{k} + b_H^2 A^2(n/k) =: \text{AVAR}(k) + \text{ABIAS}^2(k),$$

with AVAR and ABIAS standing for asymptotic variance and asymptotic bias. Then, we can easily see that $k_{0|H}$ is of the order of $n^{-2\rho/(1-2\rho)}$ due to the fact that

$$k_{A|H} = \arg\min_k \left\{ \frac{1}{k} + b_H^2 \beta^2 (n/k)^{2\rho} \right\} = \left(\frac{n^{-2\rho}}{\beta^2(-2\rho)(1-\rho)^{-2}} \right)^{1/(1-2\rho)}.$$

The adequate accommodation of this bias has recently been extensively addressed. We mention the pioneering papers [1, 11, 18, 29], among others. In these papers, authors are led to second-order reduced-bias (SORB) EVI-estimators, with asymptotic variances larger than or equal to $(\gamma (1-\rho)/\rho)^2$, where $\rho(< 0)$ is the aforementioned "shape" second-order parameter, in (4). Recently, the authors in [4, 19, 21] considered, in different ways, the problem of corrected-bias EVI-estimation, being able to reduce the bias without increasing the asymptotic variance, which was shown to be kept at γ^2, the asymptotic variance of Hill's estimator. Those estimators, called minimum-variance reduced-bias (MVRB) EVI-estimators, are all based on an adequate "external" consistent estimation of the pair of second-order parameters, $(\beta, \rho) \in (\mathbb{R}, \mathbb{R}^-)$, done through estimators denoted by $(\hat{\beta}, \hat{\rho})$. For algorithms related to such estimation, see [17]. The estimation of β has been done through the class of estimators in [15]. The estimation of ρ has been usually performed though the simplest class of estimators in [12].

We now consider the simplest class of MVRB EVI-estimators in [4], defined as

$$\overline{H}(k) \equiv \overline{H}_{\hat{\beta},\hat{\rho}}(k) := H(k)\left(1 - \frac{\hat{\beta}}{1-\hat{\rho}} \left(\frac{n}{k}\right)^{\hat{\rho}}\right). \tag{10}$$

Under the same conditions as before, i.e. if as $n \to \infty$, $\sqrt{k} A(n/k) \to \lambda$, finite, possibly non-null, $\sqrt{k} (\overline{H}(k) - \gamma)$ is asymptotically normal with variance also equal to γ^2 but with a null mean value. Indeed, from the results in [4], we know that it is possible to adequately estimate the second-order parameters β and ρ, so that we get

$$\sqrt{k} (\overline{H}(k) - \gamma) \overset{d}{=} \mathcal{N}(0, \gamma^2) + o_p\left(\sqrt{k} A(n/k)\right).$$

Fig. 1 Typical patterns of
variance, bias and MSE of H
and $\overline{H}$, as a function of the
sample fraction $r = k/n$

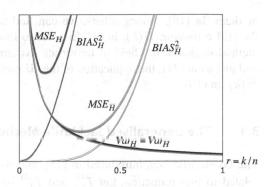

Consequently, $\overline{H}(k)$ outperforms $H(k)$ for all k. Indeed, under the validity of the aforementioned third-order condition related to the class of Pareto-type models, we can then adequately estimate the vector of second-order parameters, (β, ρ), and write [5],

$$\sqrt{k}\left(\overline{H}(k) - \gamma\right) \overset{d}{=} \mathcal{N}(0, \gamma^2) + b_{\overline{H}}\sqrt{k}\, A^2(n/k) + o_p\left(\sqrt{k}\, A^2(n/k)\right),$$

where, with ξ defined in (8), $b_{\overline{H}} = \left(\xi/(1 - 2\rho) - 1/(1 - \rho)^2\right)/\gamma$.

In Fig. 1 we picture the comparative behaviour of the bias, variance and MSE of H and $\overline{H}$, in (9) and (10), respectively.

Now, $k_{0|\overline{H}} := \arg\min_k \mathrm{MSE}(\overline{H}(k))$ can be asymptotically approximated by $k_{A|\overline{H}} = \left(n^{-4\rho}/\left(\beta^2(-2\rho)b_{\overline{H}}^2\right)\right)^{1/(1-4\rho)}$, i.e. $k_{0|\overline{H}}$ is of the order of $n^{-4\rho/(1-4\rho)}$, and depends not only on (β, ρ), as does $k_{0|H}$, but also on (γ, ξ). Recent reviews on extreme value theory and statistics of univariate extremes can be found in [2,20,31].

3   Resampling Methodologies

The use of resampling methodologies (see [10]) has revealed to be promising in the estimation of the tuning parameter k, and in the reduction of bias of any estimator of a parameter of extreme events. For a recent review on the subject, see [30].

If we ask how to choose k in the EVI-estimation, either through $H(k)$ or through $\overline{H}(k)$, we usually consider the estimation of $k_{0|H} := \arg\min_k \mathrm{MSE}(H(k))$ or $k_{0|\overline{H}} = \arg\min_k \mathrm{MSE}(\overline{H}(k))$. To obtain estimates of $k_{0|H}$ and $k_{0|\overline{H}}$ one can then use a double-bootstrap method applied to an adequate auxiliary statistic which tends to be **zero** and has an asymptotic behaviour similar to either $H(k)$ (see [6, 9, 16], among others) or $\overline{H}(k)$ (see [22, 23], also among others). Such a double-bootstrap method will be sketched in Sect. 3.2.

But at such optimal levels, we still have a non-null asymptotic bias. If we still want to remove such a bias, we can make use of the generalised jackknife (GJ). It is then enough to consider an adequate pair of estimators of the parameter of extreme events under consideration and to build a reduced-bias affine combination of them. In [18], among others, we can find an application of this technique to the Hill estimator, $H(k)$, in (9). In order to illustrate the use of these resampling methodologies in the field of univariate extremes, we shall consider, in Sect. 3.1 and just as in [24], the application of the GJ methodology to the MVRB estimators $\overline{H}(k)$, in (10).

3.1   The Generalised Jackknife Methodology and Bias Reduction

The GJ-statistic was introduced in [25], and the main objective of the method is related to bias reduction. Let $T_n^{(1)}$ and $T_n^{(2)}$ be two biased estimators of γ, with similar bias properties, i.e. $\text{Bias}(T_n^{(i)}) = \phi(\gamma)d_i(n)$, $\quad i = 1, 2$. Then, if $q = q_n = d_1(n)/d_2(n) \neq 1$, the affine combination $T_n^{GJ} := (T_n^{(1)} - qT_n^{(2)})/(1-q)$ is an unbiased estimator of γ.

Given $\overline{H}$, and with $\lfloor x \rfloor$ denoting the integer part of x, the most natural GJ r.v. is the one associated with the random pair $(\overline{H}(k), \overline{H}(\lfloor k/2 \rfloor))$, i.e.

$$\overline{H}^{GJ(q)}(k) := \frac{\overline{H}(k) - q\,\overline{H}(\lfloor k/2 \rfloor)}{1-q}, \quad q > 0,$$

with

$$q = q_n = \frac{\text{ABIAS}\left(\overline{H}(k)\right)}{\text{ABIAS}\left(\overline{H}(\lfloor k/2 \rfloor)\right)} = \frac{A^2(n/k)}{A^2(n/\lfloor k/2 \rfloor)} \xrightarrow[n/k \to \infty]{} 2^{-2\rho}.$$

It is thus sensible to consider $q = 2^{-2\rho}$, and, with $\hat{\rho}$ a consistent estimator of ρ, the approximate GJ estimator,

$$\overline{H}^{GJ}(k) := \frac{2^{2\hat{\rho}}\,\overline{H}(k) - \overline{H}(\lfloor k/2 \rfloor)}{2^{2\hat{\rho}} - 1}. \tag{11}$$

Then, and provided that $\hat{\rho} - \rho = o_p(1)$,

$$\sqrt{k}\left(\overline{H}^{GJ}(k) - \gamma\right) \overset{d}{=} \mathcal{N}(0, \sigma_{GJ}^2) + o_p(\sqrt{k}\,A^2(n/k)),$$

with $\sigma_{GJ}^2 = \gamma^2(1 + 1/(2^{-2\rho} - 1)^2)$. Further details on the estimators in (11) can be found in [24]. As expected, we have again a trade-off between variance and bias. The bias decreases, but the variance increases, and to try solving such a trade-off, an adequate estimation of third-order parameters, still an almost open topic of research in the area of statistics of extremes, would be needed. Anyway, at optimal levels, $\overline{H}^{GJ}$ can outperform $\overline{H}$, as it is theoretically illustrated in Fig. 2.

A Monte-Carlo simulation of the mean value (E) and the root MSE (RMSE) of the estimators under consideration have revealed similar patterns. On the basis of

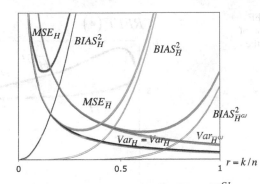

Fig. 2 Typical patterns of variance, bias and MSE of H, $\overline{H}$ and $\overline{H}^{GJ}$, as a function of the sample fraction $r = k/n$

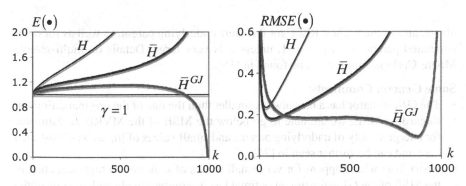

Fig. 3 Simulated mean values (*left*) and RMSEs (*right*) of the estimators under study, for a sample of size $n = 1,000$ from an underlying Burr(γ, ρ) model, with $(\gamma, \rho) = (1, -0.5)$

5,000 runs, and for a Burr(γ, ρ) parent, with d.f. $F(x) = 1 - (1 + x^{-\rho/\gamma})^{1/\rho}$, $x \geq 0$, with $\gamma = 1$ and $\rho = -0.5$, we present Fig. 3, as an illustration of the results obtained for different underlying parents and different sample sizes.

As usual, we define the relative efficiency of any EVI-estimator as the quotient between the simulated RMSE of the H-estimator and the one of any of the estimators under study, both computed at their optimal levels, i.e. for any T-statistic, consistent for the EVI-estimation,

$$\text{REFF}_{T_0|H_0} := \frac{\text{RMSE}(H_0)}{\text{RMSE}(T_0)},$$

with $T_0 := T(k_{0|T})$ and $k_{0|T} := \arg\min_k \text{MSE}(T(k))$.

The simulation of those efficiencies for the same Burr model is based on $20 \times 5,000$ replicates and, as shown in Fig. 4, the REFF-indicators as a function of n, are always larger than one, both for $\overline{H}$, in (10) and for $\overline{H}^{GJ}$, in (11). Moreover, $\overline{H}^{GJ}$, computed at its optimal level, in the sense of minimal MSE, just as mentioned

Fig. 4 Simulated REFF indicators, as a function of the sample size n, for the same Burr parent

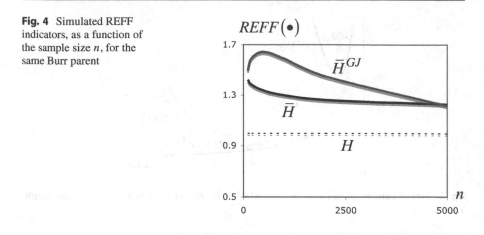

above, attains the highest REFF for this Burr underlying parent, as well as for other simulated parents with $\rho > -1$, unless n is very large. Details on multi-sample Monte-Carlo simulation can be found in [16].

Some General Comments:

- The GJ-estimator has a bias always smaller than the one of the original estimator.
- Regarding MSE, we are able to go below the MSE of the MVRB $\overline{H}$-estimator for a large variety of underlying parents and small values of $|\rho|$, as was illustrated here and can be further seen in [24].
- Apart from what happens for very small values of ρ, there is a high reduction in the MSE of the GJ-estimator, at optimal levels, comparatively with the one of the original $\overline{H}$-estimator, despite the already nice properties of the $\overline{H}$ EVI-estimator.

3.2   The Bootstrap Methodology for the Estimation of Sample Fractions

As already mentioned in Sect. 2,

$$k_{A|\overline{H}}(n) = \arg\min_k \mathrm{AMSE}(\overline{H}(k)) = \arg\min_k \left(\frac{\gamma^2}{k} + b_{\overline{H}}^2 \, A^4(n/k) \right)$$

$$= k_{0|\overline{H}}(n)(1 + o(1)).$$

The bootstrap methodology enables us to estimate the optimal sample fraction, $k_{0|\overline{H}}(n)/n$ in a way similar to the one used for the classical EVI estimation, in [6, 9, 16], now through the use of any auxiliary statistic, such as

$$T_n(k) \equiv T_n^{\overline{H}}(k) := \overline{H}(\lfloor k/2 \rfloor) - \overline{H}(k), \quad k = 2, \dots, n - 1,$$

which converges in probability to the known value **zero**, for intermediate k. Moreover, under the third-order framework, in (7), we get:

$$T_n(k) \stackrel{d}{=} \frac{\gamma \, P_k}{\sqrt{k}} + b_{\overline{H}} \, (2^{2\rho} - 1) \, A^2(n/k) + O_p\big(A(n/k)/\sqrt{k}\big),$$

with P_k asymptotically standard normal. The AMSE of $T_n(k)$ is thus minimal at a level k such that $\sqrt{k} \, A^2(n/k) \to \lambda'_A \neq 0$. Consequently, denoting

$$k_{A|T}(n) := \arg\min_k \text{AMSE}\big(T_n^{\overline{H}}(k)\big) = k_{0|T}(1 + o(1)),$$

we have

$$k_{0|\overline{H}}(n) = k_{0|T}(n) \, \big(1 - 2^{2\rho}\big)^{\frac{2}{1-4\rho}} \, (1 + o(1)). \qquad (12)$$

Note also that, with the adequate simple modifications, a similar comment applies to the GJ EVI-estimator $\overline{H}^{GJ}(k)$, in (11).

Given the sample $\underline{X}_n = (X_1, \ldots, X_n)$ from an unknown model F, and the functional $T_n(k) =: \phi_k(\underline{X}_n)$, $1 \leq k < n$, consider for any $n_1 = O(n^{1-\epsilon})$, $0 < \epsilon < 1$, the bootstrap sample $\underline{X}_{n_1}^* = (X_1^*, \ldots, X_{n_1}^*)$, from the model $F_n^*(x) = \sum_{i=1}^n I_{[X_i \leq x]}/n$, the empirical d.f. associated with our sample $\underline{X}_n$. Next, consider $T_{n_1}^*(k_1) := \phi_{k_1}(\underline{X}_{n_1}^*), 1 < k_1 < n_1$. Then, with $k_{0|T}^*(n_1) = \arg\min_{k_1} \text{MSE}\big(T_{n_1}^*(k_1)\big)$,

$$k_{0|T}^*(n_1)/k_{0|T}(n) = (n_1/n)^{\frac{4\rho}{1-4\rho}} \, (1 + o(1)), \quad \text{as } n \to \infty.$$

To get a simpler way of computing $k_{0|T}(n)$ it is then sensible to use a double bootstrap, based on another sample size n_2. Then for every $\alpha > 1$,

$$\frac{\big(k_{0|T}^*(n_1)\big)^\alpha}{k_{0|T}^*(n_2)} \left(\frac{n_1^\alpha}{n^\alpha} \frac{n}{n_2}\right)^{-\frac{4\rho}{1-4\rho}} = \{k_{0|T}(n)\}^{\alpha-1} \, (1 + o(1)).$$

It is then enough to choose $n_2 = \lfloor n \left(\frac{n_1}{n}\right)^\alpha \rfloor$, in order to have independence of ρ. If we put $n_2 = \lfloor n_1^2/n \rfloor$, i.e. $\alpha = 2$, we have

$$\big(k_{0|T}^*(n_1)\big)^2/k_{0|T}^*(n_2) = k_{0|T}(n)(1 + o(1)),$$

and the possibility of estimating $k_{0|T}(n)$ on the basis of $k_{0|T}^*(n_1)$ and $k_{0|T}^*(n_2)$ only. We are next able to estimate $k_{0|\overline{H}}(n)$, on the basis of (12) and any estimate $\hat{\rho}$ of the second-order parameter ρ. Then, with $\hat{k}_{0|T}^*$ denoting the sample counterpart of $k_{0|T}^*$, we have the estimate

$$\hat{k}_{0|\overline{H}}^*(n; n_1) := \min \left(n - 1, \left\lfloor \frac{c_{\hat{\rho}} \, (\hat{k}_{0|T}^*(n_1))^2}{\hat{k}_{0|T}^*(\lfloor n_1^2/n \rfloor + 1)} \right\rfloor + 1\right), \quad c_{\hat{\rho}} = \big(1 - 2^{2\hat{\rho}}\big)^{\frac{2}{1-4\hat{\rho}}}.$$

Fig. 5 Sample paths of the
EVI-estimators under study
and bootstrap estimates of the
$k_{0|\bullet}$-values, for a Burr
random sample with $\gamma = 1$
and $\rho = -0.5$

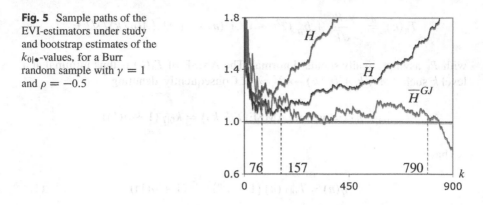

The final estimate of γ is then given by $\overline{H}^* \equiv \overline{H}^*_{n,n_1|T} := \overline{H}_{\hat{\beta},\hat{\rho}}(\hat{k}_{0|\overline{H}}(n; n_1))$. And
a similar procedure can be used to estimate any other parameter of extreme events,
as well as the EVI, either through H or through $\overline{H}^{GJ}$.

The application of the associated bootstrap algorithm, with $n_1 = n^{0.975}$ and
$B = 250$ generations, to the first randomly generated Burr(γ, ρ) sample of size
$n = 1,000$, with $\gamma = 1$ and $\rho = -0.5$ led us to $\hat{k}^*_{0|H} = 76$, $\hat{k}^*_{0|\overline{H}} = 157$, and
$\hat{k}^*_{0|\overline{H}^{GJ}} = 790$. The bootstrap EVI-estimates were $H^* = 1.259$, $\overline{H}^* = 1.108$ and
$\overline{H}^{GJ*} = 1.049$, a value indeed closer to the target value $\gamma = 1$. In Fig. 5 we present
the sample paths of the EVI-estimators under study.

4   Concluding Remarks

A few practical questions and final remarks can now be raised.

- How does the asymptotic method work for moderate sample sizes? Is the method
  strongly dependent on the choice of n_1? Although aware of the theoretical need
  of $n_1 = o(n)$, what happens if we choose $n_1 = n - 1$? Answers to these
  questions have not yet been fully given for the class of GJ EVI-estimators, in (11),
  but will surely be similar to the ones given for classical estimation and for the
  MVRB estimation. Usually, the method does not depend strongly on n_1 and
  practically we can choose $n_1 = n - 1$. And here we can mention again the
  old controversy between theoreticians and practioners: The value $n_1 = \lfloor n^{1-\epsilon} \rfloor$
  can be equal to $n - 1$ for small ϵ and a large variety of values of n, finite. Also,
  $k_n = \lceil c \ln n \rceil$ is intermediate for every constant c, and if we take, for instance,
  $c = 1/5$, we get $k_n = 1$ for every $n \leq 22,026$. And Hall's formula of the
  asymptotically optimal level for the Hill EVI-estimation (see [26]), given by
  $k_{0|H}(n) = \lfloor ((1 - \rho)^2 n^{-2\rho} / (-2 \rho \beta^2))^{1/(1-2\rho)} \rfloor$ and valid for models in (6),
  may lead, for a fixed n, and for several choices of (β, ρ), to $k_{0|H}(n)$ either equal
  to 1 or to $n - 1$ according as ρ is close to 0 or quite small, respectively.

- Note that bootstrap confidence intervals as well as asymptotic confidence intervals are easily associated with the estimates presented, and the smallest size (with a high coverage probability) is usually related to the EVI-estimator $\overline{H}^{GJ}$, in (11), as expected.

Acknowledgements Research partially supported by National Funds through **FCT**—Fundação para a Ciência e a Tecnologia, project PEst-OE/MAT/UI0006/2011, and PTDC/FEDER, EXTREMA.

References

1. Beirlant, J., Dierckx, G., Goegebeur, Y., Matthys, G.: Tail index estimation and an exponential regression model. Extremes **2**, 177–200 (1999)
2. Beirlant, J., Caeiro, F., Gomes, M.I.: An overview and open researh topics in statistics of univariate extremes. Revstat **10**(1), 1–31 (2012)
3. Bingham, N., Goldie, C.M., Teugels, J.L.: Regular Variation. Cambridge University Press, Cambridge (1987)
4. Caeiro, F., Gomes, M.I., Pestana, D.: Direct reduction of bias of the classical Hill estimator. Revstat **3**(2), 111–136 (2005)
5. Caeiro, F., Gomes, M.I., Henriques-Rodrigues, L.: Reduced-bias tail index estimators under a third order framework. Commun. Stat. Theory Methods **38**(7), 1019–1040 (2009)
6. Danielsson, J., de Haan, L., Peng, L., de Vries, C.G.: Using a bootstrap method to choose the sample fraction in the tail index estimation. J. Multivariate Anal. **76**, 226–248 (2001)
7. de Haan, L.: Slow variation and characterization of domains of attraction. In: Tiago de Oliveira, J. (ed.) Statistical Extremes and Applications, pp. 31–48. D. Reidel, Dordrecht (1984)
8. de Haan, L., Peng, L.: Comparison of tail index estimators. Stat. Neerl. **52**, 60–70 (1998)
9. Draisma, G., de Haan, L., Peng, L., Pereira, M.T.: A bootstrap-based method to achieve optimality in estimating the extreme value index. Extremes **2**(4), 367–404 (1999)
10. Efron, B.: Bootstrap methods: another look at the jackknife. Ann. Stat. **7**(1), 1–26 (1979)
11. Feuerverger, A., Hall, P.: Estimating a tail exponent by modelling departure from a Pareto distribution. Ann. Stat. **27**, 760–781 (1999)
12. Fraga Alves, M.I., Gomes, M.I., de Haan, L.: A new class of semi-parametric estimators of the second order parameter. Port. Math. **60**(2), 194–213 (2003)
13. Geluk, J., de Haan, L.: Regular Variation, Extensions and Tauberian Theorems. CWI Tract, vol. 40. Center for Mathematics and Computer Science, Amsterdam (1987)
14. Gnedenko, B.V.: Sur la distribution limite du terme maximum d'une série aléatoire. Ann. Math. **44**, 423–453 (1943)
15. Gomes, M.I., Martins, M.J.: "Asymptotically unbiased" estimators of the tail index based on external estimation of the second order parameter. Extremes **5**(1), 5–31 (2002)
16. Gomes, M.I., Oliveira, O.: The bootstrap methodology in Statistical Extremes – choice of the optimal sample fraction. Extremes **4**(4), 331–358 (2001)
17. Gomes, M.I., Pestana, D.: A sturdy reduced-bias extreme quantile (VaR) estimator. J. Am. Stat. Assoc. **102**(477), 280–292 (2007)
18. Gomes, M.I., Martins, M.J., Neves, M.: Alternatives to a semi-parametric estimator of parameters of rare events: the Jackknife methodology. Extremes **3**(3), 207–229 (2000)
19. Gomes, M.I., Martins, M.J., Neves, M.: Improving second order reduced bias extreme value index estimation. Revstat **5**(2), 177–207 (2007)
20. Gomes, M.I., Canto e Castro, L., Fraga Alves, M.I., Pestana, D.: Statistics of extremes for iid data and breakthroughs in the estimation of the extreme value index: Laurens de Haan leading contributions. Extremes **11**, 3–34 (2008)

21. Gomes, M.I., de Haan, L., Henriques-Rodrigues, L.: Tail index estimation for heavy-tailed models: accommodation of bias in weighted log-excesses. J. R. Stat. Soc. **B70**(1), 31–52 (2008)
22. Gomes, M.I., Mendonça, S., Pestana, D.: Adaptive reduced-bias tail index and VaR estimation via the bootstrap methodology. Commun. Stat. Theory Methods **40**, 2946–2968 (2011)
23. Gomes, M.I., Figueiredo, F., Neves, M.M.: Adaptive estimation of heavy right tails: the bootstrap methodology in action. Extremes (2012). doi:10.1007/s10687-011-0146-6
24. Gomes, M.I., Martins, M.J., Neves, M.M.: Generalized Jackknife-based estimators for univariate extreme-value modelling. Commun. Stat. Theory Methods **42**(7), 1227–1245 (2013)
25. Gray, H.L., Schucany, W.R.: The Generalized Jackknife Statistic. Marcel Dekker, New York (1972)
26. Hall, P.: On some simple estimates of an exponent of regular variation. J. R. Stat. Soc. B **44**, 37–42 (1982)
27. Hall, P., Welsh, A.W.: Adaptive estimates of parameters of regular variation. Ann. Stat. **13**, 331–341 (1985)
28. Hill, B.M.: A simple general approach to inference about the tail of a distribution. Ann. Stat. **3**, 1163–1174 (1975)
29. Peng, L.: Asymptotically unbiased estimator for the extreme-value index. Stat. Prob. Lett. **38**(2), 107–115 (1998)
30. Qi, Y.: Bootstrap and empirical likelihood methods in extremes. Extremes **11**, 81–97 (2008)
31. Scarrott, C.J., MacDonald, A.: A review of extreme value threshold estimation and uncertainty quantification. Revstat **10**(1), 33–60 (2012)

Robust Functional Principal Component Analysis

Juan Lucas Bali and Graciela Boente

Abstract

When dealing with multivariate data robust principal component analysis (PCA), like classical PCA, searches for directions with maximal dispersion of the data projected on it. Instead of using the variance as a measure of dispersion, a robust scale estimator s_n may be used in the maximization problem. In this paper, we review some of the proposed approaches to robust functional PCA including one which adapts the projection pursuit approach to the functional data setting.

1 Introduction

Functional data analysis provides modern analytical tools for data that are recoded as images or as a continuous phenomenon over a period of time. Because of the intrinsic nature of these data, they can be viewed as realizations of random functions often assumed to be in $L^2(\mathscr{I})$, with $\mathscr{I}$ a real interval or a finite dimensional Euclidean set.

Principal component analysis (PCA) is a standard technique in the context of multivariate analysis as a dimension-reduction technique. The goal is to search for directions with maximal dispersion of the data projected on it. The classical estimators are obtained taking as dispersion the sample variance leading to estimators which are sensitive to atypical observations. To overcome this problem, [16] proposed a procedure based on the principles of projection-pursuit to define the estimator of the first direction as

$$\hat{\mathbf{a}} = \mathrm{argmax}_{\mathbf{a}:\|\mathbf{a}\|=1} s_n(\mathbf{a}^{\mathrm{T}}\mathbf{x}_1, \cdots, \mathbf{a}^{\mathrm{T}}\mathbf{x}_n)$$

J.L. Bali (✉) • G. Boente
Facultad de Ciencias Exactas y Naturales, Universidad de Buenos Aires and CONICET, Buenos Aires, Argentina
e-mail: lbali@dm.uba.ar; gboente@dm.uba.ar

A. Pacheco et al. (eds.), *New Advances in Statistical Modeling and Applications*,
Studies in Theoretical and Applied Statistics, DOI 10.1007/978-3-319-05323-3_4,
© Springer International Publishing Switzerland 2014

where $\mathbf{x}_1, \ldots, \mathbf{x}_n$ are i.i.d. $\mathbf{x}_i \in \mathbb{R}^p$ and s_n is a robust scale estimator. The subsequent loading vectors are then obtained by imposing orthogonality conditions. When dealing with high dimensional data, the projection pursuit approach is preferable to the plug-in approach that estimates the principal components as the eigenvectors of a robust estimator of the covariance matrix. Effectively, as pointed out by Tyler [24], when the dimension is larger than the sample size, the only affine equivariant multivariate location statistic is the sample mean vector and any affine equivariant scatter matrix must be proportional to the sample covariance matrix, with the proportionality constant not being dependent on the data. Hence, in that case, any affine equivariant scatter estimator loses its robustness, so that most commonly used robust scatter estimators should be avoided for high dimensional data and projection methods become useful. Croux and Ruiz-Gazen [6] derived the influence functions of the projection-pursuit principal components, while their asymptotic distribution was studied in [7]. A maximization algorithm for obtaining $\hat{\mathbf{a}}$ was proposed in [5] and adapted for high dimensional data in [4].

When dealing with functional data, an approach to functional PCA (FPCA) is to consider the eigenvalues and eigenfunctions of the sample covariance operator. In a very general setting, [8] studied their asymptotic properties. However, this approach may produce rough principal components and in some situations, smooth ones may be preferable. One argument in favour of smoothed principal components is that smoothing might reveal more interpretable and interesting features of the modes of variation for functional data. To provide smooth estimators, [3] considered a kernel approach by regularizing the trajectories. A different approach was proposed by Rice and Silverman [20] and studied by Pezzulli and Silverman [19]. It consists in imposing an additive roughness penalty to the sample variance. On the other hand, [22] considered estimators based on penalizing the norm rather than the sample variance. More recent work on estimation of the principal components and the covariance function includes [12, 13, 25].

Not much work has been done in the area of robust functional data analysis. Of course, when $X \in L^2(\mathscr{I})$, it is always possible to reduce the functional problem to a multivariate one by evaluating the observations on a common output grid or by using the coefficients of a basis expansion, as in [17]. However, as mentioned by Gervini [9] discretizing the problem has several disadvantages which include the choice of the robust scatter estimators when the size of the grid is larger than the number of trajectories, as discussed above, the selection of the grid and the reconstruction of the functional estimators from the values over the grid. Besides, the theoretical properties of these procedures are not studied yet and they may produce an avoidable smoothing bias see, for instance, [26]. For this reason a fully functional approach to the problem is preferable. To avoid unnecessary smoothing steps, Gervini [9] considered a functional version of the estimators defined in [17] and derived their consistency and influence function. Also, [10] developed robust functional principal component estimators for sparsely and irregularly observed functional data and used it for outlier detection. Recently, [21] consider a robust approach of principal components based on a robust eigen-analysis of the coefficients of the observed data on some known basis. On the other hand, [15] gives an application of a robust projection-pursuit approach, applied to smoothed trajectories. Recently, [1] considered robust estimators of the functional principal directions using a projection-pursuit approach that may include a penalization in the scale or in the norm and derived their consistency and qualitative robustness.

In this paper, we review some notions related to robust estimation for functional data. The paper is organized as follows, Sect. 2 states some preliminary concepts and notation that will be helpful along the paper. Section 3 states the principal component problem, Sect. 4 reviews the robust proposals previously studied while a real data example is given in Sect. 5. Finally, Sect. 6 contains some final comments.

2   Preliminaries and Notation

Let us consider independent identically distributed random elements $X_1, \ldots, X_n$ in a separable Hilbert space $\mathcal{H}$ (often $L^2(\mathcal{I})$) with inner product $\langle \cdot, \cdot \rangle$ and norm $\|u\| = \langle u, u \rangle^{1/2}$ and assume that $\mathbb{E}\|X_1\|^2 < \infty$. Denote by $\mu \in \mathcal{H}$ the mean of $X \sim X_1, \mu = \mathbb{E}(X)$ and by $\Gamma_X : \mathcal{H} \to \mathcal{H}$ the covariance operator of X. Let $\otimes$ stand for the tensor product on $\mathcal{H}$, e.g., for $u, v \in \mathcal{H}$, the operator $u \otimes v : \mathcal{H} \to \mathcal{H}$ is defined as $(u \otimes v)w = \langle v, w \rangle u$. With this notation, the covariance operator Γ_X can be written as $\Gamma_X = \mathbb{E}\{(X - \mu) \otimes (X - \mu)\}$, which is just the functional version of the variance–covariance matrix in the classical multivariate analysis. The operator Γ_X is linear, self-adjoint and continuous. Moreover, it is a Hilbert–Schmidt operator having a countable number of eigenvalues, all of them real.

Let $\mathcal{F}$ denote the Hilbert space of Hilbert–Schmidt operators with inner product defined by $\langle H_1, H_2 \rangle_{\mathcal{F}} = \text{trace}(H_1 H_2) = \sum_{\ell=1}^{\infty} \langle H_1 u_\ell, H_2 u_\ell \rangle$ and norm $\|H\|_{\mathcal{F}} = \langle H, H \rangle_{\mathcal{F}}^{1/2} = \{\sum_{\ell=1}^{\infty} \|H u_\ell\|^2\}^{1/2}$, where $\{u_\ell : \ell \geq 1\}$ is any orthonormal basis of $\mathcal{H}$, while H_1, H_2 and H are Hilbert–Schmidt operators, i.e., such that $\|H\|_{\mathcal{F}} < \infty$. Choosing an orthonormal basis $\{\phi_\ell : \ell \geq 1\}$ of eigenfunctions of Γ_X related to the eigenvalues $\{\lambda_\ell : \ell \geq 1\}$ such that $\lambda_\ell \geq \lambda_{\ell+1}$, we get $\|\Gamma_X\|_{\mathcal{F}}^2 = \sum_{\ell=1}^{\infty} \lambda_\ell^2$.

The Karhunen–Loève expansion for the process leads to $X = \mu + \sum_{\ell=1}^{\infty} \lambda_\ell^{1/2} f_\ell \phi_\ell$, where the random variables $\{f_\ell : \ell \geq 1\}$ are the standardized coordinates of $X - \mu$ on the basis $\{\phi_\ell : \ell \geq 1\}$, that is, $\lambda_m^{1/2} f_m = \langle X - \mu, \phi_m \rangle$. Note that $\mathbb{E}(f_m) = 0$, while $\mathbb{E}(f_m^2) = 1$ if $\lambda_m \neq 0$, $\mathbb{E}(f_m f_s) = 0$ for $m \neq s$, since $\text{cov}(\langle u, X - \mu \rangle, \langle v, X - \mu \rangle) = \langle u, \Gamma_X v \rangle$. This expansion shows the importance of an accurate estimation of the principal components as a way to predict the observations and examine their atypicity.

3   The Problem

As in multivariate analysis, there are two major approaches to develop robust estimators of the functional principal components. The first aims at developing robust estimates of the covariance operator, which will then generate robust FPCA procedures. The second approach aims directly at robust estimates of the principal direction bypassing a robust estimate of the covariance operator. They are based, respectively, on the following properties of the principal components

*   **Property 1.** The principal component correspond to the eigenfunction of Γ_X related to the largest eigenvalues.
*   **Property 2.** The first principal component maximizes $\text{var}(\langle \alpha, X \rangle)$ over $\mathscr{S} = \{\alpha : \|\alpha\| = 1\}$. The subsequent are obtained imposing orthogonality constraints to the first ones.

Let $X_1, \cdots, X_n$, $1 \leq i \leq n$, be independent observations from $X \in \mathscr{H}$, $X \sim P$ with mean μ and covariance operator Γ_X. A natural way to estimate the covariance operators Γ_X is to consider the empirical covariance operator given by $\hat{\Gamma}_X = \sum_{j=1}^{n} (X_j - \overline{X}) \otimes (X_j - \overline{X}) / n$, where $\overline{X} = \sum_{j=1}^{n} X_j / n$. Dauxois et al. [8] proved that $\sqrt{n} \left(\hat{\Gamma}_X - \Gamma_X \right)$ converges in distribution to a zero mean Gaussian random element U of $\mathscr{F}$. Besides, they derived the asymptotic behaviour of the eigenfunctions of the empirical covariance operator, leading to a complete study on the behaviour of the classical unsmoothed estimators of the principal components. As mentioned in the Introduction, smooth estimators of the covariance operators were studied in [3] where also the asymptotic behaviour of its eigenfunctions was obtained. This approach to principal components follows the lines established by **Property 1**.

As is well known, FPCA is a data analytical tool to describe the major modes of variation of the process as a way to understand it and also to predict each curve. Once we have estimators $\hat{\phi}_\ell$ for the ℓ-th principal component, $1 \leq \ell \leq m$, one can predict each observation through $\hat{X}_i = \overline{X} + \sum_{\ell=1}^{m} \hat{\xi}_{i\ell} \hat{\phi}_\ell$, where $\hat{\xi}_{i\ell}$ are the scores of X_i in the basis of principal components, i.e., $\hat{\xi}_{i\ell} = \langle X_i - \overline{X}, \hat{\phi}_\ell \rangle$. In this sense, FPCA offers an effective way for dimension reduction.

However, FPCA based on the sample covariance operator is not robust. Hence, if one suspects that outliers may be present in the sample, robust estimators should be preferred. We recall that robust statistics seeks for reliable procedures when a small amount of atypical observations arise in the sample. In most cases, the estimators are functionals over the set of probability measures evaluated at the empirical probability measure and in this case, robustness is related to continuity of the functional with respect to the Prohorov distance.

In a functional setting influential observations may occur in several different ways. As mentioned by Locantore et al. [17] they may correspond to atypical trajectories entirely outlying, that is, with extreme values for the L^2 norm, also to isolated points within otherwise typical trajectories (corresponding to a single extreme measurement) or they can be related to an extreme on some principal components, the latter being more difficult to detect. In the functional case, these types of observations may significantly impact the empirical covariance operator, even if they are not outlying in the sense of being faraway of their centre. Detection of such observations is not easy and has been recently investigated by [23].

As an example for each type of influential observations, Fig. 1 shows $n = 100$ trajectories generated using a finite Karhunen–Loève expansion, $X_i = Z_{i1}\phi_1 + Z_{i2}\phi_2 + Z_{i3}\phi_3$ where $\phi_1(x) = \sin(4\pi x)$, $\phi_2(x) = \cos(7\pi x)$ and $\phi_3(x) =$

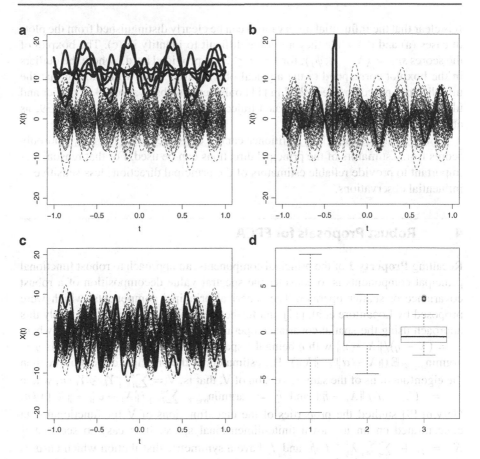

Fig. 1 Different influential trajectories with (**a**) large values on the L^2 norm, (**b**) a extreme value over a small interval and (**c**) extreme score on a principal component. Boxplot of the scores of the generated data c) over ϕ_j, $1 \leq j \leq 3$

$\cos(15\pi x)$. The uncontaminated trajectories correspond to $Z_{ij} \sim N(0, \sigma_j^2)$ with $\sigma_1 = 4$, $\sigma_2 = 2$ and $\sigma_3 = 1$, Z_{ij} independent for all $1 \leq i \leq n$ and $1 \leq j \leq 3$. The atypical observations are plotted in thick lines and they correspond in each case to

(a) add randomly to 10 % the trajectories a factor of 12,
(b) replace $X_2(t)$ by $X_2(t) + 25$ when $-0.4 < t < -0.36$,
(c) generate the random variables $Z_{i,j}$ as $Z_{i1} \sim N(0, \sigma_1^2)$,

$$\begin{pmatrix} Z_{i2} \\ Z_{i3} \end{pmatrix} \sim (1 - \epsilon)\, N\left(\begin{pmatrix} 0 \\ 0 \end{pmatrix}, \mathrm{diag}\left(\sigma_2^2, \sigma_3^2 \right) \right) + \epsilon\, N\left(\begin{pmatrix} 4 \\ 4 \end{pmatrix}, \mathrm{diag}\left(0.01, 0.01 \right) \right)$$

where $\epsilon = 0.1$, leading in this case to ten atypical observations, labelled 5, 7, 17, 32, 33, 40, 47, 69, 88 and 95.

It is clear that the influential observations can be clearly distinguished from the plots in cases (a) and (b) while they are more difficult to identify in (c). The boxplot of the scores $s_{i,j} = \langle X_i - \mu, \phi_j \rangle$, for $1 \leq j \leq 3$ are provided in (d), where the outliers in the boxplot correspond to the atypical observations. It is worth noting that the interdistance procedure described in [11] only detects observation 33 as outlier and identifies four of the uncontaminated trajectories, labelled 64, 71, 84 and 39, as atypical.

However, in practice the practitioner cannot construct the scores $s_{i,j}$ and only scores from estimators of the principal directions can be used. For that reason, it is important to provide reliable estimators of the principal directions less sensitive to influential observations.

4   Robust Proposals for FPCA

Recalling **Property 1** of the principal components, an approach to robust functional principal components is to consider the spectral value decomposition of a robust covariance or scatter operator. The spherical principal components, which were proposed by Locantore et al. [17] and further developed by Gervini [9], apply this approach using the spatial covariance operator defined as $\mathsf{V} = \mathbb{E}\,(Y \otimes Y)$, where $Y = (X - \eta)/\|X - \eta\|$ with η being the spatial median, defined in [9], that is $\eta = \mathrm{argmin}_{\alpha \in \mathscr{H}} \mathbb{E}\,(\|X - \alpha\| - \|X\|)$. The estimators of the principal directions are then the eigenfunctions of the sample version of V, that is, $\hat{\mathsf{V}} = \sum_{i=1}^{n} Y_i \otimes Y_i/n$, where $Y_i = (X_i - \hat{\eta})/\|X_i - \hat{\eta}\|$ and $\hat{\eta} = \mathrm{argmin}_{\alpha \in \mathscr{H}} \sum_{i=1}^{n} (\|X_i - \alpha\| - \|X_i\|)/n$. Gervini [9] studied the properties of the eigenfunctions of $\hat{\mathsf{V}}$ for functional data concentrated on an unknown finite-dimensional space. It is easy to see that if $X = \mu + \sum_{\ell=1}^{\infty} \lambda_\ell^{1/2} f_\ell \phi_\ell$ and f_ℓ have a symmetric distribution which ensures that $\eta = \mu$, then, the functional spherical principal components estimate the true directions since V has the same eigenfuntions as γ. Indeed, $\mathsf{V} = \sum_{\ell \geq 1} \tilde{\lambda}_\ell \phi_\ell \otimes \phi_\ell$ where $\tilde{\lambda}_\ell = \lambda_\ell \mathbb{E}\,\big(f_\ell^2 (\sum_{s \geq 1} \lambda_s f_s^2)^{-1}\big)$.

From a different point of view, taking into account **Property 2**, [1] considered a projection-pursuit approach combined with penalization to obtain robust estimators of the principal directions which provide robust alternatives to the estimators defined by Rice and Silverman [20] and Silverman [22].

To define these estimators, denote as $P[\alpha]$ for the distribution of $\langle \alpha, X \rangle$ when $X \sim P$. Given $\sigma_R(F)$ a robust univariate scale functional, define $\sigma : \mathscr{H} \to \mathbb{R}$ as the map $\sigma(\alpha) = \sigma_R(P[\alpha])$. Let $s_n^2 : \mathscr{H} \to \mathbb{R}$ be the empirical version of σ^2, that is, $s_n^2(\alpha) = \sigma_R^2(P_n[\alpha])$, where $\sigma_R(P_n[\alpha])$ stands for the functional σ_R computed at the empirical distribution of $\langle \alpha, X_1 \rangle, \ldots, \langle \alpha, X_n \rangle$.

Moreover, let us consider $\mathscr{H}_S$, the subset of "smooth elements" of $\mathscr{H}$ and $D : \mathscr{H}_S \to \mathscr{H}$ a linear operator, referred to as the "differentiator". Using D, they define the symmetric positive semidefinite bilinear form $\lceil \cdot, \cdot \rceil : \mathscr{H}_S \times \mathscr{H}_S \to \mathbb{R}$, where $\lceil \alpha, \beta \rceil = \langle D\alpha, D\beta \rangle$. The "penalization operator" is then defined as $\Psi : \mathscr{H}_S \to \mathbb{R}$, $\Psi(\alpha) = \lceil \alpha, \alpha \rceil$, and the penalized inner product as $\langle \alpha, \beta \rangle_\tau = \langle \alpha, \beta \rangle + \tau \lceil \alpha, \beta \rceil$.

Therefore, $\|\alpha\|_\tau^2 = \|\alpha\|^2 + \tau\Psi(\alpha)$. Besides, let $\{\delta_i\}_{i\geq 1}$ be a basis of $\mathcal{H}$ and denote $\mathcal{H}_{p_n}$ the linear space spanned by $\delta_1, \ldots, \delta_{p_n}$ and $\mathcal{S}_{p_n} = \{\alpha \in \mathcal{H}_{p_n} : \|\alpha\| = 1\}$.

The robust projection pursuit estimators are then defined as

$$\begin{cases} \hat{\phi}_1 = \text{argmax}_{\alpha \in \mathcal{H}_{p_n}, \|\alpha\|_\tau = 1}\, \{s_n^2(\alpha) - \rho\Psi(\alpha)\} \\ \hat{\phi}_m = \text{argmax}_{\alpha \in \hat{\mathcal{B}}_{m,\tau}}\, \{s_n^2(\alpha) - \rho\Psi(\alpha)\} \quad 2 \leq m, \end{cases} \quad (1)$$

where $\hat{\mathcal{B}}_{m,\tau} = \{\alpha \in \mathcal{H}_{p_n} : \|\alpha\|_\tau = 1, \langle\alpha, \hat{\phi}_j\rangle_\tau = 0\, , \forall\, 1 \leq j \leq m - 1\}$. In the above definition, we understand that the products $\rho\Psi(\alpha)$ or $\tau\Psi(\alpha)$ are defined as 0 when $\rho = 0$ or $\tau = 0$, respectively, even when $\alpha \notin \mathcal{H}_S$ for which case $\Psi(\alpha) = \infty$ and when $p_n = \infty$, $\mathcal{H}_{p_n} = \mathcal{H}$.

With this definition and by taking $p_n = \infty$, the robust raw estimators are obtained when $\rho = \tau = 0$, while the robust estimators penalizing the norm and scale correspond to $\rho = 0$ and $\tau = 0$, respectively. On the other hand, the basis expansion approach corresponds to a finite choice for p_n and $\tau = \rho = 0$.

Bali et al. [1] derived the qualitative robustness of these estimators and show that they turn out to be consistent with the functional principal component directions defined as

$$\begin{cases} \phi_{R,1}(P) = \text{argmax}_{\|\alpha\|=1}\sigma(\alpha) \\ \phi_{R,m}(P) = \text{argmax}_{\|\alpha\|=1, \alpha \in \mathcal{B}_m}\sigma(\alpha), \quad 2 \leq m, \end{cases}$$

where $\mathcal{B}_m = \{\alpha \in \mathcal{H} : \langle\alpha, \phi_{R,j}(P)\rangle = 0, 1 \leq j \leq m - 1\}$. To provide an explanation of what the directions $\phi_{R,m}(P)$ represent, assume that there exists a constant $c > 0$ and a self-adjoint, positive semidefinite and compact operator Γ_0, such that for any $\alpha \in \mathcal{H}$, $\sigma^2(\alpha) = c\langle\alpha, \Gamma_0\alpha\rangle$. Moreover, denote by $\lambda_1 \geq \lambda_2 \geq \ldots$ the eigenvalues of Γ_0 and by ϕ_j the eigenfunction of Γ_0 associated with λ_j. Assume that for some $q \geq 2$, and for all $1 \leq j \leq q$, $\lambda_1 > \lambda_2 > \cdots > \lambda_q > \lambda_{q+1}$, then $\phi_{R,j}(P) = \phi_j$. Conditions that guarantee that $\sigma^2(\alpha) = c\langle\alpha, \Gamma_0\alpha\rangle$ when a robust scale is used are discussed in [1] where also the results of an extensive simulation study showing the advantages of using robust procedures are reported.

As an example, we compute the robust projection-pursuit estimators for the generated data in Fig. 1c. The robust estimators correspond to an M-scale with score function, the Tukey's function $\chi_c(y) = \min\left(3\,(y/c)^2 - 3\,(y/c)^4 + (y/c)^6, 1\right)$ with tuning constant $c = 1.56$ and breakdown point $1/2$. The choice $c = 1.56$ ensures that the M-scale functional is Fisher-consistent at the normal distribution. We have also computed the classical estimators which correspond to select σ_R as the standard deviation (SD). Figure 2 reports the results corresponding to the raw estimators of each principal component. The solid line corresponds to the true direction while the broken ones to the estimators. From these plots we observe the sensitivity of the classical procedure to the influential observations introduced.

As detection rule, Fig. 3 gives parallel boxplots of the scores $\hat{s}_{i,j} = \langle X_i - \hat{\mu}, \hat{\phi}_j\rangle$ when $\hat{\phi}_j$ are the classical and robust estimators. For the classical estimators, $\hat{\mu} = \overline{X}$, while for the robust ones $\hat{\mu} = \text{argmin}_{\theta \in \mathcal{H}} \sum_{i=1}^n (\|X_i - \theta\| - \|X_i\|)/n$. Due to a

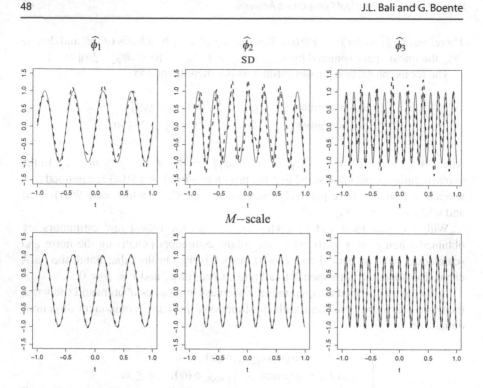

Fig. 2 Estimators of the principal directions for the generated data c). The *solid* and *broken line* correspond to the true and estimated directions, respectively

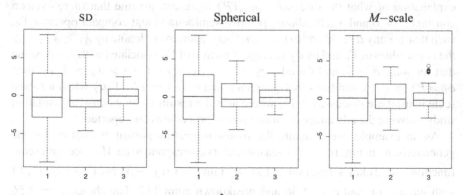

Fig. 3 Boxplots of the estimated scores $\langle X_i, \hat{\phi}_j \rangle$ for the generated data c)

masking effect, the boxplots of the scores over the classical estimators do not reveal any outlier. On the other hand, when using the robust projection-pursuit estimators the largest values of $\hat{s}_{i,3}$ correspond to the atypical observations generated. It is worth noting that the same conclusions are obtained if the plots of the residuals squared norm $\|\hat{r}_i\|^2 = \|X_i - \hat{\mu} - \sum_{j=1}^{3} \hat{s}_{i,j} \hat{\phi}_j\|^2$ are considered (see Fig. 4). The residual plot corresponding to the M-scale clearly shows that the residual squared norm of the atypical observations are out of bound. On the other hand, when

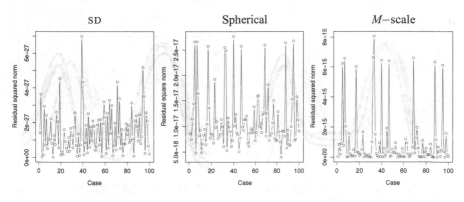

Fig. 4 Residual plots for the generated data c)

considering the eigenfunctions of the sample covariance operator, the observations with the largest residuals correspond to those labelled 19, 39, 40, 71 and 94, that is, only one of the atypical observations appears with a large residual, so leading to the wrong conclusions. It is worth noticing that, for this contamination, the boxplots of the estimated scores obtained when considering the spherical principal components considered in [9] do not detect any of the atypical observations (see Fig. 3). This is mainly due to the fact that the spherical principal components are more biased in this situation producing a masking effect on the scores. Indeed, as pointed out by Boente and Fraiman [2], spherical principal components estimators are resistant for any contamination model which preserves the property of being elliptical, while its resistance is not guaranteed when other types of contamination are involved such as the one we are considering. Even though, the residual plot corresponding to these estimators show that the atypical data have large values of the residual squared norm. Hence, highly resistant procedures should be preferred.

5   Lip Data Example

The following example was considered in [9] to show the effect of outliers on the functional principal components. A subject was asked to say the word *bob* 32 times and the position of lower lip was recorded at each time point. Lip movement data was originally analyzed by Malfait and Ramsay [18]. In Fig. 5, the plotted curves correspond to the 32 trajectories of the lower lip versus time. Three of these curves (plotted with thick lines on Fig. 5a seem to be out of line, with delayed second peaks. To determine whether or not these curves are within the normal range of variability, it is necessary to estimate accurately the principal components.

As in [9], we have estimated five principal directions using the robust projection-pursuit estimators defined in (1) related to the M-scale with Tukey's score function. The robust and classical principal components are given in Fig. 6 where the classical and robust raw estimators are plotted with a solid line and with a broken line,

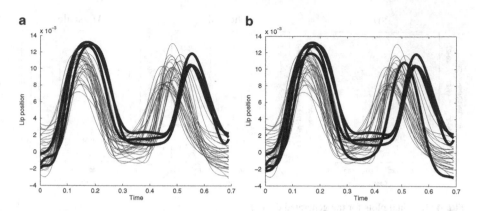

Fig. 5 Lip movement data. Smoothed lower-lip trajectories of an individual pronouncing *bob* 32 times. (**a**) The trajectories 24, 25 and 27 are indicated with *thick lines*. (**b**) The trajectories 14, 24, 25 and 27 are indicated with *thick lines*

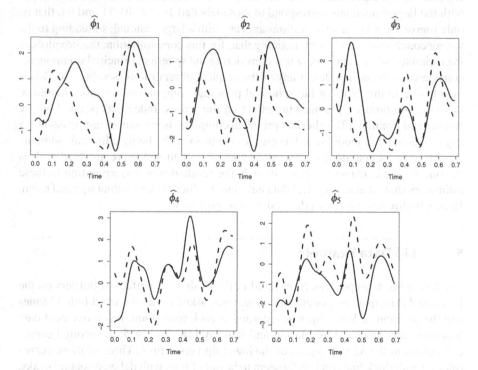

Fig. 6 Estimators of the principal directions for the lip movement data. The *solid line* correspond to the classical direction while the *broken line* to the robust ones

respectively. We refer to Gervini [9] to understand the type of variability explained by these components. Besides, as described therein a positive component score will be associated with curves that show a large first peak and a delayed second peak, as those observed in the three atypical curves.

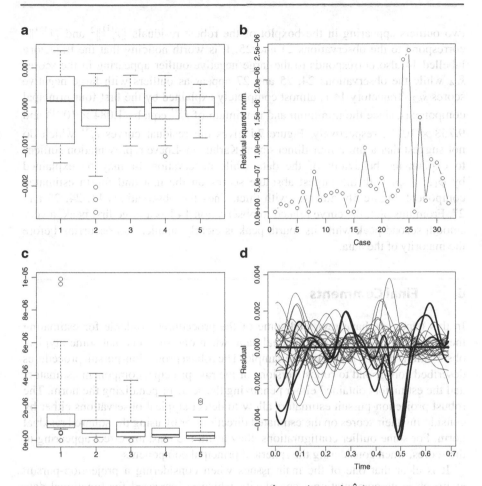

Fig. 7 Lip movement data. (**a**) Boxplot of the scores $\hat{s}_{i,j} = \langle X_i - \hat{\mu}, \hat{\phi}_j \rangle$, (**b**) residual plots of $\|\hat{r}_i^{(5)}\|^2$, (**c**) adjusted boxplots of $\|\hat{r}_i^{(q)}\|^2$, $1 \leq q \leq 5$ and (**d**) residuals plot $\hat{r}_i^{(4)}$ based on a robust fit. The *thick curves* correspond to the observations 24, 25 and 27 while the *thick horizontal line* to the trajectory 14

Figure 7 presents the parallel boxplots of the scores $\hat{s}_{i,j} = \langle X_i - \hat{\mu}, \hat{\phi}_j \rangle$ when $\hat{\phi}_j$ are the robust estimators together with the plot of the squared norm of the residuals $\|\hat{r}_i^{(q)}\|^2 = \|X_i - \hat{\mu} - \sum_{j=1}^{q} \hat{s}_{i,j} \hat{\phi}_j\|^2$ where $\hat{\mu} = \text{argmin}_{\theta \in \mathcal{H}} \sum_{i=1}^{n} (\|X_i - \theta\| - \|X_i\|) / n$. We only present the plots for the robust fit since we have already shown that when considering the classical one a masking effect may appear.

The residual plot corresponding to the M-scale shows clearly that the residual squared norm of the atypical observations are out of bound. Figure 7 also presents the boxplots of $\|\hat{r}_i^{(q)}\|^2$. Due to the skewness of the distribution of the norm, we have considered the adjusted boxplots (see [14]) instead of the usual ones. The two outliers appearing in the boxplot of the robust residuals $\|\hat{r}_i^{(1)}\|^2$ and $\|\hat{r}_i^{(5)}\|^2$ correspond to the observations 24 and 25. It is worth noticing that the trajectory labelled 14 also corresponds to the large negative outlier appearing in the scores $\hat{s}_{i,4}$ while the observations 24, 25 and 27 appear as outliers with large negative scores $\hat{s}_{i,2}$. Trajectory 14 is almost completely explained by the first four principal components, since the minimum and maximum of $\hat{r}_i^{(4)}$ equal -1.084×10^{-18} and 9.758×10^{-19}, respectively. Figure 7d gives the residual curves $\hat{r}_i^{(4)}$ which do not suggest that a finite four-dimensional Karhunen–Loève representation suffices to explain the behaviour of the data while observation 14 may be explained by $\hat{\phi}_1, \ldots, \hat{\phi}_4$ with the largest absolute scores on the first and fourth estimated component. Figure 5b indicates with thick lines the observations 14, 24, 25 and 27. From this plot, the curve related to observation 14 has a large first peak, a very smooth second peak while its fourth peak is clearly smaller and occurring before the majority of the data.

6 Final Comments

In this paper, we have reviewed some of the procedures available for estimating the principal directions for functional data, when one suspects that some atypical observations may be present in the sample. The robust projection-pursuit procedures described correspond to robust versions of the raw principal component estimators and the estimators obtained either penalizing the scale or penalizing the norm. The robust projection-pursuit estimators allow to detect atypical observations either by considering their scores on the estimated directions or by using the squared residual norm. For some outlier configurations, they avoid the masking effect appearing in the scores, when considering the spherical principal components.

It is clear that one of the main issues when considering a projection-pursuit approach is its computational complexity which is increased for functional data since in infinite-dimensional spaces the ball is sparse. In [1], an algorithm to compute the estimators is described. The numerical procedure adapts the finite-dimensional maximization algorithm proposed in [5] to the functional setting in order to deal with the penalization term included in the definition of the estimators. The consistency of the first principal direction estimators obtained from this algorithm is studied in Bali and Boente (2012, On the consistency of the projection pursuit estimators computed through the Croux–Gazen algorithm, unpublished manuscript)

Finally, it is worth noting that robust projection-pursuit methods may also be helpful to provide statistical tools less sensitive to atypical data when considering functional discrimination or functional canonical correlation.

Acknowledgements This research was partially supported by Grants 276 from the Universidad de Buenos Aires, PIP 216 from CONICET and PICT 821 from ANPCYT at Buenos Aires, Argentina. The authors wish to thank three anonymous referees for valuable comments which led to an improved version of the original paper.

References

1. Bali, J.L., Boente, G., Tyler, D.E., Wang, J.L.: Robust functional principal components: a projection-pursuit approach. Ann. Stat. **39**, 2852–2882 (2011)
2. Boente, G., Fraiman, R.: Discussion on robust principal component analysis for functional data by N. Locantore, J. Marron, D. Simpson, N. Tripoli, J. Zhang and K. Cohen. Test **8**, 28–35 (1999)
3. Boente, G., Fraiman, R.: Kernel-based functional principal components. Stat. Prob. Lett. **48**, 335–345 (2000)
4. Croux, C., Filzmoser, P., Oliveira M.R.: Algorithms for projection-pursuit robust principal component analysis. Chem. Intell. Lab. Syst. **87**, 218–225 (2007)
5. Croux, C., Ruiz-Gazen, A.: A fast algorithm for robust principal components based on projection pursuit. In: Prat, A. (ed.) Compstat: Proceedings in Computational Statistics, pp. 211–217. Physica-Verlag, Heidelberg (1996)
6. Croux, C., Ruiz-Gazen, A.: High breakdown estimators for principal components: the projection-pursuit approach revisited. J. Multivar. Anal. **95**, 206–226 (2005)
7. Cui, H., He, X., Ng, K.W.: Asymptotic distribution of principal components based on robust dispersions. Biometrika **90**, 953–966 (2003)
8. Dauxois, J., Pousse, A., Romain, Y.: Asymptotic theory for the principal component analysis of a vector random function: some applications to statistical inference. J. Multivar. Anal. **12**, 136–154 (1982)
9. Gervini, D.: Robust functional estimation using the spatial median and spherical principal components. Biometrika **95**, 587–600 (2008)
10. Gervini, D.: Detecting and handling outlying trajectories in irregularly sampled functional datasets. Ann. Appl. Stat. **3**, 1758–1775 (2009)
11. Gervini, D.: Outlier detection and trimmed estimation for general functional data. Stat. Sin. (2012). doi:10.5705/ss.2010.282
12. Hall, P., Hosseini-Nasab, M.: On properties of functional principal components analysis. J. R. Stat. Soc. Ser. B **68**, 109–126 (2006)
13. Hall, P., Müller, H.-G., Wang, J.-L.: Properties of principal component methods for functional and longitudinal data analysis. Ann. Stat. **34**, 1493–1517 (2006)
14. Hubert, M., Vandervieren, E.: An adjusted boxplot for skewed distributions. Comput. Stat. Data Anal. **52**, 5186–5201 (2008)
15. Hyndman, R.J., Ullah, S.: Robust forecasting of mortality and fertility rates: a functional data approach. Comput. Stat. Data Anal. **51**, 4942–4956 (2007)
16. Li, G., Chen, Z.: Projection-pursuit approach to robust dispersion matrices and principal components: primary theory and Monte Carlo. J. Am. Stat. Assoc. **80**, 759–766 (1985)
17. Locantore, N., Marron, J.S., Simpson, D.G., Tripoli, N., Zhang, J.T., Cohen, K.L.: Robust principal components for functional data (with discussion). Test **8**, 1–73 (1999)
18. Malfait, N., Ramsay, J.O.: The historical functional linear model. Can. J. Stat. **31**, 115–128 (2003)
19. Pezzulli, S.D., Silverman, B.W.: Some properties of smoothed principal components analysis for functional data. Comput. Stat. **8**, 1–16 (1993)
20. Rice, J., Silverman, B.W.: Estimating the mean and covariance structure nonparametrically when the data are curves. J. R. Stat. Soc. Ser. B **53**, 233–243 (1991)
21. Sawant, P., Billor, N., Shin, H.: Functional outlier detection with robust functional principal component analysis. Comput. Stat. **27**, 83–102 (2011)
22. Silverman, B.W.: Smoothed functional principal components analysis by choice of norm. Ann. Stat. **24**, 1–24 (1996)
23. Sun, Y., Genton, M.G.: Functional boxplots. J. Comput. Graph. Stat. **20**, 316–334 (2011)

24. Tyler, D.: A note on multivariate location and scatter statistics for sparse data sets. Stat. Prob. Lett. **80**, 1409–1413 (2010)
25. Yao, F., Lee, T.C.M.: Penalized spline models for functional principal component analysis. J. R. Stat. Soc. Ser. B **68**, 3–25 (2006)
26. Zhang, J.-T., Chen, J.: Statistical inferences for functional data. Ann. Stat. **35**, 1052–1079 (2007)

Testing the Maximum by the Mean in Quantitative Group Tests

João Paulo Martins, Rui Santos, and Ricardo Sousa

Abstract

Group testing, introduced by Dorfman in 1943, increases the efficiency of screening individuals for low prevalence diseases. A wider use of this kind of methodology is restricted by the loss of sensitivity inherent to the mixture of samples. Moreover, as this methodology attains greater cost reduction in the cases of lower prevalence (and, consequently, a higher optimal batch size), the phenomenon of rarefaction is crucial to understand that sensitivity reduction. Suppose, with no loss of generality, that an experimental individual test consists in determining if the amount of substance overpasses some prefixed threshold l. For a pooled sample of size n, the amount of substance of interest is represented by $(Y_1, \cdots, Y_n)$, with mean $\overline{Y}_n$ and maximum M_n. The goal is to know if any of the individual samples exceeds the threshold l, that is, $M_n > l$. It is shown that the dependence between $\overline{Y}_n$ and M_n has a crucial role in deciding the use of group testing since a higher dependence corresponds to more information about M_n given by the observed value of $\overline{Y}_n$.

J.P. Martins (✉)
School of Technology and Management, Polytechnic Institute of Leiria,
CEAUL-Center of Statistics and Applications of University of Lisbon, Lisbon, Portugal
e-mail: jpmartins@ipleiria.pt

R. Santos
CEAUL and School of Technology and Management, Polytechnic Institute of Leiria, Leiria,
Portugal
e-mail: rui.santos@ipleiria.pt

R. Sousa
Higher School of Health Technology of Lisbon, Polytechnic Institute of Lisbon,
CEAUL-Center of Statistics and Applications of University of Lisbon, Lisbon, Portugal
e-mail: ricardo.sousa@estesl.ipl.pt

A. Pacheco et al. (eds.), *New Advances in Statistical Modeling and Applications*,
Studies in Theoretical and Applied Statistics, DOI 10.1007/978-3-319-05323-3_5,
© Springer International Publishing Switzerland 2014

1   Introduction

The original idea of [1] was to use pooled samples on the screening of the defective members of a population (classification problem) in order to reduce the expected number of tests. In Dorfman's algorithm first stage, specimens are grouped for batched testing. If a pooled test is negative, all individuals in the pooled sample are declared negative. Otherwise, individual tests are performed. The optimal batch size minimizes the expected number of tests. Several extensions of this algorithm may be found in [2, 10, 11]. Alternative algorithms are presented in [5].

The seminal work of [9] deals with the problem of estimating the proportion of the defective members of a population (estimation problem).

The use of group testing schemes is usually restricted to qualitative analyses (presence or absence of the infection), without measuring any quantitative variable (antigens or antibodies or bacteria counts, or proportion of specific cells, or weight or volume of some chemical compound). If some continuous test outcome is available, it is usually transformed into a dichotomous outcome (cf. [13]). There are few works that deal with continuous outcomes (cf. [12, 13]), but even those consider only the estimation problem. In this work, we present two possible methodologies that allow the application of Dorfman's algorithm to the classification problem when the test outcome is a continuous variable.

This work outline is as follows. Section 2 presents a discussion on the group testing procedure originally defined by Dorfman and the effect of the rarefaction phenomenon or the dilution effect (cf. [4]) in the sensitivity and the specificity of individual tests. In Sect. 3, two methodologies are proposed to conduct pooled sample tests with continuous outcomes. The importance of the correlation between the sample mean and the sample maximum is discussed when rarefaction may disturb the quality of group testing. The final remarks are presented in Sect. 4, where some suggestions for further investigation are given.

2   Dorfman's Procedures and Its Extensions

Let p denote the prevalence rate of the infection and the independent Bernoulli random variables X_i, with $i = 1, \cdots, N$, represent the presence ($X_i = 1$) or absence ($X_i = 0$) of the infection in the ith population individual. Furthermore, let $+$ and $-$ represent, respectively, the result of an individual test as positive and negative. The error of an individual test is usually assessed by two probabilities: the sensitivity and the specificity. The probability of getting a correct result on one individual test performed on a healthy individual is defined as the test specificity, that is, $\varphi_e = \mathbb{P}(-|X_i = 0)$. More important is the probability of detecting an infected individual, that is, the test sensitivity $\varphi_s = \mathbb{P}(+|X_i = 1)$. These definitions could be extended to pooled sample procedures (cf. [5, 8]). Thus, using a sample of size n, the pooled sensitivity is defined by $\varphi_s^{[n]} = \mathbb{P}(+|\sum_{i=1}^{n} X_i > 0)$ and the pooled specificity by $\varphi_e^{[n]} = \mathbb{P}(-|\sum_{i=1}^{n} X_i = 0)$. The sensitivity $\varphi_s^{[n]}$ depends on the number m of infected members as a result of the dilution of the fluid and its rarefaction. Therefore, using $\varphi_s^{[m,n]} = \mathbb{P}(+\mid \sum_{i=1}^{n} X_i = m)$ and applying Bayes theorem we obtain

$$\varphi_s^{[n]} = \frac{\sum_{j=1}^{n} \mathbb{P}\left(+, \sum_{i=1}^{n} X_i = j\right)}{\mathbb{P}\left(\sum_{i=1}^{n} X_i > 0\right)} = \sum_{j=1}^{n} \frac{\binom{n}{j} p^j (1-p)^{n-j}}{1-(1-p)^n} \varphi_s^{[j,n]} = \sum_{j=1}^{n} \lambda_j \, \varphi_s^{[j,n]},$$

(1)

where $\sum_{j=1}^{n} \lambda_j = 1$.

In our problem both the healthy and the infected individual possess some substance of interest in the samples for testing. Suppose also that the amount of substance in a healthy individual follows some continuous distribution $Y \frown D_{\theta_0}$ and that the amount of substance in an infected individual is $Y^* = \beta_0 + \beta_1 Y$ (or $Y^* \frown D_{\theta_1}$ where θ_0 and θ_1 stand for distinct parameter vectors). We consider the cases where D is an exponential distribution, Gaussian distribution and Pareto distribution.

For an individual test, the hypothesis to be considered are

$$H_0 : X_i = 0 \text{ versus } H_1 : X_i = 1,$$

(2)

where the null hypothesis is equivalent to state that the amount of substance of the ith sample is described by $Y_i \frown D_{\theta_0}$. A rule to decide if a sample is to be declared positive or negative is well defined for a known D_{θ_0}. For a significance level α, the null hypothesis is to be rejected if the amount of substance Y exceeds some fixed threshold $l = F_{D_{\theta_0}}^{-1} (1 - \alpha)$ where $F_{D_{\theta_0}}^{-1} (1 - \alpha)$ stands for the generalized inverse of the distribution function D_{θ_0} at the point $1 - \alpha$ (a similar reasoning is applied if the rule is to declare a positive sample when the amount of substance is lower than some threshold l). Hence, the test significance level coincides with $1 - \varphi_e$ (i.e. $\varphi_e = 1 - \alpha$). The power of the test is given by the probability that an infected sample is declared infected, that is, $1 - \beta = 1 - F_{D_{\theta_1}} (l)$. Thus, the power of the test is equal to the sensitivity of the experimental test. Of course, if some further sources of error in the experimental test are considered, this correspondence between the two types of error of the hypothesis test and the two measures of the quality may not be valid. However, our goal is to show the association between these concepts.

As an example, consider $D_{\theta_0} \equiv \mathrm{N}(\mu_0, \sigma_0)$ and $D_{\theta_1} \equiv \mathrm{N}(\mu_1, \sigma_1)$ where $\mathrm{N}(\mu, \sigma)$ stands for a Gaussian distribution with mean μ and standard deviation σ. We will assume, with no loss in generality, $\mu_1 > \mu_0$ and let $\sigma = \sigma_1/\sigma_0$ and $\mu = \mu_1 - \mu_0 > 0$. Thus, in (2) we have $l = \mu_0 + \sigma_0 \Phi^{-1} (1 - \alpha)$ and the power of the test is $1 - \beta = 1 - \Phi\left(\frac{l-\mu_1}{\sigma_1}\right) = 1 - \Phi\left(\frac{\Phi^{-1}(1-\alpha)}{\sigma} - \frac{\mu}{\sigma_1}\right)$, where Φ denotes the cumulative distribution function of a standard Gaussian random variable. It is no surprise to verify that the test power increases with the difference of the mean values $\mu = \mu_1 - \mu_0$.

In Dorfman's methodology and its extensions to a number of stages greater than two, the decision whether a sample is or isn't infected depends firstly on the result of the pooled sample test. If the pooled sample test is considered to be positive further individual/grouped sample tests have to be conducted. The main problem here is to decide whether a pooled sample contains at least one infected individual since the substance of interest may be present both in healthy and infected individuals. This leads to the hypothesis test

$$H_0 : \sum_{i=1}^{n} X_i = 0 \; versus \; H_1 : \sum_{i=1}^{n} X_i > 0, \tag{3}$$

where the null hypothesis is equivalent to state that the amount of substance of interest is described by the $\sum_{i=1}^{n} Y_i$ of independent and identically distributed random variables to $Y \frown D_{\theta_0}$. As in the individual tests, it is necessary to establish a threshold (as a function of n) to decide if a pooled sample is or is not classified as a mixture of at least one infected individual. In Sect. 3 we propose two different methodologies in order to decide whether the null hypothesis should be rejected or maintained.

The process of getting a pooled sample is as follows. The same amount of sample is taken from n individuals and mixed (homogeneously). The new mixed sample is now tested. In a low prevalence case, a maximum of one infected sample in the pooled sample occurs with high probability. Hence, due to rarefaction the effect of this sample in the total amount of some substance in the pooled sample could be quite low. We raise this question in order to keep in mind that if the distributions D_{θ_0} and D_{θ_1} are not quite different the pooled sensitivity of the test could be seriously compromised. Some works incorporating rarefaction use some previous knowledge about this phenomenon (i.e. [12]). However, those works don't take advantage from this possibly known distributions. These pooled sample tests will be treated in detail in the next section.

3   The Pooled Sample Tests

When using pooled samples, the experimental test provides information on the batched sample as a whole although the experimenter wants to know if any of the individual samples exceeds the prefixed threshold l. The process of decision of the hypothesis test (3) isn't as obvious as the decision on individual testing represented by (2), because to classify a pooled sample it is previously needed to identify the samples in which $M_n = \max(Y_1, \cdots, Y_n)$ overpasses the threshold l using only the information of the sample mean (the only quantity observed). In this work, we discuss two different methodologies for deciding whether to reject or to maintain the null hypothesis of the pooled test (3).

Table 1 Simulation of correlation between mean $\overline{Y}_n$ and maximum M_n (1,000,000 replicates)

n	Standard Gaussian	Standard exponential	Pareto $\theta = 5$	Pareto $\theta = 3$	Pareto $\theta = 1$
2	0.8585	0.9483	0.9679	0.9801	0.9999450
3	0.7699	0.9034	0.9412	0.9646	0.9999819
5	0.6632	0.8413	0.8977	0.9378	0.9999714
50	0.3283	0.4919	0.6330	0.7607	0.9999983
100	0.2323	0.4062	0.5473	0.6889	0.9999786

3.1 T_1 Methodology: Using the Distribution of the Sample Mean

When mixing n healthy samples $Y_1, \cdots, Y_n$ and then extracting a portion $1/n$ of the total amount for batched testing, the amount of substance of interest is given by the random variable $C_{0,n}$ where $C_{m,n}$ is given by

$$C_{m,n} = \frac{\sum_{i=1}^{n-m} Y_i + \sum_{i=1}^{m} Y_i^*}{n}. \tag{4}$$

The random variable $C_{m,n}$ represents the amount of substance in a batched sample of size n with m infected individual samples. The null hypothesis of the hypothesis test (3) is rejected if $C_{0,n} > q_{1-\alpha}$ where $F_{C_{0,n}}(q_{1-\alpha}) = 1 - \alpha$ and $F_{C_{0,n}}$ stands for the distribution function of the random variable $C_{0,n}$.

If there is an infected individual in the pooled sample, the main problem is to know whether the observed value of the "mean" random variable $C_{m,n}$, with $m \geq 1$, is influenced by the presence of m infected samples.

A pooled sample that contains at least one defective individual should be screened as positive. Thus, a pooled sample is classified as defective if the sample maximum $M_n = \max(Y_1, \cdots, Y_n)$ overpasses the prefixed threshold l. However, the researcher uses only information about the mean to attain a decision that concerns only to the sample maximum. Hence, it is expected that the chance of deciding correctly increases with the dependence between the sample mean and sample maximum. For the three distributions mentioned above, the correlation between the sample mean and the sample maximum is computed for different sample sizes. All values presented in Table 1 were obtained by simulation (using software R) although we can get the same results analytically for the exponential distribution $\left[\rho_{M_n, \overline{Y}_n} = \sum_{i=1}^{n} i^{-1} \left(n \sum_{i=1}^{n} i^{-2} \right)^{-0.5} \right]$ and by numerical approximation for the Gaussian case (e.g. using $\rho_{M_n, \overline{Y}_n} = \left(\sqrt{n} \sigma_{M_n} \right)^{-1}$ with σ_{M_n} given in [7]). Thus, the simulation is an excellent resource to obtain good approximations of the theoretical value of the correlations.

The correlation decreases as n increases as expected. For the Pareto distribution, with shape parameter θ, the correlation is high even for n as high as 50. This is probably related to the heavy tails of this distribution. Otherwise, the correlation is quite moderate for the Gaussian distribution, and therefore, the power of test (3) is expected to be rather low. We simulate the Pareto(1) case to point out that the sequence of correlations converges to 1 when θ decreases (although, for $\theta = 1$, the mean does not exist and therefore the simulation cannot be interpreted as an estimate of the theoretical value of the correlations). For the exponential and Gaussian distributions the correlations are independent of the parameters values.

3.2 T_2 Methodology: Using a Simulation Method

Let α be the significance level of the hypothesis test (3). The aim is to reject the null hypothesis if at least one of the individual samples exceeds the threshold l. Therefore, under H_0,

$$\mathbb{P}(M_n \leq l) = \mathbb{P}(Y_1 \leq l, \cdots, Y_n \leq l) = F_{D_{\theta_0}}^n (l) = 1 - \alpha \Leftrightarrow l = F_{D_{\theta_0}}^{-1} \left((1 - \alpha)^{\frac{1}{n}} \right).$$

(5)

The computation of the generalized inverse distribution function $F_{D_{\theta_0}}^{-1}$ is not generally straightforward but the use of simulation provides good approximations for the value of l (quantile $\sqrt[n]{1 - \alpha}$ of distribution D_{θ_0}). Simulation is the core of this methodology. Let $(Y_{1j}, \cdots, Y_{nj})_{j=1,\cdots,N}$ be N samples of size n generated by simulation that verify $Y_{ij} \frown D_{\theta_0}$. Consider the N samples ordered by the sample maximum. Then, the k samples whose maximum is closest to l are chosen where k is an arbitrary number (in the simulations performed in Sect. 3.3 it was used $N = 10^5$ and k equals to 1 % of N). The mean sample of those k samples is computed and taken as the threshold l^* of decision for the pooled sample test, that is, if the mean sample exceeds l^* the pooled sample is declared infected.

3.3   Simulations Results

In this subsection, we compare the use of these two methodologies and their effects on sensitivity and specificity. Gaussian, exponential, and Pareto distribution are considered. The calculations are all done using simulation (via software R) although some calculus could be done analytically ($\overline{Y}_n \frown N(\mu_0, \sigma_0/\sqrt{n})$ if $Y \frown N(\mu_0, \sigma_0)$ and $\overline{Y}_n \frown \text{Gamma}(n, \frac{\lambda}{n})$ if $Y \frown \text{Exp}(\lambda)$).

Tables 2 and 3 present the specificity $\varphi_e^{[n]}$ and sensitivity $\varphi_s^{[n]}$ of a pooled sample test, applying methodologies T_1 and T_2 and assuming for each distribution that D_{θ_1} is just a translation of D_{θ_0}. The translation is chosen in order to keep both sensitivity φ_s and specificity φ_e of individual tests equal to 0.95 (case 1) and to 0.995 (case 2). In all simulations we use the most efficient value for n in Dorfman's methodology [1].

The patterns observed in each methodology were already expectable according to Liu et al. [6]. In the T_1 methodology, the loss of specificity of the test is low but it results in higher sensitivity loss than when using second methodology. Otherwise, the T_2 methodology specificity loss is very close to the one using the T_1

Table 2 Hypothesis tests simulation, Gaussian and exponential distribution (100,000 replicates)

p	$1-\alpha$	Case 1: $\varphi_s^{[1]} = \varphi_e^{[1]} = 0.95$				Case 2: $\varphi_s^{[1]} = \varphi_e^{[1]} = 0.995$			
		T_1		T_2		T_1		T_2	
		$\varphi_s^{[n]}$	$\varphi_e^{[n]}$	$\varphi_s^{[n]}$	$\varphi_e^{[n]}$	$\varphi_s^{[n]}$	$\varphi_e^{[n]}$	$\varphi_s^{[n]}$	$\varphi_e^{[n]}$
Gaussian distribution									
0.15	0.90	0.7746	0.8994	0.8573	0.8283	0.9623	0.9002	0.9830	0.8231
($n = 3$)	0.95	0.6627	0.9504	0.8012	0.8818	0.9229	0.9501	0.9728	0.8707
	0.99	0.4310	0.9900	0.6902	0.9408	0.7835	0.9901	0.9319	0.9427
0.05	0.90	0.6127	0.9003	0.7636	0.7971	0.8628	0.8992	0.9371	0.7960
($n = 5$)	0.95	0.4804	0.9501	0.7109	0.8417	0.7716	0.9500	0.9202	0.8320
	0.99	0.2534	0.9901	0.9490	0.9082	0.5431	0.9900	0.8445	0.9137
0.01	0.90	0.4066	0.8997	0.6600	0.7343	0.6261	0.9001	0.8277	0.7404
($n = 11$)	0.95	0.2759	0.9497	0.5917	0.7874	0.4880	0.9502	0.8122	0.7606
	0.99	0.1061	0.9899	0.5266	0.8338	0.2491	0.9900	0.7337	0.8347
Exponential distribution									
0.15	0.90	0.9423	0.8984	0.9923	0.8825	1.00	0.9019	1.00	0.8856
($n = 3$)	0.95	0.5496	0.9492	0.6673	0.9339	1.00	0.9554	1.00	0.9429
	0.99	0.1626	0.9894	0.2464	0.9792	0.6413	0.9895	0.9596	0.9820
0.05	0.90	0.8206	0.9005	0.9644	0.8698	1.00	0.8975	1.00	0.8667
($n = 5$)	0.95	0.4240	0.9486	0.6660	0.9198	1.00	0.9494	1.00	0.9229
	0.99	0.0792	0.9900	0.2195	0.9746	0.4019	0.9903	0.8979	0.9737
0.01	0.90	0.7668	0.9015	0.9907	0.8371	1.00	0.8957	1.00	0.8220
($n = 11$)	0.95	0.3514	0.9487	0.8710	0.8842	0.9521	0.9472	1.00	0.8896
	0.99	0.0256	0.9910	0.3292	0.9525	0.1896	0.9902	0.9367	0.9512

methodology and it performs better in what concerns to the sensitivity. We advise the use of the second methodology, since it has a better sensitivity behaviour.

4   Conclusion

The phenomenon of rarefaction can have a great effect in the quality of a pooled sample test. When the sample mean and the sample maximum correlation is high this effect is minimized and the use of batched samples is recommended.

When the correlation is low, the presence of an infected individual in the pooled sample has low effect on the amount of substance in the pooled sample. Therefore it is difficult to detect this infected sample. In this case, we have to be very careful when using pooled samples. Our recommendation is to use a pooled sample dimension lower than the optimal size obtained just by considering the relative cost of a specific methodology.

Further investigation may be conducted by considering a different null hypothesis in the test (3). As [3] points out, when the use of batched samples provides greatest savings (low prevalences), an infected pooled sample is almost certainly a pooled sample with just one infected sample ($\lambda_1 \approx 1$ for the efficient value for n in (1),

Table 3 Hypothesis tests simulation, Pareto distribution (100,000 replicates)

| | | Case 1: $\varphi_s^{[1]} = \varphi_e^{[1]} = 0.95$ | | | | Case 2: $\varphi_s^{[1]} = \varphi_e^{[1]} = 0.995$ | | | |
| | | T_1 | | T_2 | | T_1 | | T_2 | |
p	$1-\alpha$	$\varphi_s^{[n]}$	$\varphi_e^{[n]}$	$\varphi_s^{[n]}$	$\varphi_e^{[n]}$	$\varphi_s^{[n]}$	$\varphi_e^{[n]}$	$\varphi_s^{[n]}$	$\varphi_e^{[n]}$
Pareto(5)									
0.15	0.90	0.5813	0.9010	0.6300	0.8883	1.00	0.8995	1.00	0.8928
$n=3$	0.95	0.3616	0.9473	0.3994	0.9408	1.00	0.9495	1.00	0.9422
	0.99	0.0646	0.9896	0.0802	0.9874	0.4166	0.9898	0.5075	0.9873
0.05	0.90	0.4189	0.9001	0.4725	0.8827	0.9924	0.8861	0.9987	0.8861
$n=5$	0.95	0.2248	0.9507	0.2875	0.9353	0.8087	0.9514	0.9084	0.9362
	0.99	0.0406	0.9901	0.0583	0.9862	0.2526	0.9892	0.3350	0.9842
0.01	0.90	0.2585	0.9000	0.3555	0.8551	0.6771	0.9004	0.7738	0.8687
$n=11$	0.95	0.1365	0.9504	0.2113	0.9207	0.4294	0.9497	0.5940	0.9168
	0.99	0.0263	0.9900	0.0554	0.9800	0.1118	0.9902	0.2054	0.9881
Pareto(3)									
0.15	0.90	0.5691	0.9019	0.5883	0.8978	1.00	0.8994	1.00	0.8905
$(n=3)$	0.95	0.2900	0.9518	0.3216	0.9479	1.00	0.9534	1.00	0.9494
	0.99	0.0287	0.9891	0.0325	0.9886	0.3409	0.9904	0.3840	0.9893
0.05	0.90	0.3669	0.9047	0.3996	0.8966	1.00	0.8984	1.00	0.8873
$(n=5)$	0.95	0.1836	0.9529	0.2189	0.9457	0.9359	0.9529	0.9793	0.9472
	0.99	0.0286	0.9901	0.0339	0.9887	0.1924	0.9903	0.2304	0.9886
0.01	0.90	0.2265	0.9032	0.2714	0.8803	0.7715	0.9002	0.8551	0.8768
$(n=11)$	0.95	0.1030	0.9490	0.1337	0.9382	0.4418	0.9492	0.5234	0.9355
	0.99	0.0183	0.9888	0.0220	0.9863	0.0772	0.9904	0.1089	0.9875

therefore the sensitivity $\varphi_s^{[1,n]}$ is crucial in $\varphi_s^{[n]}$ determination). This means that the study of the hypothesis test

$$H_0 : \sum_{i=1}^{n} X_i = 1 \text{ versus } H_1 : \sum_{i=1}^{n} X_i = 0$$

is quite general and may be an alternative to follow up.

Acknowledgements The authors thank the referees for their very useful comments. Research partially sponsored by national funds through the Fundação Nacional para a Ciência e Tecnologia, Portugal—FCT under the project PEst-OE/MAT/UI0006/2011.

References

1. Dorfman, R.: The detection of defective members in large populations. Ann. Math. Stat. **14**, 436–440 (1943)
2. Finucan, H.M.: The blood testing problem. Appl. Stat. **13**, 43–50 (1964)
3. Gastwirth, J.L., Johnson W.O.: Screening with cost-effective quality control: potential applications to HIV and drug testing. J. Am. Stat. Assoc. **89**, 972–981 (1994)

4. Hung, M., Swallow, W.: Robustness of group testing in the estimation of proportions. Biometrics **55**, 231–237 (1999)
5. Kim, H., Hudgens, M., Dreyfuss, J., Westreich, D., Pilcher, C.: Comparison of group testing algorithms for case identification in the presence of testing errors. Biometrics **63**, 1152–1163 (2007)
6. Liu, S.C., Chiang, K.S., Lin, C.H., Chung, W.C., Lin, S.H., Yang, T.C.: Cost analysis in choosing group size when group testing for potato virus Y in the presence of classification errors. Ann. Appl. Biol. **159**, 491–502 (2011)
7. Parrish, R.S.: Computing variances and covariances of normal order statistics. Commun. Stat. Simul. Comput. **21**, 71–101 (1992)
8. Santos, R., Pestana, D., Martins, J.P.: Extensions of Dorfman's theory. In: Oliveira, P.E., et al. (eds.) Recent Developments in Modeling and Applications in Statistics, pp. 179–189. Studies in Theoretical and Applied Statistics. Selected Papers of the Statistical Societies. Springer, Berlin (2013)
9. Sobel, M., Elashoff, R.: Group testing with a new goal, estimation. Biometrika **62**, 181–193 (1975)
10. Sobel, M., Groll, P.A.: Group testing to eliminate efficiently all defectives in a binomial sample. Bell Syst. Tech. J. **38**, 1179–1252 (1959)
11. Sterret, A.: On the detection of defective members of large populations. Ann. Math. Stat. **28**, 1033–1036 (1957)
12. Wein, L.M., Zenios, S.A.: Pooled testing for HIV screening: capturing the dilution effect. Oper. Res. **44**, 543–569 (1996)
13. Zenios, S., Wein, L.: Pooled testing for HIV prevalence estimation exploiting the dilution effect. Stat. Med. **17**, 1447–1467 (1998)

Testing Serial Correlation Using the Gauss–Newton Regression

Efigénio Rebelo, Patrícia Oom do Valle, and Rui Nunes

Abstract

This paper proposes two types of autocorrelation tests based on a methodology that uses an auxiliary regression, named Gauss–Newton regression. All tests are derived considering that the regression function contains contemporary values of endogenous variables, situation in which the model is estimated using the nonlinear instrumental variables method. The first type of test intends to identify the presence of serial correlation, whether genuine or not. The second type of test is proposed to distinguish the genuine serial correlation from the non-genuine serial correlation, being the latter an evidence of misspecification. This study also shows that this second type of test, called the "Common Factor Restrictions" test, can be deduced as a χ^2 test or as a t test.

1 Introduction

The context of this study refers to the need of estimating and testing models with serial correlation by focusing on first-order autoregressive errors. There will be no loss of generality since all the results carry over to higher order processes in an obvious fashion.

So, let us consider the model

$$y_t = x_t \beta + u_t; u_t = \rho u_{t-1} + \xi_t \; ; \xi_t \; IID \sim \left(0, w^2\right) \tag{1}$$

where x_t is row t of matrix $X, t = 1, 2, \ldots, n$, and β is a k-dimensional vector. In (1), $|\rho| < 1$ is the necessary and sufficient condition for $Var(u_t) = \sigma^2 = \frac{w^2}{1-\rho^2}$.

E. Rebelo (✉) • P.O. do Valle • R. Nunes
Research Centre for Spatial and Organizational Dynamics (CIEO), University of Algarve, Algarve, Portugal
e-mail: elrebelo@ualg.pt; pvalle@ualg.pt; rnunes@ualg.pt

A. Pacheco et al. (eds.), *New Advances in Statistical Modeling and Applications*,
Studies in Theoretical and Applied Statistics, DOI 10.1007/978-3-319-05323-3__6,
© Springer International Publishing Switzerland 2014

The consequences of serial correlation in the model can be analysed in two different situations: (1) if x_t contains neither lagged dependent variables nor any current endogenous variables, $\hat{\beta}$ (OLS) is unbiased, therefore consistent, but not efficient. Worse than that, $\widehat{\text{Var}}\left(\hat{\beta}\right) = \hat{\sigma}^2 \left(X^T X\right)^{-1}$ is a biased and inconsistent estimator of the covariance matrix of the $\hat{\beta}$'s. Therefore, in this case, the statistical inference on the model is invalid because the standard errors of $\hat{\beta}_j$'s are inconsistent with bias of unknown signal; (2) if x_t contains lagged dependent variables and/or current endogenous variables, $\hat{\beta}$ is not even a consistent estimator. Therefore, in this situation, the statistical inference will also be invalid.

The Aitken's transformation is often adopted to deal with the serial correlation problem. Let us review how to obtain the Aitken's model.

Based on model (1), and lagging it an observation, we obtain the model $y_{t-1} = x_{t-1}\beta + u_{t-1}$. After multiplying both members of this equality by ρ, the model equals to $\rho y_{t-1} = \rho x_{t-1}\beta + \rho u_{t-1}$. This latter model can be subtracted from model (1), allowing to obtain $y_t - \rho y_{t-1} = (x_t - \rho x_{t-1})\beta + u_t - \rho u_{t-1}$, or finally

$$ y_t = x_t\beta + \rho\left(y_{t-1} - x_{t-1}\beta\right) + \xi_t; \xi_t \sim IID\left(0, w^2\right). \tag{2}$$

Model (2) is the Aitken's model which is spherical, that is with well-behaved disturbance terms, but nonlinear in the parameters since the regression function is $f_t(\beta, \rho) = x_t\beta + \rho(y_{t-1} - x_{t-1}\beta)$ depending on β as well as on ρ in a nonlinear fashion.

In model (2), regardless of the existence of lagged dependent variables, if x_t does not contain contemporary values of endogeneous variables, the application of nonlinear least squares (NLLS) produces consistent and asymptotically efficient estimates for β and ρ. Furthermore, the estimate of the covariance matrix of (β, ρ) will be consistent.

However, if x_t in model (2) contains also contemporaneous values of endogenous variables, only the application of the nonlinear instrumental variables method (NLIVM) to this model will produce consistent estimates for β, ρ, and for the covariance matrix of (β, ρ).

Therefore, the use of the Aitken's model is the solution to be adopted when the errors are genuinely serially correlated. However, evidence of serial correlation may result from two factors: (1) lack of dynamism in the model, transferring the missing lags (from y_t) to u_t; (2) incorrect omission of an important variable, lagged or not, transferring the phenomenon of serial correlation, always present in economic variables, to u_t. In any of these situations, misspecification of the regression function (the main part of the model) will not be resolved by the Aitken's model.

Within this framework, this study has two purposes. The first one is to propose a test for apparent serial correlation using the Gauss–Newton Regression (GNR) method, firstly proposed by Davidson and Mackinnon [1] in the context of non-nested testing. In his study, Godfrey [2] proposed tests for apparent serial correlation but using a different methodology. The second objective is to propose two tests to distinguish between genuine serial correlation and non-genuine serial correlation (evidence of misspecification). As we will explain, these latter tests can be named as Tests for Common Factor Restrictions. These tests were early proposed by Hendry and Mizon [3], Mizon and Hendry [4] and Sargan [7] but only in the context in which the estimation by OLS produces consistent estimates. In this study, Tests for Common Factor Restrictions are derived using a different methodology, based on the GNR, and assuming that x_t contains current endogenous variable so that the NLIVM must be used.

2 The Gauss–Newton Regression

The GNR is an auxiliary regression derived from the first-order Taylor expansion around one point in order to obtain a linear approximation of a nonlinear function as proposed in [1]. In a GNR the dependent variable is the residual variable of the model and the regressors are obtained by differentiation of the regression function with respect to the parameters as proposed in [5, 6].

Concerning model (2), its regression function is given by

$$f_t(\beta, \rho) = x_t \beta + \rho(y_{t-1} - x_{t-1}\beta). \tag{3}$$

Thus, the corresponding GNR will be given by

$$\xi_t = \frac{\partial f_t(\beta, \rho)}{\partial \beta} b + \frac{\partial f_t(\beta, \rho)}{\partial \rho} r + \text{error term}$$

which is equivalent to

$$y_t - x_t\beta - \rho(y_{t-1} - x_{t-1}\beta) = (x_t - \rho x_{t-1})b + (y_{t-1} - x_{t-1}\beta)r + \text{error term}$$

and, finally,

$$(y_t - \rho y_{t-1}) - (x_t - \rho x_{t-1})\beta = (x_t - \rho x_{t-1})b + r(y_{t-1} - x_{t-1}\beta) + \text{error term}. \tag{4}$$

Any restrictions on the parameters of a nonlinear regression function can be tested by estimating the corresponding GNR evaluated at restricted estimates, provided the estimation methods involved produce consistent root-n estimates under the null hypothesis.

3 Testing for Evidence of Serial Correlation

Consider again the regression function given by (3)

$$f_t(\beta, \rho) = x_t\beta + \rho(y_{t-1} - x_{t-1}\beta)$$

with corresponding GNR

$$(y_t - \rho y_{t-1}) - (x_t - \rho x_{t-1}) \beta = (x_t - \rho x_{t-1}) b + r (y_{t-1} - x_{t-1}\beta) + \text{error term}.$$

Regarding our first purpose, testing for apparent serial correlation is to test $H_0 : \rho = 0$ against $H_A : \rho \neq 0$. Consider $(\beta_R^*, 0)$ any consistent root-n restricted estimate of the parameter vector (β, ρ) under $H_0 : \rho = 0$. Due to the existence of contemporary variables in the right-hand side of the equation, this is $\hat{\beta}_R$, the instrumental variables estimator (IVE) of model (1). Evaluated at these estimates, the GNR becomes

$$y_t - x_t\hat{\beta}_R = x_t b + r \left(y_{t-1} - x_{t-1}\hat{\beta}_R\right) + \text{error term} \qquad (5)$$

and its estimation using the instrumental variables method (IVM) produces consistent estimates to b and r.

After this estimation, to test for serial correlation using the GNR approach, the GNR from the restricted model needs to be obtained. In this sense, and under the null hypothesis, the regression function itself is simply given by the linear function

$$f_t (\beta, \rho) = x_t\beta \qquad (6)$$

with corresponding GNR evaluated at $\hat{\beta}_R$ given by:

$$y_t - x_t\hat{\beta}_R = x_t b + \text{error term}. \qquad (7)$$

By comparing the GNRs (5) and (7), it is easy to see that to test the restriction $\rho = 0$ it is the same as testing the significance of the extra term $y_{t-1} - x_{t-1}\hat{\beta}_R$ in the GNR (5) by using the t statistics. Moreover, this extra term is the lagged residual of model (1), $\hat{u}_{t-1}$, previously estimated using the IVM (as proposed by Godfrey [2], using a different methodology).

It should be noted that both the estimation of GNR (5) and the estimate $\hat{\beta}_R$ that results from the estimation of β in model (1) (after imposition the null hypothesis $\rho = 0$) should be obtained using the same set of instruments used to create the instrumental variables for x_t. This procedure ensures that the test will be based on the Lagrange multiplier principle, since only in this case $\hat{u}_R$, the left-hand side of Eq. (5) will be orthogonal to x_t in GNR (5). In other words, just in this situation the GNR (7), estimated by IVM has no asymptotic explanatory power.

This implies that, for efficiency reasons, the set of instruments should include all predetermined variables included in $\{x_t, x_{t-1}, y_{t-1}\}$ (of course without repeating any of them) and other exogenous variables that explain the endogenous variables contained in x_t.

4 Testing for Common Factor Restrictions

In their studies, Hendry and Mizon [3], Mizon and Hendry [4], and Sargan [7] proposed this type of tests, but only in the context in which the estimation by OLS produces consistent estimates. In this section, two new tests, based on the IVM and using a GNR are proposed: a χ^2 test and a t test.

4.1 χ^2 Test

Let us return to model (2),

$$y_t = x_t \beta + \rho (y_{t-1} - x_{t-1}\beta) + \xi_t \; ; \; \xi_t \sim IID\left(0, w^2\right)$$

and consider, as alternative, model (8)

$$y_t = x_t \beta + \rho y_{t-1} + x_{t-1}\gamma + \xi_t \; ; \; \xi_t \sim IID\left(0, w^2\right). \tag{8}$$

The former can be rewritten as

$$(1 - \rho L) \, y_t = (1 - \rho L) \, x_t \beta + \xi_t \; ; \; \xi_t \sim IID\left(0, w^2\right) \tag{9}$$

where L is the lag operator, and the latter as

$$(1 - \rho L) \, y_t = x_t \beta + L x_t \gamma + \xi_t \; ; \; \xi_t \sim IID\left(0, w^2\right). \tag{10}$$

The absence of a common factor in the right side of Eq. (10) justifies the name given to the test. It is easily seen that model (9) is the restricted version of model (10) when imposing the nonlinear restrictions $\gamma = -\rho\beta$.

Consider now the regression function of the alternative (unrestricted) model (8),

$$f_t \left(\beta, \rho, \gamma \right) = x_t \beta + \rho y_{t-1} + x_{t-1}\gamma \tag{11}$$

which is a linear function in their parameters. The corresponding GNR will be given by

$$(y_t - \rho y_{t-1}) - (x_t \beta + x_{t-1}\gamma) = x_t b + c y_{t-1} + x_{t-1} d + \text{error term}. \tag{12}$$

Consider now $\left(\beta_R^*, \rho_R^*, -\rho_R^*\beta_R^*\right)$, a restricted consistent root-n estimate of the vector (β, ρ, γ) under the null hypothesis $H_0 : \gamma = -\rho\beta$. In that case, $\beta_R^* = \tilde{\beta}_R$ and $\rho_R^* = \tilde{\rho}_R$ [nonlinear IV estimates of model (2)].

Evaluating GNR (10) at this point, it becomes

$$(y_t - \tilde{\rho}_R y_{t-1}) - (x_t - \tilde{\rho}_R x_{t-1}) \tilde{\beta}_R = x_t b + c y_{t-1} + x_{t-1} d + \text{error term} \tag{13}$$

and its estimation by the IVM produces consistent estimates for b, c and d.

The next step is to clarify how to test $H_0 : \gamma = -\rho\beta$ after this estimation. With this aim, let us return to GNR (4), associated with the restricted model,

$$(y_t - \rho y_{t-1}) - (x_t - \rho x_{t-1})\beta = (x_t - \rho x_{t-1})b + r(y_{t-1} - x_{t-1}\beta) + \text{error term.}$$

This GNR assessed under the same nonlinear IV estimates becomes

$$(y_t - \tilde{\rho}_R y_{t-1}) - (x_t - \tilde{\rho}_R x_{t-1})\tilde{\beta}_R = (x_t - \tilde{\rho}_R x_{t-1})b + r\left(y_{t-1} - x_{t-1}\tilde{\beta}_R\right)$$

$$+\text{error term} \tag{14}$$

or, taking into account the regression function in (3)

$$\tilde{\xi} = \frac{\partial f_t(\beta,\rho)}{\partial \beta}\bigg|_{(\beta,\rho)=(\tilde{\beta}_R,\tilde{\rho}_R)} b + r\frac{\partial f_t(\beta,\rho)}{\partial \rho}\bigg|_{(\beta,\rho)=(\tilde{\beta}_R,\tilde{\rho}_R)} + \text{error term} \tag{15}$$

which allows us to conclude that this GNR has no explanatory asymptotic power, because all regressor vectors are orthogonal (asymptotically) with $\tilde{\xi}$, the vector of residuals of model (2), obtained by nonlinear IV (first order conditions of the IV minimization process).

Under these circumstances, the comparison of the sum of squared residuals (SSR) of the GNR's (11) and (12) to test the restriction (using an F test) becomes unnecessary. Instead, it is enough to perform a χ^2 test using the statistics $(n - k - 1) R^2(g)$ calculated from (11). In fact, GNR (12), the restricted one, is composed of $k + 1$ parameters and would be estimated based on n observations (where n represents the number of observations actually used). The adjustment to finite samples is thus given by $n - k - 1$.

It is also important to note that the number of degrees of freedom (g) is not equal to the number of non-redundant regressors in (11). It is rather given by the difference between that number and $k + 1$ [number of parameters in GNR (12)]. For example, if $x_t = \{1, z_{1,t}, z_{2,t}, y_{t-1}\}$, $x_{t-1} = \{1, z_{1,t-1}, z_{2,t-1}, y_{t-2}\}$. Therefore, the number of non-redundant regressors in (11) seven. Since in this case Eq. (12) contains five parameters (four b's plus one r), the number of degrees of freedom is two ($g = 2$).

4.2 T Test

To complete the "Common Factor Restrictions" family of tests there is still another test that can be proposed. So, let us return to model (2)

$$y_t = x_t\beta + \rho(y_{t-1} - x_{t-1}\beta) + \xi_t$$

and consider, as alternative,

$$y_t = x_t\beta + \rho y_{t-1} - \delta x_{t-1}\beta + \xi_t. \tag{16}$$

It is easily seen to model (2) is the restricted version of model (13) when imposing the linear restriction $\delta = \rho$.

As an alternative, models (2) and (13) can be rewritten as

$$(1 - \rho L)\, y_t = (1 - \rho L)\, x_t \beta + \xi_t \tag{17}$$

and

$$(1 - \rho L)\, y_t = (1 - \delta L)\, x_t \beta + \xi_t \tag{18}$$

to clarify, once again, the name given to the test.

The regression function of this alternative model (non-restricted) is

$$f_t\,(\beta, \rho, \delta) = x_t \beta + \rho y_{t-1} - \delta x_{t-1}\beta \tag{19}$$

which is nonlinear in the parameters. The corresponding GNR will be

$$(y_t - \rho y_{t-1}) - (x_t - \delta x_{t-1})\,\beta = (x_t - \delta x_{t-1})\,b + r y_{t-1} - d x_{t-1}\beta + \text{error term.} \tag{20}$$

Consider now $\left(\beta_R^*, \rho_R^*, \delta_R^*\right)$, a restricted consistent root-n estimates of the vector (β, ρ, δ) under the null hypothesis $H_0 : \delta = \rho$. Using, once again, the estimates given by the NLIV applied to model (2), when evaluated at those estimates the GNR becomes

$$(y_t - \tilde{\rho}_R y_{t-1}) - (x_t - \tilde{\rho}_R x_{t-1})\,\tilde{\beta}_R = (x_t - \tilde{\rho}_R x_{t-1})\,b + r y_{t-1} - d x_{t-1}\tilde{\beta}_R$$

$$+\text{error term} \tag{21}$$

or, after subtracting and adding up the term $r x_{t-1}\beta_R^*$ in the right-hand side of the equation, and collecting terms,

$$(y_t - \tilde{\rho}_R y_{t-1}) - (x_t - \tilde{\rho}_R x_{t-1})\,\tilde{\beta}_R =$$
$$(x_t - \tilde{\rho}_R x_{t-1})\,b + r\left(y_{t-1} - x_{t-1}\tilde{\beta}_R\right) - (r - d)\,x_{t-1}\tilde{\beta}_R + \text{error term.} \tag{22}$$

Comparing (22) and (4), once evaluated at the same point,

$$(y_t - \tilde{\rho}_R y_{t-1}) - (x_t - \tilde{\rho}_R x_{t-1})\,\tilde{\beta}_R = (x_t - \tilde{\rho}_R x_{t-1})\,b + r\left(y_{t-1} - x_{t-1}\tilde{\beta}_R\right)$$

$$+\text{error term}$$

it is straightforward to conclude that, for testing the restriction $\delta = \rho$ it is enough to use a simple t test to test the significance of the extra term $x_{t-1}\tilde{\beta}_R$ in regression (22), after its estimation with IV using the same set of instruments (the previously used in the serial correlation test, in the previous section).

5   Conclusions

In this work, two types of tests have been deduced using the same methodology, based on the GNR. The first type allows identifying evidence of serial correlation (if the null hypothesis is rejected). The second type, designated as "Common Factor Restrictions" tests, intends to discriminate between genuine and non-genuine serial correlation. Both the common factor tests should be carried out, as any one may have better performance than the other (as reported by Mizon and Hendry [4] in an OLS context).

Aitken's model must be chosen if and only if both common factor tests do not reject the corresponding null hypotheses. This is the case of genuine serial correlation. Otherwise, if at least one of the common factor tests rejects the null hypothesis, the Aitken's model must be seen as a restricted version of more general models, where false restrictions have been imposed. This is the case of non-genuine serial correlation. In this situation, the NLIV estimates applied to the Aitken's model will be inconsistent and the inference based on those estimates will be invalid.

Acknowledgements Supported by FCT under the Annual Financial Support Program for Research Units granted to CEC and CMUP.

References

1. Davidson, R., Mackinnon, J.G.: Several tests for model specification in the presence of alternative hypotheses. Econometrica **49**, 781–793 (1981)
2. Godfrey, L.G.: Misspecification Tests in Econometrics. Cambridge University Press, Cambridge (1988)
3. Hendry, D.F., Mizon, G.E.: Serial correlation as a convenient simplification not a nuisance: a comment on a study of the demand for money by the bank of England. Econ. J. **88**, 63 (1978)
4. Mizon, G.E., Hendry, D.F.: An empirical and Monte Carlo analysis of tests of dynamic specification. Rev. Econ. Stud. **47**, 21–45 (1980)
5. Rebelo, E., Gonçalves, I.: Testes ao Modelo de Aitken. In: Proceedings of the IX Congress of the Statistics Portuguese Society, pp. 319–326 (2002)
6. Rebelo, E., Valle, P., Nunes, R.: Nonnested testing for competing autoregressive dynamic models estimated by instrumental variables. Commun. Stat. Theory Methods **41**(20), 3799–3812 (2012)
7. Sargan, J.D.: Some tests of dynamic specification for a single equation. Econometrica **48**, 879–897 (1980)

Part II

Probability and Stochastic Processes

Cantor Sets with Random Repair

M. Fátima Brilhante, Dinis Pestana, and M. Luísa Rocha

Abstract

The effect of random repair in each step of the construction of Cantor-like sets, defined by the union of segments determined by the minimum and maximum of two independent observations from a population with support on $[0, 1]$, is investigated here. Independence between the samples used in the damage and repair stages is also assumed. The final assessment of the repair benefits is done in terms of the mean diameter and mean total length of the set obtained after a small number of iterations.

1 Introduction

In many important biological and industrial issues such as in the recovery of patients with cerebral lesions, or in the recovery of information in damaged storage units, the damage extent or the ability to repair what is damaged is often random. These kinds of issues motivated us to evaluate the random repair benefits in Cantor-like sets (a much simpler setting).

M.F. Brilhante (✉)
Universidade dos Açores (DM) and CEAUL, Campus de Ponta Delgada, Apartado 1422, 9501-801 Ponta Delgada, Portugal
e-mail: fbrilhante@uac.pt

D. Pestana
Faculdade de Ciências (DEIO) and CEAUL, Universidade de Lisboa, Bloco C6, Piso 4, Campo Grande, 1749-016 Lisboa, Portugal
e-mail: dinis.pestana@fc.ul.pt

M.L. Rocha
Universidade dos Açores (DEG) and CEEAplA, Campus de Ponta Delgada, Apartado 1422, 9501-801 Ponta Delgada, Portugal
e-mail: lrocha@uac.pt

A. Pacheco et al. (eds.), *New Advances in Statistical Modeling and Applications*, Studies in Theoretical and Applied Statistics, DOI 10.1007/978-3-319-05323-3_7, © Springer International Publishing Switzerland 2014

A huge variety of deterministic and random Cantor-like sets can be constructed using different iterative procedures. In most common constructions some part of the set is deleted using a deterministic or stochastic rule in each step. Pestana and Aleixo [1] introduced a stuttering procedure in which each deletion (damage) stage is followed by a partial random reconstruction (repair) stage, and where redundancy is allowed in the sense that repair can operate on non-damaged areas. In their study they used order statistics from some Beta populations to model the segment to delete or to repair. The assessment of the random repair benefits was done in terms of the Hausdorff dimension of the fractal obtained as limit of the damage/repair iterative procedure. (Note that the Hausdorff dimension allows to measure complicated geometric forms such as fractals.)

In Aleixo et al. [2] a wider range for the beta parameters was considered in order to reveal the asymptotic effect of different combinations of damage/repair. These authors observed two important features: (1) the Hausdorff dimension is always significantly lower for the fractals that only suffered damage stages, when compared to the corresponding damage/repair fractals; (2) the Hausdorff dimension decreases under randomness in the damage stage. For practical purposes Aleixo et al. [2] also presented plots for some combinations of Beta damage/repair and of Beta or BetaBoop damage/repair, for steps 2 and 3 of the construction procedure. (Note that the BetaBoop family, introduced by Brilhante et al. [3], generalizes the Beta family.)

One of the nasty effects of random repair is obviously the possibility of repairing what is not damaged—this is actually common in many aspects of our everyday life. For example, in scheduled car maintenances, the automaker protocols often establish that some auto parts (e.g., spark plugs and fan belt) should be replaced even if in perfect conditions. Since it is important to observe what happens with Cantor-like sets with random repair after a moderate number of steps, we decided to compare the mean diameter and mean total length of each set obtained after a small number of iterations, for some combinations of Beta damage/repair in order to assess the actual repair benefits.

This paper is organized as follows. In Sect. 2 we describe how to construct a stuttering Cantor-like random set and revisit some of its important features. In Sect. 3 we tabulate the simulated mean diameter and mean total length, based on 5,000 runs, for a small number of iterations and for some combinations of Beta damage/repair.

2 Stuttering Cantor-Like Random Sets Construction Procedure

Let F_k, $k = 1, 2, \ldots$, denote the set that we obtain after k steps of damage/repair. Starting with $F_0 = [0, 1]$, the set F_1 is constructed in the following way:

Damaging Stage: Generate two independent random points X_1 and X_2 from a parent population X with support on $[0, 1]$ and delete from F_0 the set $(X_{1:2}, X_{2:2})$, where $X_{j:2}$, $j = 1, 2$, represents the jth order statistic of the sample (X_1, X_2);

Table 1 Summary statistics for the estimated mean diameters and estimated mean total lengths under deterministic damage stages

	D_1	D_2	D_3	D_4	D_5	D_6	D_7
No repair							
Min	0.2974	0.0884	0.0263	0.0078	0.0023	0.0007	0.0002
Max	0.6429	0.4133	0.2657	0.1708	0.1098	0.0706	0.0454
Mean	0.4718	0.2358	0.1239	0.0677	0.0383	0.0222	0.0131
SD	0.1214	0.1161	0.0871	0.0604	0.0406	0.0268	0.0176
With repair							
Min	0.3897	0.1686	0.0746	0.0340	0.0157	0.0073	0.0034
Max	0.7697	0.5915	0.4569	0.3516	0.2700	0.2088	0.1611
Mean	0.5899	0.3676	0.2352	0.1536	0.1018	0.0684	0.0465
SD	0.0937	0.1024	0.0909	0.0748	0.0594	0.0463	0.0358
	L_1	L_2	L_3	L_4	L_5	L_6	L_7
No repair							
Min	0.5947	0.3537	0.2103	0.1251	0.0744	0.0442	0.0263
Max	0.7857	0.6173	0.4851	0.3811	0.2994	0.2353	0.1849
Mean	0.7369	0.5464	0.4075	0.3053	0.2297	0.1735	0.1314
SD	0.0619	0.0863	0.0909	0.0856	0.0761	0.0654	0.0549
With repair							
Min	0.6898	0.4761	0.3272	0.2253	0.1553	0.1069	0.0736
Max	0.8911	0.7925	0.7052	0.6289	0.5598	0.4995	0.4448
Mean	0.8314	0.6933	0.5795	0.4859	0.4084	0.3441	0.2905
SD	0.0456	0.0740	0.0905	0.0987	0.1012	0.1000	0.0964

Repair Stage: Generate two other independent random points Y_1 and Y_2, independent of X_1 and X_2, from a parent population Y with support on $[0, 1]$, and set
$F_1 = [F_0 - (X_{1:2}, X_{2:2})] \cup (Y_{1:2}, Y_{2:2})$.
Construction of F_k, $k = 2, 3, \ldots$:

Damaging Stage: For each segment $S_{i,k-1}$ of $F_{k-1} = \bigcup_i S_{i,k-1}$, generate two independent random points $X_{1;i,k-1}$ and $X_{2;i,k-1}$ from the parent population X truncated on $S_{i,k-1}$ and delete from $S_{i,k-1}$ the set $(X_{1:2;i,k-1}, X_{2:2;i,k-1})$, where $X_{j:2;i,k-1}$, $j = 1, 2$, denotes the jth order statistic of the sample $(X_{1;i,k-1}, X_{2;i,k-1})$;

Repair Stage: For each segment $S_{i,k-1}$, generate two other random points $Y_{1;i,k-1}$ and $Y_{2;i,k-1}$ from the parent population Y truncated on $S_{i,k-1}$, independent of $X_{1;i,k-1}$ and $X_{2;i,k-1}$, and set

$$F_k = \bigcup_i \{[S_{i,k-1} - (X_{1:2;i,k-1}, X_{2:2;i,k-1})] \cup (Y_{1:2;i,k-1}, Y_{2:2;i,k-1})\}.$$

The fractal $\mathscr{F} = \bigcap_{k=1}^{\infty} F_k$ is the stuttering Cantor-like random set.

Table 2 Summary statistics for the estimated mean diameters and estimated mean total lengths under random damage stages

	D_1	D_2	D_3	D_4	D_5	D_6	D_7
No repair							
Min	0.4964	0.2521	0.1310	0.0691	0.0369	0.0198	0.0106
Max	0.6540	0.4284	0.2807	0.1858	0.1228	0.0821	0.0546
Mean	0.5489	0.3094	0.1785	0.1050	0.0624	0.0377	0.0230
SD	0.0598	0.0668	0.0577	0.0450	0.0333	0.0241	0.0172
With repair							
Min	0.5624	0.3233	0.1894	0.1133	0.0684	0.0411	0.0249
Max	0.7745	0.6024	0.4689	0.3653	0.2847	0.2227	0.1750
Mean	0.6452	0.4250	0.2847	0.1932	0.1325	0.0917	0.0640
SD	0.0536	0.0684	0.0675	0.0601	0.0509	0.0419	0.0339
	L_1	L_2	L_3	L_4	L_5	L_6	L_7
No repair							
Min	0.5959	0.3585	0.2140	0.1277	0.0757	0.0450	0.0267
Max	0.7898	0.6207	0.4884	0.3845	0.3018	0.2371	0.1862
Mean	0.7369	0.5454	0.4074	0.3056	0.2297	0.1734	0.1312
SD	0.0621	0.0852	0.0904	0.0853	0.0759	0.0652	0.0548
With repair							
Min	0.6883	0.4732	0.3243	0.2239	0.1536	0.1057	0.0726
Max	0.8860	0.7841	0.6966	0.6162	0.5462	0.4843	0.4281
Mean	0.8202	0.6744	0.5568	0.4610	0.3827	0.3185	0.2656
SD	0.0480	0.0765	0.0913	0.0976	0.0980	0.0949	0.0897

If in the previous construction procedure we establish a deterministic damaging stage, the set to delete from each segment $S_{i,k-1}$ is $(\mathbb{E}(X_{1:2;i,k-1}), \mathbb{E}(X_{2:2;i,k-1}))$. For example, the classical Cantor set only has deterministic damaging stages and uses the expected value of the order statistics of samples of size 2 from the standard uniform distribution.

Aleixo and Pestana [1] showed that when X and Y are both standard uniform random variables,

$$
F_1 = \begin{cases}
[0,1] & \text{if } Y_{1:2} < X_{1:2} \text{ and } Y_{2:2} > X_{2:2} \text{ (with probability } \tfrac{1}{6}) \\
[0, X_{1:2}] \cup [X_{2:2}, 1] & \text{if } Y_{2:2} < X_{1:2} \text{ or } Y_{1:2} > X_{2:2} \text{ (with probability } \tfrac{1}{3}) \\
[0, Y_{2:2}] \cup [X_{2:2}, 1] & \text{if } Y_{1:2} < X_{1:2} < Y_{2:2} < X_{2:2} \text{ (with probability } \tfrac{1}{6}) \\
[0, X_{1:2}] \cup [Y_{1:2}, 1] & \text{if } X_{1:2} < Y_{1:2} < X_{2:2} < Y_{2:2} \text{ (with probability } \tfrac{1}{6}) \\
[0, X_{1:2}] \cup [Y_{1:2}, Y_{2:2}] \cup [X_{2:2}, 1] & \text{if } X_{1:2} < Y_{1:2} < Y_{2:2} < X_{2:2} \text{ (with probability } \tfrac{1}{6})
\end{cases}
\tag{1}
$$

and hence the random set $F_1 = \bigcup_{i=1}^{N_1} S_{i,1}$, where N_1, which counts the number of segments forming F_1, has probability mass function $\mathbb{P}(N_1 = 1) = \mathbb{P}(N_1 = 3) = \tfrac{1}{6}$ and $\mathbb{P}(N_1 = 2) = \tfrac{2}{3}$. For other parent populations X and Y the random set F_1 is also defined by (1), but with N_1 having a different distribution. From the self-

Table 3 Mean diameter under deterministic damage stages

$(p_d, q_d)/(p_r, q_r)$	D_1	D_2	D_3	D_4	D_5	D_6	D_7
(1,1)/no repair	0.33333	0.11111	0.03704	0.01235	0.00412	0.00137	0.00046
(1,1)/(0.5,0.5)	0.59340	0.38360	0.24784	0.16701	0.11020	0.07365	0.05015
(1,1)/(1,1)	0.55480	0.33483	0.20442	0.12791	0.08099	0.05207	0.03311
(1,1)/(1,2)	0.48028	0.25641	0.14067	0.07846	0.04426	0.02525	0.01461
(1,1)/(1,3)	0.41680	0.19225	0.09339	0.04614	0.02291	0.01150	0.00576
(1,1)/(2,1)	0.48807	0.26295	0.14336	0.08070	0.04568	0.02634	0.01500
(1,1)/(2,2)	0.50679	0.27920	0.15923	0.09189	0.05356	0.03123	0.01870
(1,1)/(2,3)	0.46801	0.24359	0.12979	0.07092	0.0385	0.02121	0.01179
(1,1)/(3,1)	0.41921	0.19348	0.09477	0.04657	0.02317	0.01164	0.00595
(1,1)/(3,2)	0.47533	0.24690	0.13119	0.07117	0.03939	0.02179	0.01219
(1,1)/(3,3)	0.47689	0.24342	0.12983	0.07155	0.03925	0.02184	0.01223
(3,2)/no repair	0.48571	0.23592	0.11459	0.05566	0.02703	0.01313	0.00638
(3,2)/(0.5,0.5)	0.68094	0.47254	0.33566	0.2387	0.17243	0.12507	0.09181
(3,2)/(1,1)	0.65354	0.4316	0.29214	0.20013	0.13832	0.09671	0.06748
(3,2)/(1,2)	0.57362	0.33386	0.19576	0.11727	0.07045	0.04275	0.02614
(3,2)/(1,3)	0.52282	0.2746	0.14578	0.07813	0.04209	0.02289	0.01246
(3,2)/(2,1)	0.62097	0.39492	0.25764	0.17022	0.11398	0.07628	0.05145
(3,2)/(2,2)	0.62089	0.38799	0.24599	0.15857	0.10294	0.06744	0.04424
(3,2)/(2,3)	0.57275	0.33179	0.19366	0.11434	0.06786	0.04058	0.0244
(3,2)/(3,1)	0.56763	0.33281	0.19765	0.12013	0.07321	0.04543	0.02821
(3,2)/(3,2)	0.61036	0.38155	0.24096	0.15436	0.10110	0.06645	0.04343
(3,2)/(3,3)	0.59508	0.36039	0.22041	0.13680	0.08554	0.05394	0.03434

similarity property of fractals it follows that $F_k = \bigcup_{i=1}^{N_k} S_{i,k}$, $k = 2, 3, \ldots$, where $N_k = \sum_{i=1}^{N_{k-1}} N_{1,i}$ and the $N_{1,i}$'s are independent replicas of N_1. Note that if (X_1, X_2) and (Y_1, Y_2) in the construction procedure above are not identically distributed, the mathematical analysis of the problem becomes more complex, since we no longer can consider (X_1, X_2, Y_1, Y_2) as a random sample from the same population, which ultimately affects the distribution of the counting variables N_k.

3 Random Repair Benefits for Cantor-Like Sets

Since in reality no item undergoes infinite repair, it is crucial to see what happens after a moderate number of cycles of damage/repair in the construction of Cantor-like sets. In order to evaluate the effective random repair benefit, we shall compare the mean diameter and mean total length of each set obtained after a moderate number of steps for some combinations of Beta damage/repair. (The diversity of

Table 4 Mean diameter under random damage stages

$(p_d, q_d)/(p_r, q_r)$	D_1	D_2	D_3	D_4	D_5	D_6	D_7
(1,1)/no repair	0.49947	0.25477	0.13401	0.07174	0.03861	0.02094	0.01135
(1,1)/(0.5,0.5)	0.67626	0.46719	0.32629	0.22826	0.16096	0.11579	0.08259
(1,1)/(1,1)	0.64736	0.42635	0.28304	0.18745	0.12570	0.08520	0.05771
(1,1)/(1,2)	0.59742	0.36590	0.22579	0.14037	0.08926	0.05715	0.03677
(1,1)/(1,3)	0.56406	0.32795	0.19294	0.11477	0.06835	0.04107	0.02486
(1,1)/(2,1)	0.60070	0.36815	0.22778	0.14451	0.09046	0.05750	0.03681
(1,1)/(2,2)	0.59895	0.36705	0.23001	0.1456	0.09201	0.05807	0.03772
(1,1)/(2,3)	0.58491	0.34727	0.21099	0.12907	0.07994	0.04929	0.03036
(1,1)/(3,1)	0.56244	0.32325	0.18941	0.11327	0.06860	0.04141	0.02492
(1,1)/(3,2)	0.58489	0.35171	0.21369	0.13171	0.08077	0.05089	0.03193
(1,1)/(3,3)	0.58239	0.34385	0.20649	0.12549	0.07694	0.04770	0.02987
(3,2)/no repair	0.52870	0.28481	0.15641	0.08659	0.04827	0.02699	0.01524
(3,2)/(0.5,0.5)	0.70856	0.51136	0.37223	0.27176	0.20045	0.14893	0.10952
(3,2)/(1,1)	0.68533	0.47660	0.33334	0.23497	0.16701	0.12009	0.08565
(3,2)/(1,2)	0.61797	0.38679	0.24678	0.15894	0.10386	0.06811	0.04494
(3,2)/(1,3)	0.57612	0.33651	0.20228	0.12120	0.07374	0.04514	0.02781
(3,2)/(2,1)	0.64779	0.43075	0.28931	0.19708	0.13486	0.09292	0.06474
(3,2)/(2,2)	0.64636	0.42672	0.28220	0.18854	0.12754	0.08752	0.05951
(3,2)/(2,3)	0.61491	0.38465	0.24352	0.15654	0.10108	0.06493	0.04232
(3,2)/(3,1)	0.60775	0.37737	0.24067	0.15501	0.10141	0.06554	0.04359
(3,2)/(3,2)	0.63287	0.41263	0.26968	0.17918	0.11899	0.08035	0.05429
(3,2)/(3,3)	0.62521	0.39594	0.25522	0.16389	0.10732	0.07062	0.04656

forms in the Beta family makes it an interesting candidate to model randomness patterns in [0, 1].)

Let $L_{i,k}$ denote the length of the segment $S_{i,k}$ in $F_k = \bigcup_i S_{i,k}$, and let also D_k and L_k denote the mean diameter and the mean total length of F_k, respectively, i.e.

$$D_k = \mathbb{E}\left(\max_i L_{i,k}\right) \quad \text{and} \quad L_k = \mathbb{E}\left(\sum_i L_{i,k}\right), \quad k = 0, 1, 2, \dots$$

For the classical Cantor set we have $D_k = \left(\frac{1}{3}\right)^k$ and $L_k = \left(\frac{2}{3}\right)^k$, $k = 0, 1, 2, \dots$ It is quite straightforward to obtain the exact values of D_1 and L_1 for the Cantor set under deterministic damaging stages and random repair stages, both modeled by the standard uniform distribution (i.e., Beta(1,1) distribution). For this particular case we know that

Table 5 Mean total length under deterministic damage stages

$(p_d, q_d)/(p_r, q_r)$	L_1	L_2	L_3	L_4	L_5	L_6	L_7
(1,1)/no repair	0.66667	0.44444	0.29630	0.19753	0.13169	0.08779	0.05853
(1,1)/(0.5,0.5)	0.83147	0.69074	0.57210	0.47671	0.39562	0.32834	0.27271
(1,1)/(1,1)	0.82764	0.68559	0.56572	0.46823	0.38709	0.32058	0.26532
(1,1)/(1,2)	0.78708	0.62082	0.48911	0.38559	0.30392	0.23982	0.18919
(1,1)/(1,3)	0.74112	0.55107	0.40935	0.30351	0.22543	0.16747	0.12436
(1,1)/(2,1)	0.79220	0.62658	0.49332	0.38939	0.30669	0.24184	0.19055
(1,1)/(2,2)	0.81924	0.67168	0.55109	0.45133	0.37042	0.30382	0.24922
(1,1)/(2,3)	0.79916	0.64010	0.51208	0.40985	0.32777	0.26218	0.20968
(1,1)/(3,1)	0.74296	0.55192	0.41030	0.30483	0.22636	0.16816	0.12491
(1,1)/(3,2)	0.80113	0.64217	0.51425	0.41127	0.32906	0.26327	0.21058
(1,1)/(3,3)	0.81521	0.66237	0.53821	0.43805	0.35656	0.28992	0.2358
(3,2)/no repair	0.77143	0.59510	0.45908	0.35415	0.27320	0.21075	0.16258
(3,2)/(0.5,0.5)	0.88185	0.77799	0.68791	0.60720	0.53612	0.47340	0.41836
(3,2)/(1,1)	0.87879	0.77196	0.67809	0.59639	0.52451	0.46121	0.40539
(3,2)/(1,2)	0.83248	0.69339	0.57747	0.48115	0.40067	0.33370	0.27805
(3,2)/(1,3)	0.79882	0.63894	0.51139	0.40941	0.32789	0.26247	0.21011
(3,2)/(2,1)	0.87521	0.76524	0.66919	0.58531	0.51194	0.44754	0.39178
(3,2)/(2,2)	0.87377	0.76086	0.66260	0.57770	0.50345	0.43872	0.38226
(3,2)/(2,3)	0.83930	0.70431	0.59118	0.49611	0.41583	0.34864	0.29217
(3,2)/(3,1)	0.84708	0.71889	0.60965	0.51789	0.43935	0.37249	0.31618
(3,2)/(3,2)	0.87986	0.77486	0.68159	0.59859	0.52692	0.46368	0.4079
(3,2)/(3,3)	0.86593	0.75044	0.64996	0.56350	0.48751	0.42233	0.36562

$$
F_1 = \begin{cases}
[0,1] & \text{if } Y_{1:2} < \frac{1}{3} \text{ and } Y_{2:2} > \frac{2}{3} \text{ (with probability } \frac{2}{9}) \\
[0,\frac{1}{3}] \cup [\frac{2}{3},1] & \text{if } Y_{2:2} < \frac{1}{3} \text{ or } Y_{1:2} > \frac{2}{3} \text{ (with probability } \frac{2}{9}) \\
[0,Y_{2:2}] \cup [\frac{2}{3},1] & \text{if } Y_{1:2} < \frac{1}{3} < Y_{2:2} < \frac{2}{3} \text{ (with probability } \frac{2}{9}) \\
[0,\frac{1}{3}] \cup [Y_{1:2},1] & \text{if } \frac{1}{3} < Y_{1:2} < \frac{2}{3} < Y_{2:2} \text{ (with probability } \frac{2}{9}) \\
[0,\frac{1}{3}] \cup [Y_{1:2},Y_{2:2}] \cup [\frac{2}{3},1] & \text{if } \frac{1}{3} < Y_{1:2} < Y_{2:2} < \frac{2}{3} \text{ (with probability } \frac{1}{9})
\end{cases}
\tag{2}
$$

(see Aleixo and Pestana [1]). If we denote by $D_1^* = \max_i L_{i,1}$ and $L_1^* = \sum_i L_{i,1}$ the diameter and total length of F_1, respectively, it follows from (2) that D_1^* can be $\frac{1}{3}$, $\frac{1}{2}$ and 1 with probability $\frac{3}{9}$, $\frac{4}{9}$ and $\frac{2}{9}$, respectively, and L_1^* can be $\frac{2}{3}$, $\frac{7}{9}$, $\frac{5}{6}$ and 1 with probability $\frac{2}{9}$, $\frac{1}{9}$, $\frac{4}{9}$ and $\frac{2}{9}$, respectively. Therefore, $D_1 = \mathbb{E}(D_1^*) = \frac{5}{9}$ and $L_1 = \mathbb{E}(L_1^*) = \frac{67}{81}$.

Comparing these values with the homologous values for the classical Cantor set, we see that there is a significant repair benefit just after the first repair intervention (an improvement of approximately 66.7 % for D_1 and 24.1 % for L_1). For $k > 1$ it becomes unfeasible to obtain exact values for D_k and L_k, and even messier if we try to use other beta models. However, this obstacle can be bypassed if we estimate D_k and L_k through Monte Carlo methods.

Table 6 Mean total length under random damage stages

$(p_d, q_d)/(p_r, q_r)$	L_1	L_2	L_3	L_4	L_5	L_6	L_7
(1,1)/no repair	0.66700	0.44241	0.29551	0.19751	0.13162	0.08786	0.05854
(1,1)/(0.5,0.5)	0.81462	0.66487	0.54424	0.44361	0.36198	0.29683	0.24213
(1,1)/(1,1)	0.80354	0.64421	0.51653	0.41308	0.33027	0.26421	0.21139
(1,1)/(1,2)	0.77185	0.59437	0.45774	0.35237	0.27157	0.20971	0.16192
(1,1)/(1,3)	0.74046	0.55071	0.41129	0.30672	0.22797	0.16967	0.12603
(1,1)/(2,1)	0.77376	0.59467	0.45832	0.35288	0.27190	0.20959	0.1615
(1,1)/(2,2)	0.77406	0.60272	0.46885	0.36482	0.28383	0.22022	0.17185
(1,1)/(2,3)	0.76488	0.58319	0.44518	0.33954	0.25957	0.19815	0.15135
(1,1)/(3,1)	0.74154	0.54932	0.40959	0.30550	0.22768	0.16926	0.12581
(1,1)/(3,2)	0.76445	0.58586	0.44724	0.34222	0.26118	0.20062	0.15323
(1,1)/(3,3)	0.76479	0.58202	0.44448	0.33938	0.25903	0.19796	0.15168
(3,2)/no repair	0.77248	0.59464	0.45883	0.35457	0.27329	0.21094	0.16274
(3,2)/(0.5,0.5)	0.87827	0.77066	0.67671	0.59450	0.52172	0.45820	0.40160
(3,2)/(1,1)	0.87190	0.75859	0.66134	0.57549	0.50156	0.43603	0.37935
(3,2)/(1,2)	0.83258	0.69136	0.57661	0.48104	0.40247	0.33582	0.28062
(3,2)/(1,3)	0.80760	0.65155	0.52705	0.42611	0.34473	0.27925	0.22575
(3,2)/(2,1)	0.85970	0.73975	0.63569	0.54758	0.47075	0.40544	0.34928
(3,2)/(2,2)	0.85848	0.73706	0.63148	0.54057	0.46307	0.39714	0.34021
(3,2)/(2,3)	0.83533	0.69846	0.58352	0.48893	0.40895	0.34211	0.28637
(3,2)/(3,1)	0.84099	0.70566	0.59342	0.49920	0.42057	0.35294	0.29724
(3,2)/(3,2)	0.85589	0.73376	0.62809	0.53793	0.46005	0.39424	0.33757
(3,2)/(3,3)	0.84829	0.71908	0.60992	0.51661	0.43735	0.37050	0.31397

The beta parameters considered in our simulation study are (0.5,0.5), (1,1), (1,2), (1,3), (2,1), (2,2), (2,3), (3,1), (3,2), and (3,3), and the maximum of seven iterations was performed. Due to space restrictions, we only present here a small subset of the results obtained, which are indicated in Tables 3, 4, 5, and 6, and do not indicate the standard errors associated with each estimate. However, we inform that the maximum standard error of all estimates under deterministic damage stages is less than 0.0041 for the mean diameters and less than 0.0025 for the mean total lengths. Under random damage stages the standard errors are less than 0.0039 for the mean diameter estimates and less than 0.0042 for the mean total length estimates. In Tables 1 and 2 we indicate some summary statistics for the estimated values under each type of damage stage, taking into consideration all combinations of Beta damage/repair studied (Note that these summaries are based on samples of means, where each value is obtained in a different setting.)

Analyzing individually each Tables 3, 4, 5, 6 we observe that higher values are always attained when a damaging stage (deterministic or random) is followed by a (random) repair stage. Comparing the homologous mean diameter values of Tables 3 and 4, we see that the mean value is slightly higher under randomness in the damage stage (with or without repair). On the other hand, if we compare the homologous mean total length values of Tables 5 and 6, these are, in general,

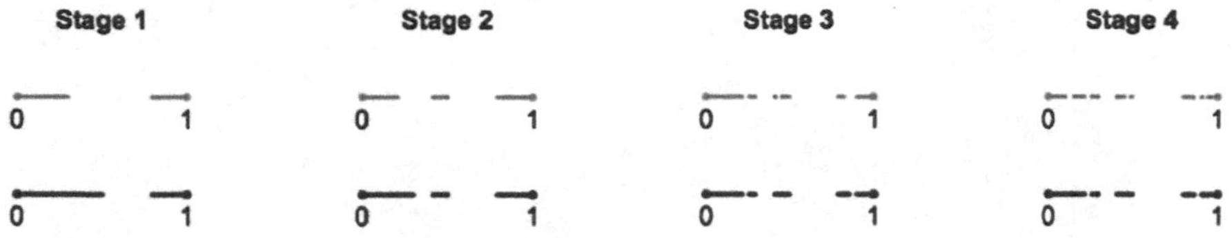

Stage 1	Stage 2	Stage 3	Stage 4
0　　　1	0　　　1	0　　　1	0　　　1
0　　　1	0　　　1	0　　　1	0　　　1

Fig. 1 Plots for the first four stages of a stuttering Cantor-like random set construction

higher under deterministic damage stages (with or without repair), although the difference between some values can be quite small. We also notice that the lowest D_k and L_k values are achieved for combinations of damage/repair involving the Beta(1,3) or Beta(3,1) models (note that these two models have the highest and lowest skewness coefficients of the group). The previous conclusions remain true for all other cases which are not shown here. Comparing now values in Tables 1 and 2 with $D_0 = L_0 = 1$ for the initial set $[0, 1]$, we also see that there seems to be some repair benefit even after seven iterations, although the range of possible values within each case can be considerable, specially for the mean total length under random repair stages.

In order to actually observe what happens in the first steps of a stuttering Cantor-like random set, we show in Fig. 1 the first four stages of the construction using (1,1) and (2,3) as Beta damage and Beta repair parameters, respectively. The plots in gray represent the set obtained after damage is inflicted and the plots in black after repair. In this case, there is always a partial reconstruction.

Acknowledgments This research has been supported by National Funds through FCT—Fundação para a Ciência e a Tecnologia, project PEst-OE/MAT/UI0006/2011.

References

1. Aleixo, S., Pestana, D.: Stuttering Cantor-like random sets. In: Luzar-Stiffler, V., et al. (eds.) Proceedings of the 33rd International Conference on Information Technology Interfaces, pp. 29–34 (2011)
2. Aleixo, S., Brilhante, M.F., Pestana, D.: General stuttering $Beta(p, q)$ Cantor-like random sets. In: Proceedings of the 58th International Statistical Institute Congress, Session CPS008, pp. 6076–6081, Dublin, Ireland (2011)
3. Brilhante, M.F., Gomes, M.I., Pestana, D.: BetaBoop brings in chaos. In: Proceedings of the 4th Chaotic Modeling and Simulation International Conference, pp. 45–52. Crete, Greece (2011)

Nearest Neighbor Connectivity in Two-Dimensional Multihop MANETs

Gonçalo Jacinto, Nelson Antunes, and António Pacheco

Abstract

A Mobile Ad Hoc Network (MANET) is characterized to be a network with free, cooperative, and dynamic nodes, self-organized in a random topology, without any kind of infrastructure, where the communication between two nodes usually occurs using multihop paths. The number of hops used in the multihop path is an important metric for the design and performance analysis of routing protocols in MANETs. In this paper, we derive the probability distribution of the hop count of a multihop path between a source node and a destination node, fixed at a known distance from each other, and when a fixed number of nodes are uniformly distributed in a region of interest. This distribution is obtained by the Poisson randomization method. To obtain the multihop path, we propose a novel routing model in which the nearest distance routing protocol (NR) is analyzed. Numerical results are obtained to evaluate the performance of the NR.

1   Introduction

When the source and destination nodes of a Mobile Ad Hoc Network (MANET) are at a distance greater than the transmission range, the communication between them is made via a multiple hop path that is determined by the routing protocol (cf., e.g.,

G. Jacinto (✉)
CIMA-UE and ECT of University of Évora, Evora, Portugal
e-mail: gjcj@uevora.pt

N. Antunes
FCT of University of Algarve and CEMAT, Faro, Portugal
e-mail: nantunes@ualg.pt

A. Pacheco
CEMAT and Departamento de Matemática, Instituto Superior Técnico, Universidade de Lisboa, Lisboa, Portugal
e-mail: apacheco@math.tecnico.ulisboa.pt

A. Pacheco et al. (eds.), *New Advances in Statistical Modeling and Applications*,
Studies in Theoretical and Applied Statistics, DOI 10.1007/978-3-319-05323-3_8,
© Springer International Publishing Switzerland 2014

[8]). One of the most popular strategies a node can use to decide to which neighbor node it should forward a given packet is the nearest distance routing protocol (NR), for which the packet is forwarded to the nearest relay node in the direction of the destination node.

As stated in [7] and references therein, one of the most important metrics to evaluate the performance of routing protocols is the number of hops of the multihop path. In [1], we have derived the hop count distribution for the one-dimensional scenario with relay nodes uniformly distributed between the source and destination nodes. However, the derivation of the hop count distribution in a two-dimensional scenario must take into account, among other factors, the transmission range and the routing protocol, aside from the node spatial distribution. The interaction of these characteristics turns the derivation of the hop count distribution a difficult task. This is the reason why, despite its importance, there are few analytical studies on the subject and most of them just consider single link models (cf. [5, 10]) and/or approximation results (cf. [4, 7]).

In [5] relay nodes are assumed to be distributed according to a Poisson process and the distribution of the distance from the source to the furthest neighbor node within transmission range is derived. The analysis was extended in [10] to a model where a finite number of relay nodes are uniformly distributed in a region of interest, but again only assuming a single link model. Few papers focus their analysis in more than a single link. In [4], an approximation for the relationship between the number of hops and the distance between the source and the destination nodes is derived, and an approximation for the probability of existence of a multihop path between the source and destination nodes is derived in [7].

In [9] one of the few closed-form results on the hop count distribution is derived for the case in which nodes are randomly distributed according to a Poisson process, for both one-dimensional and two-dimensional networks, and using three routing protocols: the nearest, the furthest, and the random routing protocol. However, the average hop length has to be used and estimated, turning the obtained results approximations of the exact hop count distribution.

In this paper, we derive the exact hop count probability distribution with an arbitrary number of hops, when the source and destination nodes are fixed, at a known distance form each other, and a known and fixed number of relay nodes are uniformly distributed in a region of interest. To obtain the multihop path, we propose a novel propagation model where the routing region of each relay node is defined by a given angular span and a radius equal to the transmission range. Since the angular span depends on the distance between the emitter and destination nodes, we call this model the dynamic propagation model. Inside each routing region, we use the NR protocol to choose the relay node to forward the packet.

The mathematical analysis of the problem of an existing path on a random set of points, with the source and destination nodes at known locations, is often called a navigation problem. Within this literature, the paper [2] proposes a model with the nearest routing protocol using routing regions with a fixed angular span. The authors proved that when the number of random nodes is large enough, almost surely exists a path between the source and the destination nodes.

As far as we know, our results are the first exact analytical results for the hop count distribution with an arbitrary number of hops in a two-dimensional scenario, when a finite number of relay nodes are uniformly distributed in an area of interest. These results are suitable to use when the number of hops is not too large, because the dynamic angular span decreases when the source or relay nodes are far way from the destination node. However, in MANETs the number of hops between the source and destination nodes cannot be large due to the small duration of multihop paths with a large number of hops [6]. In dense networks that does not constitute a problem since the multihop path is similar to a path on a straight line. Note that the usage of the position-based protocols requires that a node knows its own geographical position and the geographical position of the destination node, but the localization problem of the nodes are not focused in this paper. We also should note that we consider the transmission range of each node constant, not taking into account the SINR (signal-to-interference noise ratio), which will be the scope of future research.

The outline of this paper is the following. In Sect. 2 we describe the dynamic propagation model. In Sect. 3 we derive the hop count distribution for the NR protocol. In Sect. 4 we present some numerical results to evaluate the performance of the NR protocol. Finally, in Sect. 5, we conclude the paper.

2 Model Description

We consider an ad hoc network with the source node fixed at the origin and the destination node fixed at a distance L from the source node. A multihop path with m hops is defined as an existing path from the source to the destination node using exactly m relay nodes. Denote by $X_i, 1 \leq i \leq m$, the location of the relay node i of a multihop path, with these nodes ordered according to their distance to the origin, and let $X_0 = (0,0)$ and $X_{m+1} = (L,0)$ denote the locations of the source and destination nodes, respectively. Note that, without loss of generality, we have assumed that the destination node is located in the x-axis. Given a fixed transmission range R, $0 < R < L$, equal for all nodes, nodes i and j are connected with zero hops if $\|X_i - X_j\| < R$.

We assume that the locations of the source node, the destination node, and all relay nodes of the multihop path belong to a compact set $\Omega \subset \mathbb{R}^2$, with area B. The set Ω is defined by an isosceles triangle with one vertice at the origin $(0,0)$ with associated angle $\phi_0 = 2 \arctan(R/L)$, and the height of the triangle lies on the horizontal axis and is equal to L. The definition of the set Ω is needed to avoid analytical intractability and preclude that a given multihop path loops around the destination, see [8]. For efficient routing progress towards the destination, we consider that each relay node transmits within a routing region limited by the transmission radius R and an angular span oriented to the destination node. The angular span ϕ_i of relay node i is chosen in a dynamic way, being dependent on the location X_i of the relay node, and is such that it originates a triangle with vertices at points (L, R), $(L, -R)$ and X_i, increasing when it gets closer to the destination node and decreasing when the relay node gets further away from the

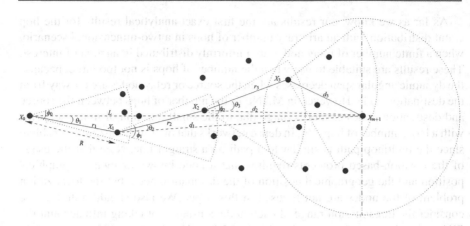

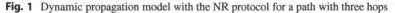

Fig. 1 Dynamic propagation model with the NR protocol for a path with three hops

destination node. This is the reason why we denominate the model as the dynamic propagation model. Within each routing region the relay node chosen to forward the packet will be the nearest relay node from the emitter node. The polar coordinates of the location of relay node i relative to the location of relay node $i-1$ are denoted by (r_i, θ_i), assuming that $-\pi \leq \theta_i \leq \pi$. In Fig. 1 we can observe a multihop path with three hops using the NR protocol and the dynamic propagation model. Note that if a given node is in the range of the destination node, they will connect directly.

3   Hop Count Distribution

To describe the routing regions of each relay node, we make a translation and rotation of the plane to locate the origin of the new plane at the current emitter node (in this case at relay node i), with horizontal axis being the line drawn from the emitter node to the destination node. For a relay node i located at X_i, the routing region relative to X_i is denoted by $\mathscr{A}_i \equiv \mathscr{A}(X_i, X_{m+1}, \phi_i)$ and, at each hop, an angular slice of a circular disk with radius R and with area $\frac{\phi_i}{2} R^2$ is covered (see Fig. 2). More precisely, the routing region of relay node i relative to X_i is defined by

$$\mathscr{A}_i \equiv \mathscr{A}(X_i, X_{m+1}, \phi_i) = \{(r, \theta) : 0 < r < R, -\phi_i^- \leq \theta \leq \phi_i^+\}$$

The angular span ϕ_i is dynamic and depends on the location of the relay node. Given (r_i, θ_i) and the distance from relay node $i-1$ to the destination node, d_{i-1}, the distance from relay node i to the destination node, d_i, is given by the function

$$d_i \equiv f(d_{i-1}, r_i, \theta_i) = \sqrt{(d_{i-1} - r_i \cos \theta_i)^2 + (r_i \sin \theta_i)^2}, \qquad 1 \leq i \leq m,$$

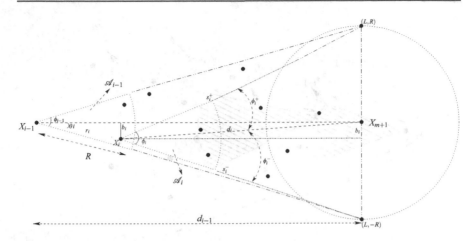

Fig. 2 Routing regions and angular spans of relay nodes $i - 1$ and i

with $d_0 = L$. The angle ϕ_i of relay node i can then be written as a function of d_{i-1} and (r_i, θ_i), $\phi_i \equiv \phi(d_{i-1}, r_i, \theta_i)$, and is given by

$$\phi_i = \arcsin\left(\frac{R - sign(\theta_i)b_i}{s_i^+}\right) + \arcsin\left(\frac{R + sign(\theta_i)b_i}{s_i^-}\right),$$

where $b_i = r_i \sin \theta_i$, so that $|b_i|$ is the minimum distance between X_i and the axis that goes from X_{i-1} to X_{m+1}, and $s_i^\pm = \sqrt{(d_{i-1} - r_i \cos \theta_i)^2 + (R \mp sign(\theta_i)b_i)^2}$ is the distance between X_i and $(L, \pm R)$; see Fig. 2. Using geometric arguments, we can show that $\phi_i = \phi_i^+ + \phi_i^-$, where ϕ_i^+ is the angle formed by the points (L, R), X_i and X_{m+1}, being given by $\phi_i^+ = \arcsin\left(\frac{R - sign(\theta_i)b_i}{s_i^+}\right) + sign(\theta_i) \arcsin\left(\frac{b_i}{d_i}\right)$, and ϕ_i^- is the angle formed by the points $(L, -R)$, X_i and X_{m+1}, being given by $\phi_i^- = \arcsin\left(\frac{R + sign(\theta_i)b_i}{s_i^-}\right) - sign(\theta_i) \arcsin\left(\frac{b_i}{d_i}\right)$.

Denote by $\mathscr{V}_i$ the vacant region of relay node i, defined to be the subset of the routing region of relay node i that has no relay nodes. That is, since the relay node selected is the closest one from the emitter node, the vacant region of relay node i is given by the set of points that are closer to i than relay node $i + 1$, having an area $V_i = \frac{\phi_i}{2} r_{i+1}^2$; see Fig. 3.

The hop count probability distribution is obtained by using Poisson randomization, [3], consisting in randomizing the number of relay nodes by assuming that relay nodes are distributed in Ω according to a Poisson process with rate λ. A precise argument for the spatial Markov property in more general spaces can be found in [11]. By conditioning in the number of relay nodes that lie in Ω, the results for the case in which a fixed and known number of relay nodes are uniformly distributed in Ω pops up. Denote by $\mathbf{l}_m = (l_1, l_2, \ldots, l_m)$ the vector of relative locations of the m relay nodes, with $l_i = (r_i, \theta_i)$, and let $d\mathbf{l}_m = d\theta_m dr_m d\theta_{m-1} dr_{m-1} \ldots d\theta_1 dr_1$. Recall that B denotes the area of Ω.

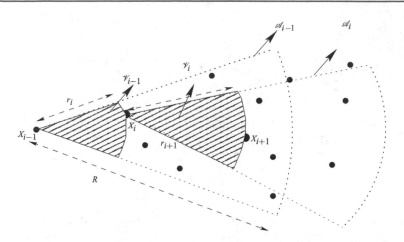

Fig. 3 Routing regions and vacant regions of relay nodes $i - 1$ and i

Theorem 1. *Given that there are n relay nodes uniformly distributed on Ω, the probability that the hop count is equal to m, for a multihop path selected by the dynamic propagation model with the NR protocol, is given by*

$$P(M = m | N = n) = \int_{N_m} \frac{n!}{(n-m)!} \frac{1}{B^m} \left(1 - \frac{1}{B} \sum_{i=0}^{m-1} V_i \right)^{n-m} \prod_{i=1}^{m} r_i \, d\mathbf{l}_m \quad (1)$$

with $K \leq m \leq n$ and $N_m = \left\{ \mathbf{l}_m : l_i = (r_i, \theta_i) \in \mathscr{A}_{i-1}, i = 1, 2, \ldots, m, d_m < R \leq d_{m-1} \right\}$.

Proof. We first derive the joint location density of the m relay nodes of the multihop path. For that, fix $(r_1, \theta_1) \in \mathscr{A}_0 = \left\{ (r_1', \theta_1') : 0 < r_1' < R, -\frac{\phi_0}{2} < \theta_1' < \frac{\phi_0}{2} \right\}$ and define $\mathscr{V}_0 = \left\{ (r_1', \theta_1') : 0 < r_1' < r_1, -\frac{\phi_0}{2} < \theta_1' < \frac{\phi_0}{2} \right\}$ and $\mathscr{V}_0^\epsilon = \{ (r_1', \theta_1') : r_1 \leq r_1' < r_1 + \epsilon_1, \theta_1 \leq \theta_1' < \theta_1 + \epsilon_2 \}$. Denote by $N(A)$ the number of points of the Poisson process in A. By the independent increment property of a Poisson process, we have

$$P\left(N(\mathscr{V}_0) = 0, N(\mathscr{V}_0^\epsilon) > 0 \right) = P\left(N(\mathscr{V}_0) = 0 \right) P\left(N(\mathscr{V}_0^\epsilon) > 0 \right)$$

$$= e^{-\lambda \frac{\phi_0}{2} r_1^2} \left(1 - \exp\left(-\lambda \int_{r_1}^{r_1+\epsilon_1} \int_{\theta_1}^{\theta_1+\epsilon_2} r \, dr d\theta \right) \right)$$

$$= e^{-\lambda \frac{\phi_0}{2} r_1^2} \lambda \int_{r_1}^{r_1+\epsilon_1} \int_{\theta_1}^{\theta_1+\epsilon_2} r \, dr d\theta + o(\epsilon_1 \epsilon_2).$$

The density of the location of the first relay node being at (r_1, θ_1) is given by

$$h(r_1, \theta_1) = \lim_{\epsilon_1, \epsilon_2 \to 0+} \frac{P\left(N(\mathcal{V}_0)=0, N(\mathcal{V}_0^\epsilon)>0\right)}{\epsilon_1 \epsilon_2} = \lambda r_1 e^{-\lambda \frac{\phi_0}{2} r_1^2}.$$

To derive the density location of the first two relay nodes, we make a rotation and translation of the plane in order to place the origin of the new plane at $(r_1 + \epsilon, \theta_1)$ with horizontal axis being the line drawn from $(r_1 + \epsilon, \theta_1)$ to the destination node. Proceeding in a similar way to the one used to derive the density of the location of the first relay node, one may conclude (see [6]) that the density of the locations of the first two relay nodes being (r_1, θ_1) and (r_2, θ_2) is $h(r_1, \theta_1, r_2, \theta_2) = \lambda^2 r_1 r_2 e^{-\lambda \frac{\phi_0}{2} r_1^2} e^{-\lambda \frac{\phi_1}{2} r_2^2}$.

Proceeding in the same manner until the m-th relay node is connected with no hops with the destination node, we obtain the joint density of the locations of the m relay nodes of the multihop path, $h(\mathbf{l}_m) = \lambda^m e^{-\lambda \sum_{i=1}^{m} \frac{\phi_{i-1}}{2} r_i^2} \prod_{i=1}^{m} r_i$, where the node locations are in N_m and the last relay node is m because $d_m < R \leq d_{m-1}$. Integrating $h(\mathbf{l}_m)$ over the set $N_m = \{\mathbf{l}_m : l_i = (r_i, \theta_i) \in \mathscr{A}_{i-1}, i = 1, 2, \ldots, m, d_m < R \leq d_{m-1}\}$, we obtain the probability that the hop count is m for the NR, when the relay nodes are randomly distributed according to a Poisson process:

$$P(M = m) = \int_{N_m} \lambda^m e^{-\lambda \sum_{i=1}^{m} \frac{\phi_{i-1}}{2} r_i^2} \prod_{i=1}^{m} \, d\mathbf{l}_m. \tag{2}$$

Multiplying equation (2) by $e^{\lambda B}$, where B is the area of Ω, we obtain

$$
\begin{aligned}
e^{\lambda B} P(M = m) &= e^{\lambda B} \int_{N_m} \lambda^m e^{-\lambda \sum_{i=0}^{m-1} V_i} \prod_{i=1}^{m} r_i \, d\mathbf{l}_m \\
&= \int_{N_m} \lambda^m \sum_{n=0}^{\infty} \frac{(\lambda B)^n}{n!} \left(1 - \frac{1}{B} \sum_{i=0}^{m-1} V_i\right)^n \prod_{i=1}^{m} r_i \, d\mathbf{l}_m \\
&= \sum_{n=m}^{\infty} \frac{(\lambda B)^n}{n!} \int_{N_m} \frac{n!}{(n-m)!} \frac{1}{B^m} \left(1 - \frac{1}{B} \sum_{i=0}^{m-1} V_i\right)^{n-m} \prod_{i=1}^{m} r_i \, d\mathbf{l}_m
\end{aligned}
$$

where the change between the sum and the integral follows by the dominated convergence theorem. On the other hand, conditioning on the value of N, which is Poisson distributed with mean λB, by the total probability law $e^{\lambda B} P(M = m) = \sum_{n=m}^{\infty} P(M = m | N = n) \frac{(\lambda B)^n}{n!}$. Since the coefficients of $\frac{(\lambda B)^n}{n!}$ in the previous two expressions for $e^{\lambda B} P(M = m)$ must match, the result follows. $\qquad \square$

Fig. 4 Connectivity
probability with the minimum
number of hops

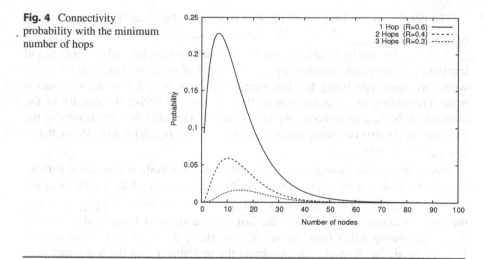

4   Numerical Results

In this section we evaluate the performance of the dynamic propagation model for
the NR protocol. We scale all parameters with respect to the distance between the
source and destination nodes assuming that $L = 1$, leading the set Ω to have area
$B = RL$. Therefore, depending on the value of R, for $1/(K + 1) < R \leq 1/K$,
$K \in \mathbb{N}$, we have multihop paths with a minimum number of hops equal to K.
The results were obtained by numerical integration using a Monte Carlo algorithm.
Despite the multi-dimensional integration, it is relatively simple and not too much
time consuming the calculation over six hops, which is a very large number of hops
for a MANET [6].

Figure 4 shows the connectivity probability with the minimum number of hops
K, $K = 1, 2, 3$, with the NR protocol and for different values of the number of
nodes. We can observe that when the number of nodes increases the minimum hop
count probability decreases and approaches the value 0, and so the NR protocol is
ineffective in a dense network because it cannot transmit with a high probability
with the minimum number of hops. For the same number of relay nodes, the hop
count probability with the minimum number of hops decreases as K increases.

In Fig. 5, we obtain the hop count probability with different values of the number
of hops. We consider $R = 0.3$, and $K = 3, 4, 5, 6$, and observe that, when there
is a small number of nodes, the NR protocol with $K + 1 = 4$ hops has the
highest probability, whereas when there is a large number of nodes, the hop count
probability with $K + 3 = 6$ has the highest probability. Again the probability with
the minimum number of hops K with the NR protocol is very ineffective, since it
has the smallest probability. Despite that, all probabilities ($K = 3, 4, 5, 6$) approach
zero with the increase of the number of nodes, and the probabilities obtained for
paths with a large number of hops are generally larger than the ones obtained for
paths with a smaller number of hops.

Fig. 5 Connectivity
probability with hop count
equal to $K = 3, 4, 5, 6$

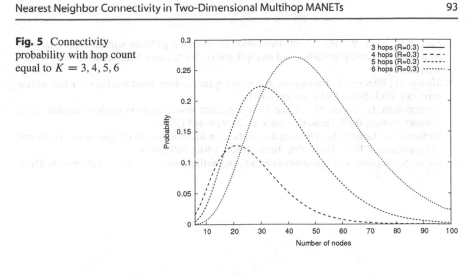

5 Conclusion

In this paper we focused on the connectivity in two-dimensional wireless ad-hoc networks. We have assumed that the source and the destination nodes are fixed, at a known distance from each other, and that a fixed and known number of relay nodes are uniformly distributed in a region of interest. To find a multihop path, we proposed a novel model called the dynamic propagation model. Using this model, we derived the hop count probability distribution when the multihop path chosen follows the NR. As far as we know, these are the first exact analytical results for the hop count probability distribution. The numerical results derived allowed us to conclude that the NR protocol is not suitable for dense networks.

Acknowledgments This work received financial support from Portuguese National Funds through FCT (Fundação para a Ciência e a Tecnologia) under the scope of projects: PTDC/EIA-EIA/115988/2009, PEstOE/MAT/UI0822/2011, and PTDC/MAT-STA/3169/2012.

References

1. Antunes, N., Jacinto, G., Pacheco, A.: On the minimum hop count and connectivity in one-dimensional ad hoc wireless networks. Telecommun. Syst. **39**(3), 366–376 (2008)
2. Bonichon, N., Marckert, J-F.: Asymptotics of geometrical navigation on a random set of points in the plane. Adv. Appl. Probab. **43**(4), 899–942 (2011)
3. Domb, C.: On the use of a random parameter in combinatorial problems. Proc. Phys. Soc. Sec. A **65**(1), 305–309 (1952)
4. Dulman, S., Rossi, M., Havinga, P., Zorzi, M.: On the hop count statistics for randomly deployed wireless sensor networks. Int. J. Sen. Netw. **1**(1/2), 89–102 (2006)
5. Haenggi, M., Puccinelli, D.: Routing in ad hoc networks: a case for long hops. IEEE Commun. Mag. **43**(10), 93–101 (2005)
6. Jacinto, G.: Modeling and performance evaluation of mobile ad hoc networks. Ph.D. thesis, Instituto Superior Técnico, Technical University of Lisbon, Lisbon (2011)

7. Kuo, J.-C., Liao, W.: Hop count distribution of multihop paths in wireless networks with arbitrary node density: modeling and its applications. IEEE Trans. Veh. Technol. **56**(4), 2321–2331 (2007)
8. Mauve, M., Widmer, A., Hartenstein, H.: A survey on position-based routing in mobile ad hoc networks. IEEE Netw. **15**(6), 30–39 (2001)
9. Rahmatollahi, G., Abreu, G.: Closed-form hop-count distributions in random networks with arbitrary routing. IEEE Trans. Commun. **60**(2), 429–444 (2012)
10. Srinivasa, S., Haenggi, M.: Distance distributions in finite uniformly random networks: theory and applications. IEEE Trans. Veh. Technol. **59**(2), 940–949 (2010)
11. Zuyev, S.: Stopping sets: gamma-type results and hitting properties. Adv. Appl. Probab. **31**(2), 355–366 (1999)

Modeling Human Population Death Rates: A Bi-Dimensional Stochastic Gompertz Model with Correlated Wiener Processes

Sandra Lagarto and Carlos A. Braumann

Abstract

This study presents an innovative approach to human mortality data analysis, namely a transversal analysis across time using stochastic differential equation models, as a form of considering random environmental oscillations on the death rates. For each age between 0 and 99, we use a bi-dimensional stochastic Gompertz model with correlated Wiener processes to model the dynamics of female (first component of the stochastic process) and male (second component) crude death rates of the Portuguese population over the period 1940–2009. We test the complete model, with correlation between the unidimensional Wiener processes associated with males and with females, against the model without correlation effects. Results show significant correlations for most ages, particularly on ages below 5 and above 50.

1 Introduction

Population aging is becoming a very pertinent issue. In several and different contexts, like continuous health care and retirement funds, longevity is becoming a challenging issue. In the last decades, mortality has been exhaustively studied through both deterministic and stochastic models [1, 3, 8]. The most used of all, currently with many variations, is the Lee–Carter model [2, 5].

So why doing a transversal analysis of mortality data and why modeling death rates with stochastic differential equations (SDE) models? When analyzing simultaneously death rates of distinct ages across time, instead of doing a cohort (or longitudinal) data analysis, there is evidence that mortality at all ages is,

S. Lagarto (✉) • C.A. Braumann
Colégio Luís Verney, CIMA-University of Évora, Rua Romão Ramalho 59,
7000-671 Évora, Portugal
e-mail: smdl@uevora.pt; braumann@uevora.pt

A. Pacheco et al. (eds.), *New Advances in Statistical Modeling and Applications*,
Studies in Theoretical and Applied Statistics, DOI 10.1007/978-3-319-05323-3_9,
© Springer International Publishing Switzerland 2014

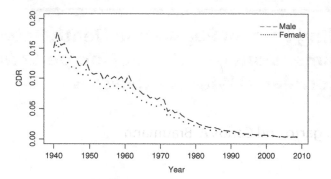

Fig. 1 Crude death rates of Portuguese population (in number of deaths per 10,000 inhabitants): 1940–2009, age 0, by sex

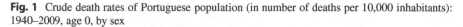

for both males and females, influenced by random environmental fluctuations (this variability overwhelmingly exceeds sampling errors, which are not considered in this paper). Also, the death rates for all ages have a strong decreasing trend during the last century. Considering these facts, we start by using a simple SDE model to describe the dynamics over time of the crude death rates (CDR) of individuals of a certain age and sex, namely a generalized stochastic Gompertz model (GSGM). This model, with only three parameters, has previously been proposed to describe animal growth [4].

This approach is quite different from the cohort approach, which uses cumbersome models in order to cover the whole life span. This unidimensional model was initially applied to each age and sex. Then, we noticed that, frequently, increases in mortality and decreases in mortality occur for males and females of the same age at the same time (see Fig. 1). This similar death behavior between the two sexes suggests that the associated SDE driving Wiener processes must also be correlated and this leads to a bi-dimensional (two-sexes) model.

In Sect. 2, we briefly describe and analyse the models. In Sect. 3, we present an application of a bi-dimensional stochastic Gompertz model (BSGM) to the CDR of the Portuguese population. Section 4 compares this correlated model with the non-correlated one using a likelihood ratio test. Section 5 presents the conclusions.

2 The Stochastic Mortality Model

The GSGM consists in applying a transformation to the data and assume that the transformed data satisfies the classical stochastic Gompertz model. In our case, the logarithmic transformation is appropriate and we assume that $X(t)$ (CDR at time t, of a certain age and sex) follows an unidimensional stochastic Gompertz model, that is, $Y(t) = \ln(X(t))$ follows the SDE

$$dY(t) = b(A - Y(t))dt + \sigma dW(t), \qquad (1)$$

with $Y(t_0) = y_{t_0}$, $W(t)$ a standard Wiener process and parameters $A = \ln(a)$ (a, the asymptotic death rate), $b > 0$ (rate of approach to the asymptotic regime) and $\sigma > 0$ (measures the effect of the environmental fluctuations on the mortality dynamics).

The solution of Eq. (1) is easily obtained by Itô calculus (see, for example, [7]):

$$Y(t) = A + (y_{t_0} - A)\exp\{-b(t - t_0)\} + \sigma \exp\{-bt\}\int_{t_0}^{t} \exp\{bs\}\, dW(s).$$

Therefore $Y(t)$ has a normal distribution:

$$Y(t) \frown \mathcal{N}(A + (y_{t_0} - A)\exp\{-b(t - t_0)\}, \sigma^2(1 - \exp\{-2b(t - t_0)\})/2b).$$

2.1 The Bi-Dimensional Stochastic Gompertz Model with Correlated Wiener Processes

Considering that there is a random environmental variability which affects both males and females, we coupled, for each age, the models for the two sexes, obtaining the bi-dimensional SDE system

$$\begin{cases} dY_1(t) = b_1(A_1 - Y_1(t))dt + \sigma_1\, dW_1^*(t) \\ dY_2(t) = b_2(A_2 - Y_2(t))dt + \sigma_2\, dW_2^*(t) \end{cases}$$

with $Y_i(t) = \ln(X_i(t))$, $Y_i(t_0) = y_{i,t_0}$, $X_i(t)$ the CDR at time t for the age under study ($i = 1$ for females; $i = 2$ for males), and $W_i^*(t)$ standard correlated Wiener processes. We call this the BSGM.

Let ρ be the correlation coefficient between $W_1^*(t)$ and $W_2^*(t)$. To avoid collapsing to the one-dimensional case, we assume $|\rho| \neq 1$. Using two independent standard Wiener processes $W_1(t)$, $W_2(t) \frown \mathcal{N}(0, t)$, we put

$$W_1^*(t) = \alpha\, W_1(t) + \beta\, W_2(t)$$

$$W_2^*(t) = \beta\, W_1(t) + \alpha\, W_2(t),$$

with

$$\alpha = ((1 + (1 - \rho^2)^{1/2})/2)^{1/2},$$

$$\beta = sign(\rho)((1 - (1 - \rho^2)^{1/2})/2)^{1/2} = sign(\rho)(1 - \alpha^2)^{1/2},$$

so that we have $E[W_i^*(t)^2] = t$ and $E[W_1^*(t)W_2^*(t)] = \rho t$.

The solution of the SDE system, for the age under study, at time t, is

$$\begin{cases} Y_1(t) = A_1 + (y_{1,t_0} - A_1)\exp\{-b_1(t - t_0)\} + \sigma_1 \exp\{-b_1 t\}\int_{t_0}^{t}\exp\{b_1 s\}\, dW_1^*(s) \\ Y_2(t) = A_2 + (y_{2,t_0} - A_2)\exp\{-b_2(t - t_0)\} + \sigma_2 \exp\{-b_2 t\}\int_{t_0}^{t}\exp\{b_2 s\}\, dW_2^*(s), \end{cases}$$

with transient distributions

$$
\begin{cases}
Y_1(t) \frown \mathcal{N}(A_1+(y_{1,t_0}-A_1)\exp\{-b_1(t-t_0)\}, \sigma_1^2(1-\exp\{-2b_1(t-t_0)\})/2b_1) \\
Y_2(t) \frown \mathcal{N}(A_2+(y_{2,t_0}-A_2)\exp\{-b_2(t-t_0)\}, \sigma_2^2(1-\exp\{-2b_2(t-t_0)\})/2b_2).
\end{cases}
$$

The joint (Y_1, Y_2) distribution is bivariate normal, with correlation coefficient

$$
r(t_0,t) = \rho \frac{(1-\exp\{-(b_1+b_2)(t-t_0)\})}{(1-\exp\{-2b_1(t-t_0)\})^{1/2}(1-\exp\{-2b_2(t-t_0)\})^{1/2}} \frac{2(b_1 b_2)^{1/2}}{b_1+b_2}.
$$

We use Maximum Likelihood (ML) estimation to obtain the parameters' estimates for the age under study.

Let $t_k = t_0 + k$ $(k = 0, 1, 2, \ldots, n)$ be the years where CDR were observed and let $Y_{i,k} = Y_i(t_k) = \ln(X_i(t_k))$. The transition p.d.f. of $(Y_1(t), Y_2(t))$ between t_{k-1} and t_k is

$$
f(y_1, y_2|Y_{1,k-1} = y_{1,k-1}, Y_{2,k-1} = y_{2,k-1}) =
$$

$$
\frac{1}{2\pi s_{Y_{1,k}} s_{Y_{2,k}} \sqrt{1-r^2}} \exp\{-\frac{1}{2} Q_k(y_1, y_2)\},
$$

with

$$
Q_k(y_1, y_2) = \frac{1}{1-r^2}\left[\left(\frac{y_1 - \mu_{Y_{1,k}}}{s_{Y_1}}\right)^2 - 2r\frac{(y_1 - \mu_{Y_{1,k}})(y_2 - \mu_{Y_{2,k}})}{s_{Y_{1,k}} s_{Y_{2,k}}}\right.
$$
$$
\left. + \left(\frac{y_2 - \mu_{Y_{2,k}}}{s_{Y_{2,k}}}\right)^2\right],
$$

where

$$
\mu_{Y_{i,k}} = A_i + (y_{i,k-1} - A_i)\exp\{-b_i(t_k - t_{k-1})\},
$$
$$
s^2_{Y_{i,k}} = \sigma_i^2(1-\exp\{-2b_i(t_k - t_{k-1})\})/2b_i)
$$

and

$$
r = r(t_{k-1}, t_k) = \rho \frac{(1-\exp\{-(b_1+b_2)(t_k - t_{k-1})\})}{(1-\exp\{-2b_1(t_k - t_{k-1})\})^{1/2}(1-\exp\{-2b_2(t_k - t_{k-1})\})^{1/2}} \frac{2(b_1 b_2)^{1/2}}{b_1+b_2}.
$$

Let $(y_{1,k}, y_{2,k})$ be the observed values of $(Y_{1,k}, Y_{2,k})$. Using the Markov properties of the SDE solution, the log-likelihood function is given, for each age, by:

$$L(A_1, b_1, \sigma_1, A_2, b_2, \sigma_2, r) =$$

$$\sum_{k=1}^{n} \ln(f(y_{1,k}, y_{2,k} | Y_{1,k-1} = y_{1,k-1}, Y_{2,k-1} = y_{2,k-1})). \tag{2}$$

The ML parameters' estimates are obtained by maximization of Eq. (2), which we did numerically. Obviously, for the model without correlation between sexes, we can estimate the parameters the same way with $r = 0$.

If we have observations up to time t_k and want to make predictions for a future time $t > t_k$, the Markov property gives

$$E[Y_i(t) | Y_{1,0}, Y_{2,0}, Y_{1,1}, Y_{2,1}, \ldots, Y_{1,k}, Y_{2,k}] = E[Y_{i,t} | Y_{1,k}, Y_{2,k}].$$

Since

$$Y_i(t) | Y_{1,k}(t) = y_{1,k}, Y_{2,k}(t) = y_{2,k} \frown$$

$$\mathcal{N}(A_i + (y_{i,k} - A_i) \exp\{-b_i(t - t_k)\}, \sigma_i^2(1 - \exp\{-2b_i(t - t_k)\})/2b_i),$$

we can use, for the age under study, the long-term predictions

$$\hat{Y}_i(t) = \hat{E}[Y_i(t) | Y_{1,k} = y_{1,k}, Y_{2,k} = y_{2,k}] = \hat{A}_i + (y_{i,k} - \hat{A}_i) \exp\{-\hat{b}_i(t - t_k)\}.$$

The "one-step" (year-by-year) predictions are estimated the same way as mentioned but updating t_k and the parameter estimates each time we get an estimate.

3   Application to Human Portuguese Population Death Rates

We fitted the BSGM to the Portuguese population CDR (death counts divided by the estimate of midyear exposure-to-risk population). In Fig. 2 we illustrate the female and male surfaces of the dataset. Annual data (from 1940 to 2009) were obtained from the Human Mortality Database [6], by age (1-year age groups: 0,1,...,99) and sex.

For the time series of the age under study, we have used 60 "observations" to fit the model (1940–1999) and have left 10 "observations" for prediction (2000–2009). Computations were performed with R free software.

In Fig. 3, we represent estimated parameters, for all ages, for the complete BSGM. We also estimate confidence intervals for the parameters using the Gaussian asymptotic approximation of ML estimates and approximating the variance–covariance matrix by the inverse of the empirical Fisher information matrix. Table 1 shows, as an example, the estimated values for age 80.

We chose age 0 (individuals with less than one complete year) to illustrate fitted and predicted values of CDR (Fig. 4).

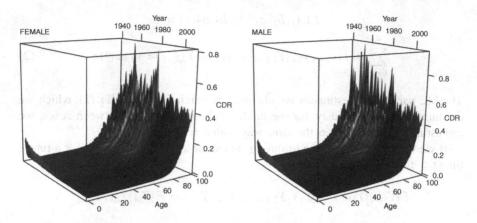

Fig. 2 Surface of female (*left*) and male (*right*) mortality data from 1940 to 1999, ages 0–99

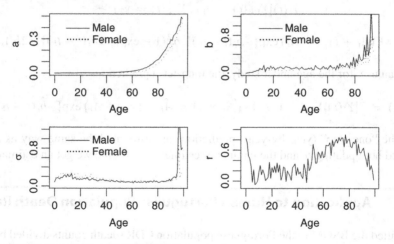

Fig. 3 Parameters estimates for the BSGM, by age and sex

Table 1 Example (for age 80) of the ML estimates and semi-amplitude of the 95 % confidence interval

Age	A_1	$b_1 year^{-1}$	$\sigma_1 year^{-1/2}$	r	A_2	$b_2 year^{-1}$	$\sigma_2 year^{-1/2}$
...	...	...	...	...	...	...	...
80	−2.5327	0.0857	0.0770	0.8180	−2.1405	0.1333	0.0854
	±0.0150	±0.0879	±0.1841	±0.1105	±0.0168	±0.0168	±0.0168
...	...	...	...	...	...	...	...

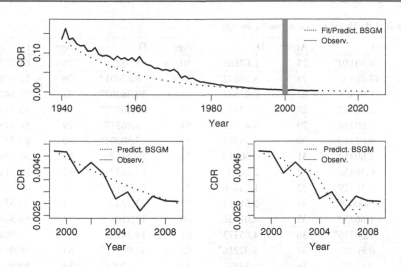

Fig. 4 On *top*: Fitted (years 1940–1999) and (long-term) predicted values (years 2000–2009) for CDR of age 0 females. *Bottom left*: Amplification of the prediction section of the top figure. *Bottom right*: Corresponding one-step predictions

4   Testing for Correlations Between Sexes

Finally, we have used a likelihood ratio test to compare the complete BSGM with the model without correlations. To test

$$H_0 : r = 0 \; vs \; H_1 : r \neq 0$$

we use the test statistic

$$D = -2\ln\left(\frac{\text{max. likelihood for model with } r = 0}{\text{max. likelihood for model with } r \neq 0}\right),$$

which, under the null hypothesis, is approximately chi-squared distributed with one degree of freedom. In Table 2, we report the values of the D test statistics for all ages. Results suggest that the very young and the older individuals have mortality rates significantly correlated with the rates of the other gender.

5   Conclusions/Future Work

The BSGM seems to be appropriate to this type of data, presenting satisfactory fit and prediction for all ages. The likelihood ratio test suggests that the complete BSGM (with correlations between the Wiener processes that measure the effect of environmental fluctuations) has significant differences when compared to the model without correlation effects, for almost every age after 50 and also for the very young.

Table 2 Likelihood ratio test results, by age

Age	D	Age	D	Age	D	Age	D
0	68.33150*	25	1.67865	50	3.53230	75	68.47583*
1	47.86390*	26	5.88632*	51	8.85401*	76	32.92903*
2	15.98596*	27	0.02848	52	39.76502*	77	92.84271*
3	12.15687*	28	0.62421	53	22.76972*	78	43.81416*
4	5.00230*	29	4.43612*	54	6.96348*	79	56.39553*
5	0.02140	30	3.79528	55	7.50503*	80	63.33385*
6	2.80414	31	0.00019	56	21.43279*	81	30.82691*
7	0.26830	32	2.26401	57	17.48574*	82	29.31415*
8	0.10529	33	1.23678	58	12.78173*	83	44.10562*
9	0.00000	34	0.51013	59	22.33432*	84	40.63140*
10	1.66353	35	0.01452	60	13.31082*	85	59.34599*
11	5.34736*	36	4.05373*	61	14.36165*	86	15.86499*
12	0.86281	37	5.02216*	62	37.07712*	87	36.90376*
13	2.67163	38	6.73935*	63	35.64075*	88	19.50755*
14	1.14750	39	0.71664	64	36.53741*	89	17.42499*
15	0.00688	40	4.02279*	65	46.17152*	90	23.87252*
16	0.09950	41	1.02278	66	16.15175*	91	27.90838*
17	0.33019	42	1.88173	67	56.88815*	92	15.31689*
18	8.12858*	43	0.48002	68	34.06079*	93	10.46706*
19	0.17147	44	0.13951	69	46.69097*	94	3.86429*
20	4.09134*	45	6.39170*	70	61.06492*	95	20.12945*
21	1.62040	46	2.63369	71	44.56891*	96	2.68888
22	5.15558*	47	0.04081	72	72.83185*	97	0.81996
23	1.54344	48	11.49340*	73	53.77353*	98	2.06987
24	7.55196*	49	2.34264	74	68.18138*	99	0.03821

$^* p < 0.05$

For future work, we will continue to explore the correlation effect on mortality, by sex, namely with different SDE models; additionally, we will consider correlations among ages.

Acknowledgements The authors are members of the CIMA (Centro de Investigação em Matemática e Aplicações), a research center of the Universidade de Évora financed by FCT (Fundação para a Ciência e Tecnologia, Portugal). Braumann gratefully acknowledges the financial support from the FCT grant PTDC/MAT/115168/2009.

References

1. Alho, J.M., Spencer, B.D.: Statistical Demography and Forecasting. Springer Series in Statistics. Springer, New York (2005)
2. Bravo, J.: Tábuas de mortalidade contemporâneas e prospectivas: modelos estocásticos, aplicações actuariais e cobertura do risco de longevidade–Tese. Universidade de Évora, Évora (2007)

3. Cairns, A.J.G., Blake, D., Dowd, K., Coughlan, G.D., Epstein, D.: Mortality density forecasts: an analysis of six stochastic mortality models. Insur. Math. Econ. **48**, 355–367 (2011)
4. Filipe, P.A., Braumann, C.A., Roquete, C.J.: Multiphasic individual growth models in random environments. Methodol. Comput. Appl. Probab. **14**, 49–56 (2010)
5. Giacometti, R., Bertocchi, M., Rachev, S.T., Fabozzi, F.J.: A comparison of the Lee-Carter model and AR–ARCH model for forecasting mortality rates. Insur. Math. Econ. **50**, 85–93 (2012)
6. Human Mortality Database. University of California, Berkeley (USA) and Max Planck Institute for Demographic Research (Germany). http://www.mortality.org (2011)
7. Øksendal, B.: Stochastic Differential Equations. An Introduction with Applications, 6th edn. Springer, New York (1998)
8. Yashin, A.I., Arbeev, K.G., Akushevich, I., Kulminski, A., Akushevich, L., Ukraintseva, S.V.: Stochastic model for analysis of longitudinal data on aging and mortality. Math. Biosci. **208**, 538–551 (2007)

Consequences of an Incorrect Model Specification on Population Growth

Clara Carlos and Carlos A. Braumann

Abstract

We consider stochastic differential equations to model the growth of a population in a randomly varying environment. These growth models are usually based on classical deterministic models, such as the logistic or the Gompertz models, taken as approximate models of the "true" (usually unknown) growth rate. We study the effect of the gap between the approximate and the "true" model on model predictions, particularly on asymptotic behavior and mean and variance of the time to extinction of the population.

1   Introduction

In [4, 5] we study the extinction of population growth in a random environment for the classical logistic and Gompertz stochastic models. These and other similar models have been frequently proposed in the literature (see [2] for detailed references). Braumann et al. [3] and Carlos et al. [6] study the first passage times for generalized stochastic Gompertz models of individual growth in a random environment.

However, we often do not know the exact form of the average growth rate and so we assume that the "true" unknown rate differs from the one in the classical logistic or Gompertz "incorrect" stochastic models by a small amount. The "true" stochastic model will be called near-logistic or near-Gompertz, respectively, and its

C. Carlos (✉)
Escola Superior de Tecnologia do Barreiro, Instituto Politécnico de Setúbal, Rua Américo da Silva Marinho, 2890-001 Lavradio, Portugal
e-mail: clara.carlos@estbarreiro.ips.pt

C.A. Braumann
Universidade de Évora, Rua Romão Ramalho 59, 7000-671 Évora, Portugal
e-mail: braumann@uevora.pt

A. Pacheco et al. (eds.), *New Advances in Statistical Modeling and Applications*,
Studies in Theoretical and Applied Statistics, DOI 10.1007/978-3-319-05323-3_10,
© Springer International Publishing Switzerland 2014

properties will be studied here (Sect. 2), as well as the population extinction times
(Sect. 3). In Sect. 3, we will also see how wrong we will be if we use the extinction
times of the "incorrect" logistic or Gompertz models instead of the extinction time
of the "true" near-logistic or near-Gompertz models. We illustrate the results with a
numerical example. Section 4 contains the conclusions.

2 Model

In the classical deterministic population growth models, the *per capita* growth rate
can be written in the form

$$\frac{1}{X}\frac{dX}{dt} = f(X), \quad X(0) = x, \tag{1}$$

where $X = X(t)$ is the population size at time $t \geq 0$. For instante, when
$f(X) = r\left(1 - \frac{X}{K}\right)$ we obtain the logistic model and, for $f(X) = r \ln \frac{K}{X}$ we obtain
the Gompertz model, having as parameters the intrinsic growth rate $r > 0$ and the
carrying capacity of the environment $K > 0$.

Suppose that the "true" *per capita* growth rate is only approximated by the
logistic model or the Gompertz model and suppose that the deviation is $\alpha(X) =
f(X) - r\left(1 - \frac{X}{K}\right)$ or $\alpha(X) = f(X) - r \ln \frac{K}{X}$, respectively, i.e.,

$$f(X) = r\left(1 - \frac{X}{K}\right) + \alpha(X), \tag{2}$$

or

$$f(X) = r \ln \frac{K}{X} + \alpha(X). \tag{3}$$

Suppose α is a C^1 function and $\frac{|\alpha(X)|}{r} < \delta$, where $0 < \delta < 1$ is a kind of relative
error committed when we use the logistic or the Gompertz models instead of the
"true" model.

The stochastic differential equation (SDE) we present are generalizations of the
classical deterministic model but incorporate a random dynamical term, describing
the effects of environmental random fluctuations on the growth process. Now $f(X)$
should fluctuate randomly over time and, since growth is a multiplicative type
process, it should be taken as the geometric average *per capita* growth rate to which
we should add the random environmental fluctuations, assumed to be uncorrelated.
We obtain

$$\frac{1}{X}\frac{dX}{dt} = f(X) + \sigma\epsilon(t), \quad X(0) = x, \tag{4}$$

where $\epsilon(t)$ is a continuous-time standard white noise and $\sigma > 0$ measures the
noise intensity. Since we use for f the geometric average growth rate, we will use
Sratonovich calculus, which is the appropriate one (see [2]).

In this paper we use the near-logistic (2) or the near-Gompertz (3) models. For both, the solution exists and is unique up to an explosion time (see, for instance, [1]). We will show later that there is no explosion and therefore the solution exists and is unique for all $t \geq 0$. The solution $X(t)$ is a homogeneous diffusion process with drift coefficient

$$a(x) = x \left(f(x) + \frac{\sigma^2}{2} \right) \tag{5}$$

and diffusion coefficient

$$b^2(x) = \sigma^2 x^2. \tag{6}$$

For the model (4) the state space is $(0, +\infty)$. We now define, in the interior of the state space, the scale and speed measures of $X(t)$ (see, for instance, p. 194 of [7]). The scale density is

$$s(y) := \exp \left(-\int_n^y \frac{2a(\theta)}{b^2(\theta)} d\theta \right) = \frac{C}{y} \exp \left(-\frac{2}{\sigma^2} \int_n^y \frac{f(\theta)}{\theta} d\theta \right) \tag{7}$$

and the speed density is

$$m(y) := \frac{1}{b^2(y)s(y)} = \frac{1}{\sigma^2 y^2 s(y)}, \tag{8}$$

where n is an arbitrary (but fixed) point in the interior of the state space and $C > 0$ is a constant. Different choices of n correspond to scale densities differing by a multiplicative constant, which does not affect their relevant properties. The "distribution" functions of these measures are $S(z) = \int_c^z s(y) dy$ and $M(z) = \int_c^z m(y) dy$, the scale function and speed function, respectively, where c is an arbitrary (but fixed) point in the interior of the state space. The scale and speed measures of an interval (a, b) are given by $S(a, b) = S(b) - S(a)$ and $M(a, b) = M(b) - M(a)$, respectively.

The state space has boundaries $X = 0$ and $X = +\infty$.

One can see that $X = 0$ is non-attracting and therefore, "mathematical" extinction has zero probability of occurring. It suffices to show that, for some $x_0 > 0$, $S(0, x_0] = \int_0^{x_0} s(y) dy = +\infty$ (see, for instance, p. 228 of [7]). Let $0 < x_0 < n$, $g(\theta) = \left(1 - \frac{\theta}{K}\right)$ or $g(\theta) = \ln \frac{K}{\theta}$. Choose n such that $0 < n < K$ and sufficiently small to insure that $g(n) - \delta > 0$. Then, since g is a decreasing function, for $y \in (0, x_0]$ we have

$$s(y) \geq \frac{C}{y} \exp \left(\frac{2r}{\sigma^2} (g(n) - \delta) \int_y^n \frac{1}{\theta} d\theta \right) = c_1 y^{-\frac{2r}{\sigma^2}(g(n)-\delta)-1} \tag{9}$$

and $S(0, x_0] \geq +\infty$, with $c_1 > 0$ constant.

One can see that $X = +\infty$ is non-attracting and therefore, explosion cannot occur and the solution exists and is unique for all $t > 0$. It suffices to show that, for some $x_0 > 0$, $S[x_0, +\infty) = \int_{x_0}^{+\infty} s(y)dy = +\infty$ (see, for instance, p. 236 of [7]). Let $x_0 > K$. Then, for $y \in [x_0, +\infty)$,

$$s(y) \geq c_2 y^{-\frac{2r}{\sigma^2}\delta - 1} \exp\left(-\frac{2r}{\sigma^2} \int_n^y \frac{g(\theta)}{\theta} d\theta\right) = s^-(y) \tag{10}$$

with c_2 positive constant, and so $S[x_0, +\infty) = +\infty$ because $s^-(y) \to +\infty$ as $y \to +\infty$.

Contrary to the deterministic model (1), the stochastic model (4) does not have an equilibrium point, but there may exist an equilibrium probability distribution for the population size, called the stationary distribution, with a probability density function $p(y)$, known as stationary density. Indeed since the boundaries are non-attracting, the stationary density exists if $M = \int_{0^+}^{+\infty} m(y)dy < +\infty$ (see p. 241 of [7]) and is given by $p_y(y) = \frac{m(y)}{M}$, with $0 < y < +\infty$.

We will now prove that, in model (4), the population size has a stationary density. We need to show that $M < +\infty$. Let $y_1 < K < y_2$ be such that $0 < y_1 < n < y_2 < +\infty$ and $g(y_2) + \delta < 0$. Break the integration interval, $M = M_1 + M_2 + M_3 = \int_{0^+}^{y_1} m(y)dy + \int_{y_1}^{y_2} m(y)dy + \int_{y_2}^{+\infty} m(y)dy$.

We first show that M_1 is finite. Let $y \in (0, y_1]$ and $\theta \in [y, n]$. Let $h(y) = -\frac{2r}{\sigma^2}(g(n) - \delta)\int_y^n \frac{1}{\theta}d\theta$, note that $h(0^+) = -\infty$. We have $m(y) \leq c_3 \frac{1}{y} e^{h(y)} = c_4 \frac{d(e^{h(y)})}{dy}$, with c_3 and c_4 are positive finites constants. Therefore $M_1 \leq c_4\left(e^{h(y_1)} - e^{h(0^+)}\right) < +\infty$.

We now prove that $M_3 < +\infty$. Let $y \in [y_2, +\infty)$ and $\theta \in [n, y]$. Decompose $\frac{2}{\sigma^2}\int_n^y \frac{f(\theta)}{\theta}d\theta = \frac{2}{\sigma^2}\int_n^{y_2} \frac{f(\theta)}{\theta}d\theta + \frac{2}{\sigma^2}\int_{y_2}^y \frac{f(\theta)}{\theta}d\theta = A + B(y)$. Obviously $A < +\infty$ and $B(y) \leq \frac{2r}{\sigma^2}(g(y_2) + \delta)\int_{y_2}^y \frac{1}{\theta}d\theta = \frac{2r}{\sigma^2}(g(y_2) + \delta)\ln\frac{y}{y_2}$. Then $m(y) \leq c_3 e^{(A+B(y))} = c_5 y^{\frac{2r}{\sigma^2}(g(y_2)+\delta)-1}$, where $c_5 > 0$ is a finite constant. Therefore, we get $M_3 < +\infty$.

Finally, we show that $M_2 < +\infty$. Put $M_2 = M_2' + M_2'' = \int_{y_1}^n m(y)dy + \int_n^{y_2} m(y)dy$. We will show that M_2' is finite, the proof that $M_2'' < +\infty$ is similar. Let $y \in [y_1, n]$ and $\theta \in [y, n]$. Therefore $m(y) \leq \frac{c_3}{y_1}\left(\frac{y}{n}\right)^{\frac{2r}{\sigma^2}(g(n)-\delta)} \leq c_6\left(\frac{n}{y_1}\right)^{\frac{2r}{\sigma^2}(g(n)-\delta)}$, where c_6 is positive finite constant. Therefore M_2' is finite.

The stationary density is given by $p_y(y) = \frac{D}{y}\exp\left(-\frac{2}{\sigma^2}\int_y^n \frac{f(\theta)}{\theta}d\theta\right)$, where D is a constant such that $\int_0^{+\infty} p_y(y)dy = 1$.

3 Extinction Times

"Mathematical" extinction has zero probability of occurring, but we prefer the concept of "realistic" extinction, meaning the population dropping below an extinction threshold $a > 0$. We are interested in the time required for the population to reach the extinction threshold a for the first time. Let us denote this extinction time by $T_a = \inf\{t > 0 : X(t) = a\}$. Assuming that the initial population x is above a and the boundary $X = +\infty$ is non-attracting, we present expressions for the mean $E_x[T_a]$ and variance $V_x[T_a]$ of the first passage time T_a.

In [3] or [6] one finds recursive expressions for the nth order moment of T_a. In particular, the expression for the mean of the first passage time is

$$E_x[T_a] = 2 \int_a^x s(\zeta) \int_\zeta^{+\infty} m(\psi)d\psi d\zeta \tag{11}$$

and for the variance is

$$V_x[T_a] = 8 \int_a^x s(\zeta) \int_\zeta^{+\infty} s(\xi) \left(\int_\xi^{+\infty} m(\psi)d\psi \right)^2 d\xi d\zeta. \tag{12}$$

The expressions obtained are valid for the sufficiently regular homogeneous ergodic diffusion processes with drift coefficient a and diffusion coefficient b^2.

As usual in dynamical systems, to reduce the number of parameters and work with adimensional quantities, let us consider $\beta(X) = \frac{\alpha(X)}{r}$, $R = \frac{r}{\sigma^2}$, $d = \frac{a}{K}$ and $z = \frac{x}{a}$. For the same reason, we will obtain expressions for the adimensional quantities $rE_x[T_a]$ and $r^2V_x[T_a]$.

In [4, 5] we obtain the expression for the mean and the variance for the standard models. For the logistic model, the mean (multiplied by r) and variance (multiplied by r^2) of the first passage time T_a are given by

$$rE_x^{L(R,d)}[T_a] = 2R \int_{2Rd}^{2Rdz} y^{-2R-1} e^y \Gamma(2R, y) \, dy, \tag{13}$$

and

$$r^2 V_x^{L(R,d)}[T_a] = 8R^2 \int_{2Rd}^{2Rdz} y^{-2R-1} e^y \int_y^{+\infty} u^{-2R-1} e^u \left(\Gamma(2R, u) \right)^2 du \, dy, \tag{14}$$

with $\Gamma(c, x) = \int_x^{+\infty} t^{c-1} e^{-t} dt$. For the Gompertz model, the mean (multiplied by r) and variance (multiplied by r^2) of the first passage time T_a are given by

$$rE_x^{G(R,d)}[T_a] = 2\sqrt{\pi} \int_{\sqrt{R}\ln(d)}^{\sqrt{R}\ln(dz)} e^{y^2} \left(1 - \Phi(\sqrt{2}y) \right) dy \tag{15}$$

and

$$r^2 V_x^{G(R,d)}[T_a] = 8\pi \int_{\sqrt{R}\ln(d)}^{\sqrt{R}\ln(dz)} e^{y^2} \int_y^{+\infty} e^{u^2} \left(1 - \Phi(\sqrt{2}u) \right)^2 du \, dy, \tag{16}$$

with $\Phi(z) = \frac{1}{\sqrt{2\pi}} \int_{-\infty}^z \exp\left(-t^2/2\right) dt$.

For the near-logistic model, using (11) and the expressions (7) and (8), we obtain, making the change of variables $v = \frac{2r}{\sigma^2 K}\theta$, $t = \frac{2r}{\sigma^2 K}\psi$ and $y = \frac{2r}{\sigma^2 K}\zeta$,

$$
rE_x[T_a] = 2R \int_{2Rd}^{2Rdz} y^{-2R-1} e^y \int_y^{+\infty} t^{2R-1} e^{-t} \exp\left(2R \int_y^t \frac{\beta\left(\frac{Kv}{2R}\right)}{v} dv\right) dt\, dy.
$$

$$(17)$$

Using the condition $\beta(X) > -\delta$, we obtain the following inequalities

$$
rE_x[T_a] \geq \frac{2R^*}{1-\delta} \int_{2R^*d^*}^{2R^*d^*z} y^{-2R^*-1} e^y\, \Gamma\left(2R^*, y\right) dy = \frac{1}{1-\delta} rE_x^{L(R^*,d^*)}[T_a],
$$

$$(18)$$

with $R^* = R(1-\delta)$ and $d^* = \frac{d}{1-\delta}$. Therefore, this lower bound can be obtained using the expression for the logistic model with K replaced by $K^* = K(1-\delta)$. Similarly, since $\beta(X) < \delta$, one has

$$
rE_x[T_a] \leq \frac{2R^{**}}{1+\delta} \int_{2R^{**}d^{**}}^{2R^{**}d^{**}z} y^{-2R^{**}-1} e^y\, \Gamma\left(2R^{**}, y\right) dy = \frac{1}{1+\delta} rE_x^{L(R^{**},d^{**})}[T_a],
$$

$$(19)$$

with $R^{**} = R(1+\delta)$ and $d^{**} = \frac{d}{(1+\delta)}$.

Using (12) and the expressions (7) and (8), with the same change of variables, we obtain for the near-logistic model

$$
r^2 V_x[T_a] = 8R^2 \int_{2Rd}^{2Rdz} y^{-2R-1} e^y \int_y^{+\infty} u^{-2R-1} e^u \int_u^{+\infty} t^{2R-1} e^{-t} e^{2R \int_y^t \frac{\beta\left(\frac{Kv}{2R}\right)}{v} dv} dt
$$

$$
\int_u^{+\infty} t^{2R-1} e^{-t} e^{2R \int_u^t \frac{\beta\left(\frac{Kv}{2R}\right)}{v} dv} dt\, du\, dy.
$$

$$(20)$$

Using the condition $|\beta(X)| > \delta$ we obtain the following inequalities

$$
r^2 V_x[T_a] \geq \frac{8R^{*2}}{(1-\delta)^2} \int_{2R^*d^*}^{2R^*d^*z} y^{-2R^*-1} e^y \int_y^{+\infty} u^{-2R^*-1} e^u \left(\Gamma\left(2R^*, u\right)\right)^2 du\, dy
$$

$$
= \frac{1}{(1-\delta)^2} r^2 V_x^{L(R^*,d^*)}[T_a]
$$

$$(21)$$

and

$$
r^2 V_x[T_a] \leq \frac{8R^{**2}}{(1+\delta)^2} \int_{2R^{**}d^{**}}^{2R^{**}d^{**}z} y^{-2R^{**}-1} e^y \int_y^{+\infty} u^{-2R^{**}-1} e^u \left(\Gamma\left(2R^{**}, u\right)\right)^2 du\, dy
$$

$$= \frac{1}{(1+\delta)^2} r^2 V_x^{L(R^{**},d^{**})}[T_a]. \tag{22}$$

For the near-Gompertz model, using (11) and the expressions (7) and (8), making the change of variables $v = \frac{\sqrt{r}}{\sigma} \ln \frac{\theta}{K}, t = \frac{\sqrt{r}}{\sigma} \ln \frac{\psi}{K}$ e $y = \frac{\sqrt{r}}{\sigma} \ln \frac{\zeta}{K}$, we obtain

$$rE_x[T_a] = 2 \int_{\sqrt{R}\ln d}^{\sqrt{R}\ln(dz)} e^{y^2} \int_y^{+\infty} e^{-t^2} \exp\left(2\sqrt{R} \int_y^t \beta\left(Ke^{\frac{v}{\sqrt{R}}}\right) dv\right) dt\, dy. \tag{23}$$

Since $|\beta(X)| < \delta$ and putting $d^* = de^\delta$ and $d^{**} = de^{-\delta}$, we obtain the following inequalities

$$rE_x[T_a] \geq 2\sqrt{\pi} \int_{\sqrt{R}\ln(d^*)}^{\sqrt{R}\ln(d^*z)} e^{y^2} \left(1 - \Phi(\sqrt{2}y)\right) dy = rE_x^{G(R,d^*)}[T_a] \tag{24}$$

and

$$rE_x[T_a] \leq 2\sqrt{\pi} \int_{\sqrt{R}\ln(d^{**})}^{\sqrt{R}\ln(d^{**}z)} e^{y^2} \left(1 - \Phi(\sqrt{2}y)\right) dy = rE_x^{G(R,d^{**})}[T_a]. \tag{25}$$

Using (12) and the expressions (7) and (8), with the same change of variables, we obtain for the near-Gompertz model

$$r^2 V_x[T_a] = 8 \int_{\sqrt{R}\ln d}^{\sqrt{R}\ln(dz)} e^{y^2} \int_y^{+\infty} e^{u^2} \int_u^{+\infty} e^{-t^2} \exp\left(2\sqrt{R} \int_y^t \beta\left(Ke^{\frac{v}{\sqrt{R}}}\right) dv\right) dt$$

$$\int_u^{+\infty} e^{-t^2} \exp\left(2\sqrt{R} \int_u^t \beta\left(Ke^{\frac{v}{\sqrt{R}}}\right) dv\right) dt\, du\, dy. \tag{26}$$

Similarly we get the following bounds for the variance,

$$r^2 V_x[T_a] \geq 8\pi \int_{\sqrt{R}\ln(d^*)}^{\sqrt{R}\ln(d^*z)} e^{y^2} \int_y^{+\infty} e^{u^2} \left(1 - \Phi(\sqrt{2}u)\right)^2 du\, dy = r^2 V_x^{G(R,d^*)}[T_a] \tag{27}$$

and

$$r^2 V_x[T_a] \leq 8\pi \int_{\sqrt{R}\ln(d^{**})}^{\sqrt{R}\ln(d^{**}z)} e^{y^2} \int_y^{+\infty} e^{u^2} \left(1 - \Phi(\sqrt{2}u)\right)^2 du\, dy = r^2 V_x^{G(R,d^{**})}[T_a]. \tag{28}$$

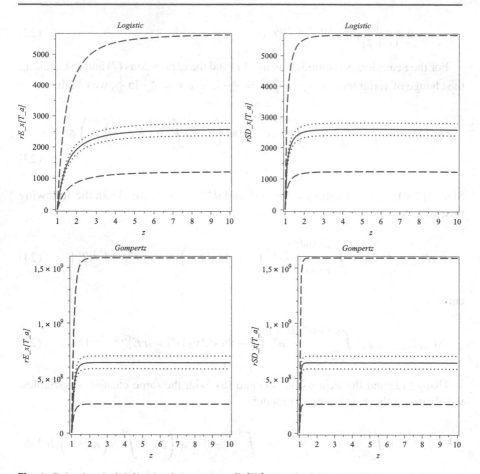

Fig. 1 Behavior (*solid lines*) of the mean $rE_x[T_a]$ (*on the left*) and the standard deviation $rSD_x[T_a]$ (*on the right*) for the extinction time of the population as a function of $z = \frac{x}{a}$ for the logistic (*top figures*) and Gompertz (*bottom figures*) models. The figure also shows the upper and lower bounds for the near-logistic and near-Gompertz models, respectively, with relative errors in f of $\delta = 0.01$ (*dotted lines*) and $\delta = 0.1$ (*dashed lines*). We consider $R = \frac{r}{\sigma^2} = 1$ and $d = \frac{a}{K} = 0.01$

To give an example, consider $R = 1$, equivalent to $r = \sigma^2$, i.e., strong relative noise intensity, and $d = 0.01$, equivalent to $a = \frac{K}{100}$. In Fig. 1, we show the behavior of the mean and the standard deviation (both multiplied by r) of the time required for the population to reach size a for the first time (the extinction times) as a function of z for the standard models (logistic and Gompertz models) and its upper and lower bounds for the near-standard models (near-logistic and near-Gompertz models), for the cases of relative errors in f of $\delta = 0.01$ and $\delta = 0.1$. The standard deviation of the population extinction time is of the same order of magnitude as the mean extinction time. So, use of means alone is not very informative.

4 Conclusions

The qualitative behavior of the "true" models coincides with the behavior of the near-standard models (near-logistic or near-Gompertz). "Mathematical" extinction has zero probability of occurring and there is a stationary density. "Realistic" extinction occurs with probability one.

If the true average growth rate is very close to the standard model (logistic or Gompertz), the mean and standard deviation of the population extinction time are close to the ones of the standard model. One can then use a standard model, which is much simpler to deal with, as a convenient approximation. One can even have bounds for the error committed. Otherwise, the use of the standard model may lead to values quite different from the true ones.

Acknowledgments The authors are members of the CIMA (Centro de Investigação em Matemática e Aplicações), a research center of the Universidade de Évora financed by FCT (Fundação para a Ciência e Tecnologia, Portugal). Clara Carlos benefits from a PROTEC doctoral scholarship financed by IPS (Instituto Politécnico de Setúbal). Braumann gratefully acknowledges the financial support from the FCT grant PTDC/MAT/115168/2009.

References

1. Arnold, L.: Stochastic Differential Equations. Theory and Applications. Wiley, New York (1974)
2. Braumann, C.A.: Itô versus Stratonovich calculus in random population growth. Math. Biosci. **206**, 81–107 (2007)
3. Braumann, C.A., Filipe, P.A., Carlos C., Roquete, C.J.: Growth of individuals in randomly fluctuating environments. In: Vigo-Aguiar, J., Alonso, P., Oharu, S., Venturino, E., Wade, B. (eds.) Proceedings of the International Conference in Computational and Mathematical Methods in Science and Engineering, Gijon, pp. 201–212 (2009)
4. Carlos, C., Braumann, C.A.: Tempos de extinção para populações em ambiente aleatório. In: Braumann, C.A., Infante, P., Oliveira, M.M., Alpízar-Jara, R., Rosado, F. (eds.), Estatística Jubilar, pp. 133–142. Edições SPE (2005)
5. Carlos, C., Braumann, C.A.: Tempos de extinção para populações em ambiente aleatório e cálculos de Itô e Stratonovich. In: Canto e Castro, L., Martins, E.G. Rocha, C., Oliveira, M.F., Leal, M.M., Rosado, F. (eds.) Ciência Estatística, pp. 229–238. Edições SPE (2006)
6. Carlos, C., Braumann, C.A., Filipe, P.A.: Models of individual growth in a random environment: study and application of first passage times. In: Lita da Silva, J., et al. (eds.), Advances in Regression, Survival Analysis, Extreme Values, Markov Processes and Other Statistical Applications, pp. 103–111. Studies in Theoretical and Applied Statistics. Selected Papers of the Statistical Societies. Springer, Berlin (2013)
7. Karlin, S., Taylor, H.M.: A Second Course in Stochastic Processes. Academic, New York (1981)

Individual Growth in a Random Environment: An Optimization Problem

Patrícia A. Filipe, Carlos A. Braumann, Clara Carlos, and Carlos J. Roquete

Abstract

We consider a class of stochastic differential equations model to describe individual growth in a random environment. Applying these models to the weight of mertolengo cattle, we compute the mean profit obtained from selling an animal to the meat market at different ages and, in particular, determine which is the optimal selling age. Using first passage time theory we can characterize the time taken for an animal to achieve a certain weight of market interest for the first time. In particular, expressions for the mean and standard deviation of these times are presented and applied to real data. These last results can be used to determine the optimal selling weight in terms of mean profit.

1 Introduction

In previous work (see, for instance, [4, 5]) we have presented a class of stochastic differential equation (SDE) models for individual growth in randomly fluctuating environments and we have applied such models using real data on the evolution of bovine weight. The Gompertz and the Bertalanffy–Richards stochastic models are particular cases of this more general class of SDE models. The work we present here is dedicated to the optimization of the mean profit obtained by selling an animal.

P.A. Filipe (✉) • C.A. Braumann • C. Carlos
Centro de Investigação em Matemática e Aplicações, Universidade de Évora,
Colégio Luís Verney, Rua Romão Ramalho 59, 7000-671 Évora, Portugal
e-mail: pasf@uevora.pt; braumann@uevora.pt; clara.carlos@estbarreiro.ips.pt

C.J. Roquete
Instituto de Ciências Agrárias e Ambientais Mediterrânicas, Universidade de Évora,
Núcleo da Mitra, Apartado 94, 7002-774 Évora, Portugal
e-mail: croquete@uevora.pt

A. Pacheco et al. (eds.), *New Advances in Statistical Modeling and Applications*,
Studies in Theoretical and Applied Statistics, DOI 10.1007/978-3-319-05323-3_11,
© Springer International Publishing Switzerland 2014

On the one hand, based on our models, we can compute the mean profit obtained by selling an animal at different ages and, in particular, we can determine the optimal age at which we should sell the animal in order to maximize the mean profit. We can also obtain the probability distribution of the selling profit and then compute probabilities involving that profit. On the other hand, knowing which animal weight is demanded by the market, we can study the properties of the time required for an animal to reach such weight for the first time. We present expressions for the mean and variance of these times, known from first passage time theory (see, for instance, [2]), and use these results to determine the optimal weight in order to obtain the maximum mean profit. A comparison between these two approaches is presented.

2 SDE Model for Individual Growth

Denoting by X_t the size of the individual (organism) at age t, the individual growth in a random environment can be described by stochastic differential equations of the form

$$dY_t = \beta(\alpha - Y_t)dt + \sigma dW_t, \quad Y_{t_0} = y_0, \tag{1}$$

where $Y_t = h(X_t)$, with h a strictly increasing C^1 function, and $y_0 = h(x_0)$, x_0 being the size at age t_0 (initial age). Notice that α is the asymptotic mean value of Y_t and therefore $A = h^{-1}(\alpha)$ is the asymptotic size or size at maturity. The parameter β is a growth coefficient, σ is an environmental noise intensity parameter and W_t is the standard Wiener process. Here, we will work with the stochastic versions of the classical Gompertz model (SGM) and the Bertalanffy–Richards model (SBRM), choosing $h(x) = \ln x$ and $h(x) = x^{1/3}$, respectively, but there are other growth models proposed in the literature that corresponds to different choices of h.

The solution of (1), Y_t, is a homogeneous diffusion process with drift coefficient $a(y) = \beta(\alpha - y)$ and diffusion coefficient $b(y) = \sigma^2$. The drift coefficient is the mean speed of growth described by Y_t and the diffusion coefficient gives a measure of the local magnitude of the fluctuations. It can be seen, for instance in [1], that the explicit solution of (1) is given by $Y_t = \alpha - (\alpha - y_0)e^{-\beta(t-t_0)} + \sigma e^{-\beta t} \int_{t_0}^t e^{\beta u} dW_u$, and follows a Gaussian distribution with mean $\alpha - (\alpha - y_0)e^{-\beta(t-t_0)}$ and variance $\frac{\sigma^2}{2\beta}\left(1 - e^{-2\beta(t-t_0)}\right)$.

In [4], we have applied maximum likelihood and non-parametric estimation methods to our models. Here we will work with the maximum likelihood estimates of the parameters A, β and σ involved in model (1) (Table 1). We will apply the methods to data on the weight, in kilograms, of 97 females of mertolengo cattle (2,129 observations).

Table 1 Maximum likelihood estimates and approximate half-width of the 95 % confidence bands

	SGM	SBRM
A (kg)	411.2 ± 8.1	425.7 ± 9.5
β (year^{-1})	1.676 ± 0.057	1.181 ± 0.057
σ (year$^{-1/2}$)	0.302 ± 0.009	0.597 ± 0.019

3 Optimization

Our SDE models can be useful in financial context. In our application, by having more information on the growth of animals, growers can, for instance, optimize the average profit obtained from selling an animal. We will explore two approaches to study the problem of optimization of the mean profit. One consists in deciding to sell the animal at a fixed age t chosen by the producer irrespective of the (random) size it reaches at that age. The mean profit is a function of t and we determine the optimal value of t, i.e., the value of t the producer should choose in order to maximize the mean profit. The other approach consists in deciding to sell the animal when it first reaches a fixed size Q^* irrespective of the (random) time required to reach that size. The mean profit is now a function of Q^* and we determine the optimal value of Q^*, i.e., the value of Q^* the producer should choose in order to maximize the mean profit.

The profit obtained from selling an animal is $L = V - C$, where V represents the selling price and C the acquisition (if it is the case) and animal breeding costs.

3.1 Profit Optimization by Age

Let x_0 be the weight of the animal at age t_0 (the age, assumed known, when it is bought) and $t > t_0$ the selling age. We represent by P the sale price per kg of carcass, by R the dressing proportion (carcass weight divided by live weight; typically around 0.5), by C_1 the fixed costs (initial price + veterinary costs + transportation and commercialization costs) and by $C_2(t) = c_2(t - t_0)$ the variable costs, supposed proportional to the breeding period. The profit at age t is given by $L_t = PRX_t - C_1 - C_2(t)$. Since $X_t = h^{-1}(Y_t)$, we can write the profit as a function of Y_t. For the SGM and the SBRM, we obtain

$$L_t = l_t(Y_t) = \begin{cases} PRe^{Y_t} - C_1 - c_2(t - t_0), & \text{for SGM} \\ \\ PRY_t^3 - C_1 - c_2(t - t_0), & \text{for SBRM.} \end{cases} \quad (2)$$

Considering the Gaussian probability distribution of Y_t, we can determine the probability density function of L_t using $f_{L_t}(u) = f_{Y_t}(l_t^{-1}(u)) \left| \frac{dl_t^{-1}(u)}{du} \right|$, where

$$f_{Y_t}(y) = \frac{1}{\sqrt{2\pi \frac{\sigma^2}{2\beta}(1 - e^{-2\beta(t-t_0)})}} \exp\left(-\frac{(y - \alpha - (y_0 - \alpha)\,e^{-\beta(t-t_0)})^2}{2\frac{\sigma^2}{2\beta}(1 - e^{-2\beta(t-t_0)})}\right),$$

and

$$l_t^{-1}(u) = \begin{cases} \ln\left(\frac{u + C_1 + c_2(t-t_0)}{PR}\right), & \text{for SGM} \\[2ex] \left(\frac{u + C_1 + c_2(t-t_0)}{PR}\right)^{1/3}, & \text{for SBRM} \end{cases} \tag{3}$$

$$\frac{dl_t^{-1}(u)}{du} = \begin{cases} (u + C_1 + c_2(t-t_0))^{-1}, & \text{for SGM} \\[2ex] \frac{(u + C_1 + c_2(t-t_0))^{-2/3}}{3(PR)^{1/3}}, & \text{for SBRM}. \end{cases} \tag{4}$$

The expressions for the mean and variance of L_t are, respectively, given by

$$E[L_t] = PRE[X_t] - C_1 - c_2(t - t_0) \tag{5}$$

and

$$Var[L_t] = P^2 R^2 Var[X_t], \tag{6}$$

where the expressions for $E[X_t]$ and $Var[X_t]$ are determined, according to the model used, as follows. In the SGM case, X_t follows a log-normal distribution, and consequently

$$E[X_t] = E[e^{Y_t}] = \exp\left(E[Y_t] + \frac{Var[Y_t]}{2}\right) =$$

$$= \exp\left(\alpha + e^{-\beta(t-t_0)}(\ln x_0 - \alpha) + \frac{\sigma^2}{4\beta}(1 - e^{-2\beta(t-t_0)})\right)$$

and

$$Var[X_t] = E[X_t^2] - E^2[X_t] = E[e^{2Y_t}] - E^2[e^{Y_t}] =$$

$$= \exp(2E[Y_t] + 2Var[Y_t]) - \exp(2E[Y_t] + Var[Y_t]) =$$

$$= \exp\left(2\alpha + 2e^{-\beta(t-t_0)}(\ln x_0 - \alpha) + \frac{\sigma^2}{\beta}(1 - e^{-2\beta(t-t_0)})\right) +$$

$$- \exp\left(2\alpha + 2e^{-\beta(t-t_0)}(\ln x_0 - \alpha) + \frac{\sigma^2}{2\beta}(1 - e^{-2\beta(t-t_0)})\right).$$

Table 2 Typical costs, since weaning until slaughter (euro/animal)[a]

Transportation and commercialization	Feeding/month	Health	Other
18.85	26.68	7.25	1.55

[a]Values do not consider subsidies for slaughter or RPU (unit income)

For the SBRM case, using Stein's Lemma, we get

$$E[X_t] = E[Y_t^3] = 3E[Y_t]\,Var[Y_t] + E[Y_t]^3 =$$

$$= \frac{3\sigma^2}{2\beta}\left(\alpha + e^{-\beta(t-t_0)}(x_0^{1/3} - \alpha)\right)(1 - e^{-2\beta(t-t_0)})$$

$$+ \left(\alpha + e^{-\beta(t-t_0)}(x_0^{1/3} - \alpha)\right)^3$$

and

$$Var[X_t] = E[X_t^2] - E^2[X_t] = E[Y_t^6] - E^2[Y_t^3] =$$

$$= 45E^2[Y_t]\,Var^2[Y_t] + 15E^4[Y_t]\,Var[Y_t] + 15Var^3[Y_t] + E^6[Y_t] +$$

$$- \left(3E[Y_t]\,Var[Y_t] + E[Y_t]^3\right)^2 =$$

$$= 36E^2[Y_t]\,Var^2[Y_t] + 9E^4[Y_t]\,Var[Y_t] + 15Var^3[Y_t] =$$

$$= \frac{9\sigma^4}{\beta^2}\left(\alpha + e^{-\beta(t-t_0)}(x_0^{1/3} - \alpha)\right)^2(1 - e^{-2\beta(t-t_0)})^2 +$$

$$+ \frac{9\sigma^2}{2\beta}\left(\alpha + e^{-\beta(t-t_0)}(x_0^{1/3} - \alpha)\right)^4(1 - e^{-2\beta(t-t_0)})$$

$$+ \frac{15\sigma^6}{8\beta^3}(1 - e^{-2\beta(t-t_0)})^3.$$

For illustration, we will consider a mertolengo cow with 160 kg, raised with the mother up to the age of 7 months, that is bought, for 200 euros, by a producer to be raised for market sale. What is the expected profit of this producer when the animal is marketed at age t (usually $t = 16$ months)?

We have used the maximum likelihood estimates of the model parameters given in Table 1 and $t_0 = 0.583$ years (7 months), $x_0 = 160$ kg, $R = 0.5$, $C_1 = 200 + 18.85 + 7.25 + 1.55 = 227.45$ euros and $c_2 = 26.68$/month $\times 12 = 320.16$/year. Maximizing expression (5) with respect to age t (using the routine optimize of the R package), we have obtained the optimal age for selling the animal in order to reach a maximum mean profit (Table 2). Table 3 shows, for both SGM and SBRM, the optimal age t (A_{opt}) and correspondent weight $E[X_{A_{opt}}]$), maximum mean profit ($E[L_{A_{opt}}]$) and standard deviation of the profit ($sd[L_{A_{opt}}]$).

Table 3 Results for optimization by age

	P (euro/kg)	A_{opt}	$E[X_{A_{opt}}]$	$E[L_{A_{opt}}]$	$sd[L_{A_{opt}}]$
MGE	3.00	0.97	253	29.41	91.11
	3.25	1.05	271	62.19	109.60
	3.50	1.13	285	96.94	127.00
	3.75	1.19	296	133.26	143.72
MBRE	3.00	0.71	189	14.14	29.53
	3.25	0.86	219	39.73	47.70
	3.50	0.97	240	68.49	61.21
	3.75	1.07	257	99.63	72.99

3.2 Profit Optimization by Weight

In [2] we have determined explicit expressions (in the form of simple integrals that can be numerically computed) for the mean and standard deviation of the time required for an animal to reach a given size for the first time. When that size is the size at which the animal is supposed to be sold for the meat market, these results can be used to compute the mean and variance of the profit and optimize the mean profit by weight.

Let us consider thresholds q^* and Q^*, one low and one high, for the animal size X_t. We are interested in the time required for an animal to reach a specific size Q^* for the first time. Since Y_t and X_t are related through the strictly increasing function h, this is also the first passage time of Y_t (modified size) by $Q = h(Q^*)$. Let us denote it by $T_Q = \inf\{t > 0 : Y_t = Q\}$.

One can see in [7] the definition, in the interior of the state space, of the scale and speed measures of Y_t. The scale density is $s(y) = \exp\left(-\int_{y^*}^{y} \frac{2a(\theta)}{b(\theta)} d\theta\right)$ and the speed density is given by $m(y) = \frac{1}{s(y)b(y)}$, where y^* is an arbitrary (but fixed) point in the interior of the state space (the choice of which is irrelevant for our purposes) and where a and b are the drift and diffusion coefficients of Y_t. In our case, we get

$$s(y) = C \exp\left(-\frac{2\beta\alpha}{\sigma^2}y + \frac{\beta}{\sigma^2}y^2\right) \quad \text{and} \quad m(y) = \frac{1}{\sigma^2 s(y)}, \quad (7)$$

where C is a constant. The "distribution" functions of these measures are the scale function and speed function defined by $S(z) = \int_{x^*}^{z} s(u)du$ and $M(z) = \int_{x^*}^{z} m(u)du$, where x^* is an arbitrary (but fixed) point in the interior of the state space.

Let $q = h(q^*)$ and assume that $-\infty < q < y_0 < Q < +\infty$ (q and Q both in the interior of the state space of Y). Let T_q be the first passage time of Y_t by q and let $T_{qQ} = \min(T_q, T_Q)$ be the first passage time of Y_t through either of the thresholds q and Q. Denote the k-th order moment of T_{qQ} by $U_k(y_0) = E[(T_{qQ})^k | Y(0) = y_0]$. One can see, for instance in [6], that

$$u(y_0) := P[T_Q < T_q | Y(0) = y_0] = \frac{S(y_0) - S(q)}{S(Q) - S(q)} \tag{8}$$

and that $U_k(y_0)$ satisfies $\frac{1}{2} b(y_0) \frac{d^2 U_k(y_0)}{dy_0^2} + a(y_0) \frac{dU_k(y_0)}{dy_0} + k U_{k-1}(y_0) = 0$, which is easily seen to be equivalent to

$$\frac{1}{2} \frac{d}{dM(y_0)} \left(\frac{dU_k(y_0)}{dS(y_0)} \right) + k U_{k-1}(y_0) = 0. \tag{9}$$

Integrating with respect to $M(y_0)$ and with respect to $S(y_0)$, using the conditions $U_k(q) = U_k(Q) = 0$ $(k = 1, 2, \ldots)$ and (8), one obtains, for $k = 1, 2, \ldots$, the solution

$$U_k(y_0) = 2u(y_0) \int_{y_0}^{Q} (S(Q) - S(\xi)) k U_{k-1}(\xi) m(\xi) d\xi$$

$$+ 2(1 - u(y_0)) \int_{q}^{y_0} (S(\xi) - S(q)) k U_{k-1}(\xi) m(\xi) d\xi. \tag{10}$$

Since $U_0(y_0) \equiv 1$, one can iteratively obtain the moments of any arbitrary order of T_{qQ}.

We can apply (10) to our model (1). Since the process Y_t is ergodic (see [3]), we can obtain the distribution (and moments) of T_Q as the limiting case of the distribution (moments) of T_{qQ} when $q \downarrow -\infty$.

Let us denote by $V_k(y_0) := E[(T_Q)^k | Y(0) = y_0]$ the k-th order moment of T_Q. Taking the limit as $q \downarrow -\infty$ in expression (10), one obtains $u(y_0) = 1$ and, therefore, $V_k(y_0) = 2 \int_{y_0}^{Q} (S(Q) - S(\theta)) k V_{k-1}(\theta) m(\theta) d\theta$. Since $S(Q) - S(\theta) = \int_{\theta}^{Q} s(\xi) d\xi$ and exchanging the order of integration, we get

$$V_k(y_0) = 2 \int_{y_0}^{Q} s(\xi) \left(\int_{-\infty}^{\xi} k V_{k-1}(\theta) m(\theta) d\theta \right) d\xi. \tag{11}$$

For our model, (7) holds. Based on (11), after some algebraic work, one gets

$$E[T_Q] = V_1(y_0) = \frac{1}{\beta} \int_{\sqrt{2\beta}(y_0 - \alpha)/\sigma}^{\sqrt{2\beta}(Q - \alpha)/\sigma} \frac{\Phi(y)}{\phi(y)} dy \tag{12}$$

and

$$Var[T_Q] = V_2(y_0) - (V_1(y_0))^2 = \frac{2}{\beta^2} \int_{\sqrt{2\beta}(y_0 - \alpha)/\sigma}^{\sqrt{2\beta}(Q - \alpha)/\sigma} \int_{-\infty}^{z} \frac{\Phi^2(y)}{\phi(y)\phi(z)} dy dz, \tag{13}$$

Table 4 Results for optimization by weight

	P (euro/kg)	$E[A_{opt}]$	Q^*_{opt}	$E[L_{Q^*_{opt}}]$	$sd[L_{Q^*_{opt}}]$
MGE	3.00	1.04	271	33.14	57.03
	3.25	1.14	292	68.38	88.79
	3.50	1.23	309	105.95	79.85
	3.75	1.31	324	145.43	91.13
MBRE	3.00	0.77	200	14.71	34.37
	3.25	0.92	232	41.74	51.22
	3.50	1.05	256	72.29	64.62
	3.75	1.16	275	105.50	76.37

where Φ and ϕ are the distribution function and the probability density function of a standard normal random variable. To obtain the mean and variance of T_Q, one needs to numerically integrate in (12) and (13).

For the case in which we want to compute the profit obtained from selling the animal when a certain weight Q^* is achieved for the first time, we can use the expression $L_{Q^*} = PRQ^* - C_1 - c_2 T_Q$. Based on the above expressions (12) and (13) for the mean and variance of the first passage time by $Q = h(Q^*)$, we can easily obtain the mean and variance of the profit

$$E[L_{Q^*}] = PRQ^* - C_1 - c_2 E[T_Q | Y(0) = y_0] \qquad (14)$$

and

$$Var[L_{Q^*}] = c_2^2 Var[T_Q | Y(0) = y_0]. \qquad (15)$$

Considering the situation described in the previous subsection ($t_0 = 0.583$, $x_0 = 160$ kg, $R = 0.5$, $C_1 = 227.45$ euros and $c_2 = 26.68$/month $\times 12 = 320.16$/year), we have started by using expression (12) to compute the expected times to achieve weights from 200 to 400 kg. These results were then used in (14) and through the maximization with respect to Q^*, we have obtained the maximum mean profit ($E[L_{Q^*_{opt}}]$), and corresponding optimal selling weight (Q^*_{opt}). The computations were made using the software Maple. Table 4 shows these results for both models (SGM and SBRM), as well as the standard deviation of the profit ($sd[L_{Q^*_{opt}}]$) and expected age of the animal when the optimal weight is achieved ($E[A_{opt}]$). Since the animal was bought at 7 months of age (0.583 years), the expected age when Q^*_{opt} is achieved for the first time can be computed as $E[A_{opt}] = 0.583 + E[T_{Q_{opt}}]$.

4　　Final Remarks

With the goal of optimizing the mean profit obtained by selling an animal in the cattle market, based on an SDE growth model, two approaches were studied in Sects. 3.1 and 3.2. One consists in selling the animal at a fixed age (chosen in order to obtain an optimal mean profit), independently of the animal's weight. The other consists in selling the animal when a fixed weight (chosen in order to obtain an optimal mean profit) is achieved for the first time, independently of the animal's age.

We have observed that, for typical market values, the second methodology achieves a higher optimal mean profit compared with the first methodology, and even provides a lower standard deviation for this optimal profit in the SGM case (in the SBRM case the standard deviation was higher but only slightly).

Acknowledgements The first three authors are members of the Centro de Investigação em Matemática e Aplicações and the fourth author is member of the Instituto de Ciências Agrárias e Ambientais Mediterrânicas, both research centers of the Universidade de Évora financed by Fundação para a Ciência e Tecnologia (FCT). Braumann gratefully acknowledges the financial support from the FCT grant PTDC/MAT/115168/2009.

References

1. Braumann, C.A.: Introdução às Equações Diferenciais Estocásticas. Edições SPE, Lisboa (2005)
2. Braumann, C.A., Filipe, P.A., Carlos C., Roquete, C.J.: Growth of individuals in randomly fluctuating environments. In: Vigo-Aguiar, J., Alonso, P., Oharu, S., Venturino, E., Wade, B. (eds.) Proceedings of the International Conference in Computational and Mathematical Methods in Science e Engineering, Gijon, pp. 201–212 (2009)
3. Filipe, P.A.: Equações diferenciais estocásticas na modelação do crescimento individual em ambiente aleatório. Dissertação de Doutoramento, Universidade de Évora (2011)
4. Filipe, P.A., Braumann, C.A., Brites, N.M., Roquete, C.J.: Modelling animal growth in random environments: an application using nonparametric estimation. Biom. J. **52**(5), 653–666 (2010)
5. Filipe, P.A., Braumann C.A., Roquete, C.J.: Multiphasic individual growth models in random environments. Methodol. Comput. Appl. Probab. **14**(1), 49–56 (2012)
6. Ghikman, I.I., Skorohod, A.V.: Stochastic Differential Equations. Springer, Berlin (1991)
7. Karlin, S., Taylor, H.M.: A Second Course in Stochastic Processes. Academic, New York (1981)

Valuation of Bond Options Under the CIR Model: Some Computational Remarks

Manuela Larguinho, José Carlos Dias, and Carlos A. Braumann

Abstract

Pricing bond options under the Cox, Ingersoll and Ross (CIR) model of the term structure of interest rates requires the computation of the noncentral chi-square distribution function. In this article, we compare the performance in terms of accuracy and computational time of alternative methods for computing such probability distributions against an externally tested benchmark. All methods are generally accurate over a wide range of parameters that are frequently needed for pricing bond options, though they all present relevant differences in terms of running times. The iterative procedure of Benton and Krishnamoorthy (Comput. Stat. Data Anal. 43:249–267, 2003) is the most efficient in terms of accuracy and computational burden for determining bond option prices under the CIR assumption.

M. Larguinho (✉)
Department of Mathematics, ISCAC, Quinta Agrícola, Bencanta, 3040-316 Coimbra, Portugal
e-mail: mlarguinho@iscac.pt

J.C. Dias
BRU-UNIDE and ISCTE-IUL Business School, Av. Prof. Aníbal Bettencourt, 1600-189 Lisboa, Portugal
e-mail: jose.carlos.dias@iscte.pt

C.A. Braumann
Department of Mathematics, Centro de Investigação em Matemática e Aplicações, Universidade de Évora, Rua Romão Ramalho 59, 7000-671 Évora, Portugal
e-mail: braumann@uevora.pt

A. Pacheco et al. (eds.), *New Advances in Statistical Modeling and Applications*,
Studies in Theoretical and Applied Statistics, DOI 10.1007/978-3-319-05323-3_12,
© Springer International Publishing Switzerland 2014

1   Introduction

The CIR model is a general single-factor equilibrium model developed by Cox et al. [3] and has been used throughout the years because of its analytical tractability and the fact that the short rate is always positive, contrary to the well-known Vasicek model of [8].

The CIR model is used to price zero-coupon bonds, coupon bonds and to price options on these bonds. To compute option prices under this process we need to use the noncentral chi-square distribution function. There exists an extensive literature devoted to the efficient computation of this distribution function. In this article, we will examine the methods proposed by Schroder [7], Ding [4] and Benton and Krishnamoorthy [2]. The noncentral chi-square distribution function can also be computed using methods based on series of incomplete gamma series, which will be used as our benchmark.

2   Noncentral χ^2 Distribution and Alternative Methods

If $Z_1, Z_2, \ldots, Z_v$ are independent unit normal random variables, and $\delta_1, \delta_2, \ldots, \delta_v$ are constants, then $Y = \sum_{j=1}^{v} (Z_j + \delta_j)^2$ has a noncentral chi-square distribution with v degrees of freedom and noncentrality parameter $\lambda = \sum_{j=1}^{v} \delta_j^2$, which is denoted as $\chi_v^{'2}(\lambda)$. When $\delta_j = 0$ for all j, then Y is distributed as the central chi-square distribution with v degrees of freedom, which is denoted as χ_v^2. Hereafter, $P[\chi_v^{'2}(\lambda) \leq w] = F(w; v, \lambda)$ is the cumulative distribution function (CDF) of $\chi_v^{'2}(\lambda)$ and $P[\chi_v^2 \leq w] = F(w; v, 0)$ is the CDF of χ_v^2.

The CDF of $\chi_v^{'2}(\lambda)$ is given by:

$$F(w; v, \lambda) = e^{-\lambda/2} \sum_{j=0}^{\infty} \frac{(\lambda/2)^j}{j!\, 2^{v/2+j}\, \Gamma(v/2+j)} \int_0^w y^{v/2+j-1}\, e^{-y/2}\, dy, \quad w > 0,$$

(1)

while $F(w; v, \lambda) = 0$ for $w < 0$. Alternatively, it is possible to express $F(w; v, \lambda)$, for $w > 0$, as a weighted sum of central chi-square probabilities with weights equal to the probabilities of a Poisson distribution with expected value $\lambda/2$, that is,

$$F(w; v, \lambda) = \sum_{j=0}^{\infty} \left(\frac{(\lambda/2)^j}{j!}\, e^{-\lambda/2} \right) F(w; v + 2j, 0),$$

(2)

where the central chi-square probability function $F(w; v + 2j, 0)$ is given by Abramowitz and Stegun [1, Eq. 26.4.1].

2.1   The Gamma Series Method

It is well known that the function $F(w; v + 2n, 0)$ is related to the so-called incomplete gamma function (see, for instance, [1, Eq. 26.4.19]). Hence, we may express the function (2) using series of incomplete gamma functions as follows:

$$F(w; v, \lambda) = \sum_{i=0}^{\infty} \frac{(\lambda/2)^i \, e^{-\lambda/2}}{i!} \frac{\gamma(v/2 + i, w/2)}{\Gamma(v/2 + i)}, \tag{3}$$

with $\gamma(m, t)$ and $\Gamma(m)$ being, respectively, the incomplete gamma function and the Euler gamma function as defined by Abramowitz and Stegun [1, Eqs. 6.5.2 and 6.1.1].

While this method is accurate over a wide range of parameters, the number of terms that must be summed increases with the noncentrality parameter λ. To avoid the infinite sum of the series we use the stopping rule as proposed by Knüsel and Bablock [6] which allows the specification of a given error tolerance by the user.

For the numerical analysis of this article we will concentrate the discussion on [4, 7] methods since both are commonly used in the finance literature. We will also use the suggested approach of [2], since it is argued by the authors that their algorithm is computationally more efficient than the one suggested by Ding [4]. A detailed explanation of how to compute the noncentral chi-square distribution function using these three algorithms is presented below.

2.2   The Schroder Method

In the method proposed by Schroder [7], the noncentral chi-square distribution is expressed as an infinite double sum of gamma densities which does not require the computation of incomplete gamma functions, that is

$$F(w; v, \lambda) = \sum_{n=1}^{\infty} g(n + v/2, w/2) \sum_{i=1}^{n} g(i, \lambda/2), \tag{4}$$

where $g(m, u) = e^{-u} u^{m-1} / \Gamma(m)$ is the standard form of the gamma density function. As noted by Schroder [7], Eq. (4) allows the following simple iterative algorithm to be used for computing the infinite sum when w and λ are not too large. First, initialize the following four variables (with $n = 1$): $gA = \frac{e^{-w/2} (w/2)^{v/2}}{\Gamma(1+v/2)}$, $gB = e^{-\lambda/2}$, $Sg = gB$, and $R = gA \times Sg$. Then repeat the following loop beginning with $n = 2$ and incrementing n by one after each iteration: $gA \leftarrow gA \times \frac{w/2}{n+v/2-1}$, $gB \leftarrow gB \times \frac{\lambda/2}{n-1}$, $Sg \leftarrow Sg + gB$, and $R \leftarrow R + gA \times Sg$. The loop is terminated when the contributions to the sum, R, are declining and very small.

2.3   The Ding Method

A similar simple recursive algorithm for evaluating the noncentral chi-square distribution is provided also by Ding [4]. Let us define $t_0 = \frac{1}{\Gamma(v/2+1)} \left(\frac{w}{2}\right)^{v/2} e^{-w/2}$, $t_i = t_{i-1} \frac{w}{v+2i}$, $y_0 = u_0 = e^{-\lambda/2}$, $u_i = \frac{u_{i-1}\lambda}{2i}$, and $y_i = y_{i-1} + u_i$. Then the required probability that the variable with the noncentral chi-square distribution will take values smaller than w is

$$F(w; v, \lambda) = \sum_{i=0}^{\infty} y_i \, t_i. \tag{5}$$

By taking a sufficient number of terms in the series or using the bound as defined by Ding [4] for the error tolerance incurred by truncating the series, the required accuracy can be obtained.

2.4   The Benton and Krishnamoorthy Method

The function $F(w; v, \lambda)$ is also expressed by Benton and Krishnamoorthy [2] using series of incomplete gamma functions as given by Eq. (3), where $P(m, t) = \frac{\gamma(m,t)}{\Gamma(m)}$ is the standard gamma distribution function, with $\gamma(m, t)$ and $\Gamma(m)$ as defined in (3). To compute $F(w; v, \lambda)$, [2] makes use of the following recurrence relations obtained from [1, Eq. 6.5.21]:

$$P(a + 1, x) = P(a, x) - \frac{x^a e^{-x}}{\Gamma(a + 1)}, \tag{6}$$

$$P(a - 1, x) = P(a, x) + \frac{x^{a-1} e^{-x}}{\Gamma(a)}. \tag{7}$$

From Eq. (6) it follows that

$$P(a, x) = \frac{x^a e^{-x}}{\Gamma(a + 1)} \left(1 + \frac{x}{(a + 1)} + \frac{x^2}{(a + 1)(a + 2)} + \cdots\right), \tag{8}$$

which can be used to evaluate $P(a, x)$. The computational algorithm also differs from the others essentially because, in order to compute the noncentral chi-square distribution function $F(w; v, \lambda)$, it starts by evaluating the kth term, where k is the integer part of $\lambda/2$, and then the other terms $k \pm i$ are computed recursively. The proposed method runs in the following steps. First, evaluate $P_k = P(Y = k) = e^{-\lambda/2}(\lambda/2)^k / k!$ and $P(v/2 + k, w/2)$ using Eq. (8). Then, compute $P(Y = k + i)$ and $P(Y = k - i)$, for $i = 1, 2, \ldots$, using the initial value $P(Y = k)$, and the recursion relations for Poisson probabilities $P_{i+1} = \frac{\lambda/2}{i+1} P_i$, $P_{i-1} = \frac{i}{\lambda/2} P_i$. Finally, using recursion relations (6) and (7) compute $P(v/2 + k + i, w/2)$ and

$P(v/2 + k - i, w/2)$. By taking a sufficient number of terms in the series or by specifying a given error tolerance the required accuracy is then obtained.

3   Bond Options Under the CIR Model

Under the risk-neutral measure Q, Cox et al. [3] modeled the evolution of the interest rate, r_t, by the stochastic differential equation (SDE):

$$dr_t = \left[\kappa\theta - (\lambda + \kappa)r_t\right]dt + \sigma\sqrt{r_t}dW_t^Q, \tag{9}$$

where W_t^Q is a standard Brownian motion under Q, κ, θ and σ are positive constants representing reversion rate, asymptotic rate and volatility parameters, respectively, and λ is the market risk. The condition $2\kappa\theta > \sigma^2$ has to be imposed to ensure that the interest rate remains positive. Following [3], the price of a general interest rate claim $F(r, t)$ with cash flow rate $C(r, t)$ satisfies the following partial differential equation (PDE)

$$\frac{1}{2}\sigma^2 r \frac{\partial^2 F(r,t)}{\partial r^2} + \kappa(\theta - r)\frac{\partial F(r,t)}{\partial r} + \frac{\partial F(r,t)}{\partial t} - \lambda r \frac{\partial F(r,t)}{\partial r} - rF(r,t) + C(r,t) = 0. \tag{10}$$

3.1   Zero-Coupon and Coupon Bonds

A bond is a contract that pays its holder a known amount, the principal, at a known future date, called maturity. The bond may also pay periodically to its holder fixed cash dividends, called coupons. When it gives no dividends, it is known as a zero-coupon bond, sometimes referred to as pure discount bond. The price of a zero-coupon bond with maturity at time s, $Z(r, t, s)$, satisfies the PDE (10), with $C(r, t) = 0$, subject to the boundary condition $Z(r, s, s) = 1$, and is given by

$$Z(r,t,s) = A(t,s)e^{-B(t,s)r} \tag{11}$$

where $A(t,s) = \left(\dfrac{2\gamma\, e^{\left((\kappa+\lambda+\gamma)(s-t)\right)/2}}{(\kappa+\lambda+\gamma)\left(e^{\gamma(s-t)}-1\right)+2\gamma} \right)^{\frac{2\kappa\theta}{\sigma^2}}$, $B(t,s) = \dfrac{2\left(e^{\gamma(s-t)}-1\right)}{(\kappa+\lambda+\gamma)\left(e^{\gamma(s-t)}-1\right)+2\gamma}$, and
$\gamma = \left((\kappa+\lambda)^2 + 2\sigma^2\right)^{1/2}$.

Since a coupon bond is just a portfolio of zero-coupon bonds of different maturities, the value of any riskless coupon bond (under a one-dimensional setting) can be expressed as a weighted sum of zero-coupon bond prices

$$P(r,t,s) = \sum_{i=1}^{N} a_i Z(r,t,s_i), \tag{12}$$

where $s_1, s_2, \cdots, s_N$ represent the N dates on which payments are made, and the $a_i > 0$ terms denote the amount of the payments made.[1]

3.2   Bond Options

A bond option provides the investor with the right, but not the obligation, to buy or sell a given bond at a fixed price either or before a specific date. In this article, we analyze European-style plain-vanilla options on bonds, which confer the right to buy or sell at a known future date for a predetermined price, i.e. the exercise price. Denote by $c^{zc}(r, t, T, s, K)$ the price at time t of a European call option with maturity $T > t$, strike price K, written on a zero-coupon bond with maturity at time $s > T$ and with the instantaneous rate at time t given by r_t. K is restricted to be less than $A(T, s)$, the maximum possible bond price at time T, since otherwise the option would never be exercised and would be worthless. The option price will follow the basic valuation equation with terminal condition $c^{zc}(r, t, T, s, K) = max[Z(r, T, s) - K, 0]$ to the PDE (10), with $C(r, t) = 0$, and is given by

$$c^{zc}(r, t, T, s, K) = Z(r, t, s) \, F(x_1; a, b_1) - KZ(r, t, T) \, F(x_2; a, b_2), \qquad (13)$$

where $x_1 = 2r^*[\phi + \psi + B(T, s)]$, $x_2 = 2r^*[\phi + \psi]$, $a = \frac{2\kappa\theta}{\sigma^2}$, $b_1 = \frac{2\phi^2 r e^{\gamma(T-t)}}{\phi + \psi + B(T,s)}$, $b_2 = \frac{2\phi^2 r e^{\gamma(T-t)}}{\phi + \psi}$, $\psi = \frac{\kappa + \lambda + \gamma}{\sigma^2}$, $\phi = \frac{2\gamma}{\sigma^2 \left(e^{\gamma(T-t)} - 1\right)}$, $r^* = \left[\ln\left(\frac{A(T,s)}{K}\right)\right] / B(T, s)$, $F(.; \nu, \lambda)$ is the noncentral chi-square distribution function with ν degrees of freedom and non-centrality parameter λ, and r^* is the critical rate below which exercise will occur, this is, $K = Z(r^*, T, s)$.

To compute options on coupon bonds we will use the Jamshidian's approach, [5], which states that an option on a portfolio of zero-coupon bonds is equivalent to a portfolio of options with appropriate strike prices. The individual options all have the same maturity and are written on the individual zero-coupon bonds in the bond portfolio. Based on this result, a European call option with exercise price K and maturity T on a bond portfolio consisting of N zero-coupon bonds with distinct maturities s_i ($i = 1, 2, \cdots, N$ and $T < s_1 < s_2 < \cdots < s_N$) and $a_i (a_i > 0, i = 1, 2, \cdots, N)$ issues of each can be priced as

$$c^{cb}(r, t, T, s, K) = \sum_{i=1}^{N} a_i c^{zc}(r, t, T, s_i, K_i), \qquad (14)$$

where $K_i = Z(r^{**}, T, s_i)$ and r^{**} is the solution to $\sum_{i=1}^{N} a_i Z(r^{**}, T, s_i) = K$.

---

[1]As an example, consider a 10-year 6 % bond with a face amount of 100. In this case, $N = 20$ since the bond makes 19 semiannual coupon payments of 3 as well as a final payment of 103. That is, $a_i = 3, i = 1, 2, \cdots, 19, a_{20} = 3 + 100 = 103$, and $s_1 = 0.5, s_2 = 1, \cdots, s_{19} = 9.5, s_{20} = 10$.

4   Numerical Analysis

This section aims to present computational comparisons of the alternative methods of computing the noncentral chi-square distribution function for pricing European options on bonds under the CIR diffusion. We examine this CIR option pricing model using alternative combinations of input values over a wide range parameter space. All the calculations in this article were made using *Mathematica* 7.0 running on a Pentium IV (2.53 GHz) personal computer. We have truncated all the series with an error tolerance of 1E−10. All values are rounded to four decimal places. In order to understand the computational speed of the alternative algorithms, we have computed the CPU times for all the algorithms using the function Timing[.] available in *Mathematica*. Since the CPU time for a single evaluation is very small, we have computed the CPU time for multiple computations. Note that the difference in computation time among the alternative tested methods is clearly due to the specific definition of each algorithm and the corresponding stopping rule, and not on the particular software implementation.

4.1   Benchmark Selection

The noncentral chi-square distribution function $F(w; v, \lambda)$ requires values for w, v, and λ. Our benchmark is the noncentral chi-square distribution $F(w; v, \lambda)$ expressed as a gamma series (GS) as given by Eq. (3), with a predefined error tolerance of 1E−10, which is tested against three external benchmarks based on the *Mathematica, Matlab*, and *R* built-in-functions that are available for computing the CDF of the noncentral chi-square distribution. The set of parameters used in the benchmark selection: is $\kappa \in \{0.15, 0.25, \cdots, 0.85\}$, $\theta \in \{0.03, 0.06, \cdots, 0.15\}$, $r \in \{0.01, 0.02, \cdots, 0.15\}$, $\sigma \in \{0.03, 0.05, \cdots, 0.15\}$, and $\lambda \in \{-0.1, 0\}$. We also consider the next two set of parameters: for the bond maturity $s = 2$, we have $T \in \{1, 1.5, 1.75\}$, and in this case the strike price set is $K \in \{0.90, 0.95\}$; for the bond maturity of $s = 10$, we consider $T \in \{3, 5, 7\}$, and in this situation the strike prices are $K \in \{0.25, 0.35\}$. These combinations of parameters produce 98, 280 probabilities.[2] Table 1 reports the results obtained. The results show that the maximum absolute error (MaxAE) and root mean absolute error (RMSE) are higher for the comparison between the GS vs CDF of *Mathematica*, though the number of times the absolute difference between the two methods exceeds 1E−07 (k_1) is small in relative terms (it represents about 0.08 % of the 98,280 computed probabilities). However, the number of times a computed probability is greater than 1 (k_2) is slightly higher for the CDF of *Mathematica*[3] (about 1.80 % of computed probabilities computed). The results comparing the GS vs CDF of *Matlab* and

---

[2] We obtained these probabilities by computing the values of $F(x_1; v, b_1)$ for this set of parameters.

[3] This means that care must be taken if one wants to use the CDF built-in-function of *Mathematica* for computing the noncentral chi-square distribution function.

Table 1 Benchmark selection

Methods	MaxAE	RMSE	k_1	k_2
GS vs CDF of *Mathematica*	1.29E−04	4.13E−07	79	1,769
GS vs CDF of *Matlab*	6.46E−11	1.16E−11	0	0
GS vs CDF of *R*	6.45E−11	1.16E−11	0	0

Table 2 Differences in approximations for each method compared against a benchmark

Methods	MaxAE	MaxRE	RMSE	MeanAE	CPU time	k_1	k_3
Panel A: Differences in probabilities							
S89	3.79E−10	4.19E−01	1.21E−10	9.03E−11	9,773.37	0	–
D92	9.60E−11	1.83E−02	6.20E−11	5.94E−11	9,085.95	0	–
BK03	4.23E−11	1.71E−07	4.29E−12	1.29E−12	1,946.11	0	–
Panel B: Differences in call option prices on zero-coupon bonds							
S89	1.22E−10	5.24E+00	2.60E−11	1.63E−11	9,796.68	–	0
D92	3.96E−11	3.43E−02	1.25E−11	1.00E−11	9,013.49	–	0
BK03	6.97E−12	4.76E−05	6.61E−13	1.59E−13	1,967.84	–	0
Panel C: Differences in call option prices on coupon bonds							
S89	1.24E−08	1.52E+00	1.98E−09	1.35E−09	14,101.40	–	0
D92	7.00E−09	2.58E−01	2.06E−09	1.66E−09	13,309.00	–	0
BK03	2.03E−09	1.97E−06	1.58E−10	3.03E−11	6,274.70	–	0

GS vs CDF of R show that the corresponding differences are smaller and very similar (never exceeds 1E−07). Under the selected wide parameter space we have not obtained any probability value greater than 1 either in the gamma series method, *Matlab* or *R*. In summary, the results show that the gamma series method is an appropriate choice for the benchmark.

4.2 Bond Options with Alternative Methods

Now we want to evaluate the differences in approximations of noncentral chi-square probabilities $F(w; v, \lambda)$ and in zero-coupon and coupon bond option prices using the iterative procedures of [7] (S89), [4] (D92) and [2] (BK03) compared against the benchmark based on the gamma series approach. We will concentrate our analysis on call options, but the same line of reasoning applies also for put options. Panels A and B of Table 2 report such comparison results using the following set of parameters: $\kappa \in \{0.35, 0.65\}$, $\theta = 0.08$, $\sigma \in \{0.04, 0.10, 0.16\}$, $r \in \{0.01, 0.02, \cdots, 0.15\}$, $\lambda \in \{-0.1, 0.0\}$, $K \in \{0.25, 0.30\}$, $T \in \{2, 5\}$, and $s \in \{10, 15\}$. Panel C of Table 2 analyzes the impact of these competing methods for pricing call options on coupon bonds under the CIR diffusion. In this analysis we used the following set of parameters: $\kappa \in \{0.35, 0.65\}$, $\theta = 0.08$, $\sigma \in \{0.04, 0.10, 0.16\}$, $r \in \{0.01, 0.02, \cdots, 0.15\}$, $\lambda \in \{-0.1, 0.0\}$, $K \in \{95, 100, 105\}$, face value $= 100$, $T \in \{2, 5\}$, $s \in \{10, 15\}$, and a coupon rate $\in \{0.10, 0.12\}$. The third rightmost column of the table reports the CPU time for computing 1,000 times the 2,880 probabilities and 1,440 unique contracts of zero-coupon bond options[4] and the CPU time for determining 100 times the 4,320 unique contracts of coupon bond options. The MaxRE, MeanAE and k_3 denote, respectively, the maximum relative error, the mean absolute error, and the number of times the absolute difference between the two methods exceeds $0.01.

5   Conclusion

In this article, we compare the performance of alternative algorithms for computing the noncentral chi-square distribution function in terms of accuracy and computation time for evaluating option prices under the CIR model. We find that all algorithms are accurate over a wide range of parameters, though presenting significative differences on computational expenses. Overall, we find that the [2] algorithm is clearly the most accurate and efficient in terms of computation time needed for determining option prices under the CIR assumption. Moreover, it has a running time that does not vary significantly with the parameters w, v, and λ.

Acknowledgments Dias is member of the BRU-UNIDE and Larguinho and Braumann are members of the Research Center Centro de Investigação em Matemática e Aplicações (CIMA), both centers financed by the Fundação para a Ciência e Tecnologia (FCT). Dias gratefully acknowledges the financial support from the FCTs grant number PTDC/EGE-ECO/099255/2008.

References

1. Abramowitz, M., Stegun, I.A.: Handbook of Mathematical Functions. Dover, New York (1972)
2. Benton, D., Krishnamoorthy, K.: Compute discrete mixtures of continuous distributions: noncentral chisquare, noncentral t and the distribution of the square of the sample multiple correlation coefficient. Comput. Stat. Data Anal. **43**, 249–267 (2003)
3. Cox, J.C., Ingersoll, J.E., Ross, S.A.: A theory of the term structure of interest rate. Econometrica **53**(2), 385–408 (1985)
4. Ding, C.G.: Algorithm AS 4: computing the non-central χ^2 distribution function. Appl. Stat. **41**, 478–482 (1992)
5. Jamshidian, F.: An exact bond option formula. J. Finance **44**(1), 205–209 (1989)
6. Knüsel, L., Bablock, B.: Computation of the noncentral gamma distribution. SIAM J. Sci. Comput. **17**, 1224–1231 (1996)
7. Schroder, M.: Computing the constant elasticity of variance option pricing formula. J. Finance **44**(1), 211–219 (1989)
8. Vasicek, O.: An equilibrium characterization of the term structure. J. Financ. Econ. **5**, 177–188 (1977)

---

[4]The CPU time for the gamma series method is 3,303.23 s for probabilities, 3,340.48 s for zero-coupon bond options, and 10,714.40 for coupon bond options.

Part III

Extremes

A Semi-parametric Estimator of a Shape Second-Order Parameter

Frederico Caeiro and M. Ivette Gomes

Abstract

In extreme value theory, any second-order parameter is an important parameter that measures the speed of convergence of the sequence of maximum values, linearly normalized, towards its limit law. In this paper we study a new estimator of a shape second-order parameter under a third-order framework.

1 Introduction

Let us assume that $X_1, X_2, \ldots, X_n$ are independent and identically distributed (i.i.d.) random variables (r.v.'s), with a Pareto-type distribution function (d.f.) F satisfying

$$\lim_{t\to\infty} \frac{\overline{F}(tx)}{\overline{F}(t)} = x^{-1/\gamma} \quad \Leftrightarrow \quad \lim_{t\to\infty} \frac{U(tx)}{U(t)} = x^{\gamma}, \quad \forall\, x > 0, \tag{1}$$

where $\gamma > 0$, $\overline{F}(x) := 1 - F(x)$, and $U(t) := \inf\{x : F(x) \ge 1 - 1/t\}$. Then we are in the max-domain of attraction of the Extreme Value distribution

$$EV_\gamma(x) = \exp\{-(1+\gamma x)^{-1/\gamma}\}, \quad 1+\gamma x > 0,$$

F. Caeiro (✉)
Faculdade de Ciências e Tecnologia da Universidade Nova de Lisboa and CMA, 2829-516 Caparica, Portugal
e-mail: fac@fct.unl.pt

M.I. Gomes
Faculdade de Ciências da Universidade de Lisboa and CEAUL, Campo Grande, 1749-016 Lisboa, Portugal
e-mail: ivette.gomes@fc.ul.pt

A. Pacheco et al. (eds.), *New Advances in Statistical Modeling and Applications*, Studies in Theoretical and Applied Statistics, DOI 10.1007/978-3-319-05323-3_13, © Springer International Publishing Switzerland 2014

where γ is the extreme value index (EVI). This index measures the heaviness of the right tail $\overline{F}$, and the heavier the right tail, the larger the EVI is. Although we deal with the right tail $\overline{F}$, the results here presented are applicable to the left tail F, after the change of variable $Y = -X$.

The estimation of the EVI is an important subject in Extreme Value Theory. Many classical EVI-estimators, based on the k largest order statistics have a strong asymptotic bias for moderate up to large values of k. To improve the estimation of γ through the adaptive selection of k or through the reduction of bias of the classical EVI estimators, we usually need to know the nonpositive second-order parameter, ρ, ruling the rate of convergence of the normalized sequence of maximum values towards the limiting law EV_γ, in Eq. (1), through the limiting relation

$$\lim_{t\to\infty} \frac{\ln U(tx) - \ln U(t) - \gamma \ln x}{A(t)} = \frac{x^\rho - 1}{\rho}, \quad \forall\, x > 0, \tag{2}$$

where $|A|$ must be of regular variation with index $\rho \le 0$ (9).

In this paper, we are interested in the estimation of the second-order parameter ρ in (2). For technical reasons, we shall consider $\rho < 0$. In Sect. 2, after a brief review of some estimators in the literature, we introduce a new class of estimators for the second-order parameter ρ. In Sect. 3, we derive the asymptotic behavior of the ρ-estimators. Finally, in Sect. 4 we provide some applications to real and simulated data.

2   Estimation of the Second-Order Parameter ρ

2.1   A Review of Some Estimators in the Literature

Many estimators of the second-order parameter ρ, in Eq. (2), are based on the scaled log-spacings U_i or on the log-excesses V_{ik} defined by

$$U_i := i\left\{\ln \frac{X_{n-i+1:n}}{X_{n-i:n}}\right\} \quad \text{and} \quad V_{ik} := \ln \frac{X_{n-i+1:n}}{X_{n-k:n}}, \quad 1 \le i \le k < n, \tag{3}$$

where $X_{i:n}$ denotes the ith ascending order statistic from a sample of size n.

The first estimator of ρ appears in [12]. Under the second-order condition in (2), with $\rho < 0$ and $A(t) = \gamma \beta t^\rho$, the log-spacings U_i, $1 \le i \le k$, in Eq. (3), are approximately exponential with mean value $\gamma \exp(\beta(i/n)^{-\rho})$, $1 \le i \le k$. [7] considered the joint maximization, in order to γ, β, and ρ, of the approximate log-likelihood of the scaled log-spacings. Such a maximization led Feuerverger and Hall to an explicit expression for $\hat{\gamma}$, as a function of $\hat{\beta}$ and $\hat{\rho}$, and to implicit estimators of $\hat{\beta} = \hat{\beta}_n^{FH}(k)$ and $\hat{\rho} = \hat{\rho}_n^{FH}(k)$. More precisely,

$$(\hat{\beta}, \hat{\rho}) := \underset{(\beta,\rho)}{\arg\min} \left\{ \ln\left(\frac{1}{k}\sum_{i=1}^{k} e^{-\beta(i/n)^{-\rho}} U_i\right) + \beta\left(\frac{1}{k}\sum_{i=1}^{k}(i/n)^{-\rho}\right) \right\}. \quad (4)$$

[8, 11] worked with the log-excesses V_{ik}, in (3), to obtain new estimators of the second-order parameter ρ. As mentioned by Goegebeur et al. [10], the estimator generally considered to be the best working one in practice is a particular member of the class of estimators proposed by Fraga Alves et al. [8]. Such a class of estimators has been first parameterized in a tuning parameter $\tau \geq 0$, but more generally, τ can be considered as a real number (3). Using the notation $a^{b\tau} = b \ln a$ if $\tau = 0$, it is defined as

$$\hat{\rho}_n^{FAGH}(k) = \hat{\rho}_n^{FAGH(\tau)}(k) := \min\left\{0, 3(T_{n,k}^{(\tau)} - 1)/(T_{n,k}^{(\tau)} - 3)\right\}, \quad (5)$$

with

$$T_{n,k}^{(\tau)} := \frac{\left(M_{n,k}^{(1)}\right)^\tau - \left(M_{n,k}^{(2)}/2\right)^{\tau/2}}{\left(M_{n,k}^{(2)}/2\right)^{\tau/2} - \left(M_{n,k}^{(3)}/6\right)^{\tau/3}}, \quad \tau \in \mathbb{R}, \quad M_{n,k}^{(\alpha)} := \frac{1}{k}\sum_{i=1}^{k}(V_{ik})^\alpha, \quad \alpha > 0.$$

Remark 1. The use of the estimator $\hat{\rho}_n^{FAGH}(k)$ in several articles on reduced-bias tail index estimation has led several authors to choose $\tau = 0$, if $\rho \geq -1$ and $\tau = 1$ if $\rho < -1$. However, practitioners should not choose blindly the value of τ. It is sensible to draw a few sample paths of k vs. $\hat{\rho}_n^{FAGH}(k)$, for several values of τ, electing the one which provides the highest stability for large k.

More recently, [6] extended the estimators in [11] and [8] and [10] introduced a new class of estimators based on the scaled log-excesses U_i. Further details on this topic can be found in [1] and references within.

2.2 A New Estimator for the Second-Order Parameter ρ

We will now propose a new estimator for the shape second-order parameter ρ. First we will consider the ratio of a difference of estimators of the same parameter,

$$R_{n,k}^{(\tau)} = \frac{\left(N_{n,k}^{(1)}\right)^\tau - \left(N_{n,k}^{(3/2)}\right)^\tau}{\left(N_{n,k}^{(3/2)}\right)^\tau - \left(N_{n,k}^{(2)}\right)^\tau}, \quad \tau \in \mathbb{R}, \quad (6)$$

with $N_{n,k}^{(\alpha)} := \frac{\alpha}{k}\sum_{i=1}^{k}\left(\frac{i}{k}\right)^{\alpha-1} U_i$, $\alpha \geq 1$, consistent estimators of $\gamma > 0$, and using the notation $a^\tau = \ln a$ whenever $\tau = 0$. The values $\alpha = 1, 3/2$, and 2 could be changed. But those values allow us to have a small asymptotic variance, for the most common values of ρ. It will be shown that, under the second-order condition in (2), $R_{n,k}^{(\tau)}$ will converge to $\frac{2-\rho}{1-\rho}$. Then, by inversion, we get the new estimator for the shape second-order parameter ρ with functional expression,

$$\hat{\rho}_n^{CG}(k) = \hat{\rho}_n^{CG(\tau)}(k) := \min\left\{0, 1 + \left(1 - R_{n,k}^{(\tau)}\right)^{-1}\right\}, \quad \tau \in \mathbb{R}. \qquad (7)$$

3   Main Asymptotic Results

We shall next proceed with the study of the new class of ρ-estimator in (7). To compare the new class of estimators with others in the literature, we also present asymptotic results for the estimators in (4) and (5). We will need to work with intermediate values of k, i.e., a sequence of integers $k = k_n$, $1 \le k < n$, such that

$$k = k_n \to \infty \quad \text{and} \quad k_n = o(n), \quad \text{as} \quad n \to \infty. \qquad (8)$$

In order to establish the asymptotic normality of the second-order estimators, it is necessary to further assume a third-order condition, ruling the rate of convergence in (2), and which guarantees that, for all $x > 0$,

$$\lim_{t \to \infty} \frac{\frac{\ln U(tx) - \ln U(t) - \gamma \ln x}{A(t)} - \frac{x^\rho - 1}{\rho}}{B(t)} = \frac{x^{\rho + \rho'} - 1}{\rho + \rho'}, \qquad (9)$$

where $|B(t)|$ must then be of regular variation with index $\rho' \le 0$. There appears then this extra nonpositive third-order parameter $\rho' \le 0$. Although ρ and ρ' can also be zero, we shall consider $\rho, \rho' < 0$.

Remark 2. The third-order condition in (9) holds for models with a tail quantile function

$$U(t) = Ct^\gamma \left(1 + D_1 t^\rho + D_2 t^{\rho + \rho^*} + o\left(t^{\rho + \rho^*}\right)\right), \qquad (10)$$

as $t \to \infty$, with $C > 0$, D_1, $D_2 \neq 0$, $\rho, \rho^* < 0$. For this models $\rho' = \max(\rho, \rho^*)$.

Remark 3. Note that for most of the common heavy-tailed models ($\gamma > 0$), we have $\rho' = \rho$. Among those models we mention: the Fréchet model, with d.f. $F(x) = \exp(-x^{-1/\gamma})$, $x \ge 0$, for which $\rho' = \rho = -1$ and the generalized Pareto (GP) model, with d.f. $F(x) = 1 - (1 + \gamma x)^{-1/\gamma}$, $x \ge 0$, for which $\rho' = \rho = -\gamma$.

Theorem 1. *Let • designate FH, FAGH, or CG and assume that $\hat{\rho}_n^\bullet(k)$ denotes any of the ρ-estimators defined in (4), (5) or (7). If the second-order condition (2) holds, with $\rho < 0$, k is intermediate and such that $\sqrt{k} A(n/k) \to \infty$, as $n \to \infty$, then $\hat{\rho}_n^\bullet(k)$ converges in probability to ρ. Under the third-order framework in (9),*

$$\hat{\rho}_n^{\bullet}(k) \stackrel{d}{=} \rho + \left(\frac{\sigma_{\rho}^{\bullet} W_k^{\bullet}}{\sqrt{k} A(n/k)} + b_1^{\bullet} A(n/k) + b_2^{\bullet} B(n/k) \right) (1 + o_p(1)), \quad (11)$$

with $W_k^{\bullet}$ an asymptotically standard normal r.v.,

$$\sigma_{\rho}^{FH} = \frac{\gamma (1 - \rho)(1 - 2\rho)\sqrt{1 - 2\rho}}{|\rho|},$$

$$\sigma_{\rho}^{FAGH} = \frac{\gamma(1 - \rho)^3 \sqrt{2\rho^2 - 2\rho + 1}}{|\rho|},$$

$$\sigma_{\rho}^{CG} = \frac{\gamma (1 - \rho)(2 - \rho)(3 - 2\rho)\sqrt{4\rho^2 - 4\rho + 7}}{\sqrt{120}|\rho|},$$

and

$$b_1^{FH} = -\frac{(1 - \rho)(1 - 2\rho)^2}{\gamma\rho(1 - 3\rho)^2}, \qquad b_2^{FH} = \frac{(1 - \rho)(1 - 2\rho)(\rho + \rho')\rho'}{\rho(1 - \rho - \rho')(1 - 2\rho - \rho')},$$

$$b_1^{FAGH} = \frac{\rho\left[\tau(1 - 2\rho)^2(3 - \rho)(3 - 2\rho) + 6\rho\left(4(2 - \rho)(1 - \rho)^2 - 1\right)\right]}{12\gamma(1 - \rho)^2(1 - 2\rho)^2},$$

$$b_2^{FAGH} = \frac{\rho'(\rho + \rho')(1 - \rho)^3}{\rho(1 - \rho - \rho')^3},$$

$$b_1^{CG} = \frac{(\tau - 1)\rho}{2\gamma}, \qquad b_2^{CG} = \frac{(1 - \rho)(2 - \rho)(3 - 2\rho)\rho'(\rho + \rho')}{\rho(1 - \rho - \rho')(2 - \rho - \rho')(3 - 2\rho - 2\rho')}.$$

Moreover, if $\sqrt{k} A^2(n/k) \to \lambda_A$ and $\sqrt{k} A(n/k)B(n/k) \to \lambda_B$ (both finite), as $n \to \infty$, then $\sqrt{k}A(n/k)(\hat{\rho}_n^{\bullet}(k) - \rho)$ is asymptotically normal with mean value $\lambda_A b_1^{\bullet} + \lambda_B b_2^{\bullet}$ and variance $(\sigma_{\rho}^{\bullet})^2$.

Proof. For the estimators $\hat{\rho}_n^{FAGH}(k)$ and $\hat{\rho}_n^{FH}(k)$ the proof can be found in [8] and [4], respectively. Regarding the new estimator, $\hat{\rho}_n^{CG}(k)$, we only need to prove the asymptotic representation in (11). Then, consistency and asymptotic normality follows straightforward. Notice that under the third-order condition, in (9), and for intermediate k we have [5]

$$N_{n,k}^{(\alpha)} \stackrel{d}{=} \gamma + \frac{\gamma \alpha Z_k^{(\alpha)}}{\sqrt{(2\alpha - 1)k}} + \frac{\alpha A(n/k)}{\alpha - \rho} + O_p\left(\frac{A(n/k)}{\sqrt{k}}\right) + \frac{\alpha A(n/k)B(n/k)}{\alpha - \rho - \rho'}(1 + o_p(1)),$$

where $Z_k^{(\alpha)}$ is asymptotically standard normal. Then, the use of Taylor expansion $(1 + x)^{-1} = 1 - x + o(x^2)$, $x \to 0$ and after some cumbersome calculations we get

Fig. 1 Pattern of the
quotients of the asymptotic
standard deviations
$\sigma_\rho^{FAGH}/\sigma_\rho^{FH}$ and $\sigma_\rho^{CH}/\sigma_\rho^{FH}$

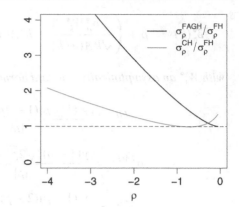

$$R_{n,k}^{(\tau)} \stackrel{d}{=} \frac{2-\rho}{1-\rho} \left\{ 1 + \frac{\sigma_\rho^R Z_k^R}{\sqrt{k} A(n/n)} + \frac{(\tau-1)\rho}{2\gamma(1-\rho)(2-\rho)} A(n/k) \left(1 + O_p\left(\frac{1}{\sqrt{k}}\right) \right) \right.$$
$$\left. + \frac{(3-2\rho)\rho'(\rho+\rho')}{\rho(1-\rho-\rho')(2-\rho-\rho')(3-2\rho-2\rho')} B(n/k)(1 + o_p(1)) \right\},$$

with Z_k^R an asymptotically standard normal r.v. and

$$\sigma_\rho^R = \frac{\gamma(3-2\rho)\sqrt{4\rho^2 - 4\rho + 7}}{\sqrt{120}|\rho|}.$$

Using again Taylor expansion for $(1 + x)^{-1}$, (11) follows.

Remark 4. From Theorem 1, we conclude that the tuning parameter τ affects the
asymptotic bias of $\hat{\rho}_n^{FAGH}$ and $\hat{\rho}_n^{CG}$, but not the asymptotic variance. Consequently
if $\rho = \rho'$ $(B(n/k) = O(A(n/k)))$, we can always choose $\tau = \tau_0$ such that the
asymptotic bias is null, even when $\sqrt{k} A(n/k) \to \infty$ and $\sqrt{k} A^2(n/k) \to \lambda_A$.

Figure 1 show us the values of the quotients of the asymptotic standard deviations
$\sigma_\rho^{FAGH}/\sigma_\rho^{FH}$ and $\sigma_\rho^{CH}/\sigma_\rho^{FH}$, for $-4 \le \rho < 0$. The patterns allow us to conclude that
$\hat{\rho}_n^{FH}$ has the smallest asymptotic variance. Also $\sigma_\rho^{CG} < \sigma_\rho^{FAGH}$ if $\rho < -0.2821$.

4   Applications to Simulated and Real Data

4.1   A Case Study in the Field of Insurance

We shall next consider an illustration through the analysis of automobile claim
amounts exceeding 1,200,000 Euro over the period 1988–2001, gathered from
several European insurance companies cooperating with the same re-insurer (Secura
Belgian Re). This data set is available in [2].

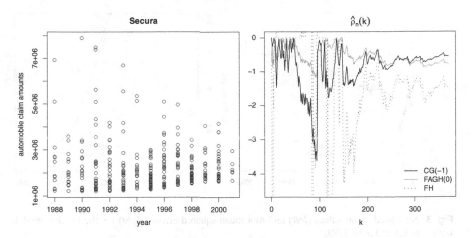

Fig. 2 Secura data (*left*) and estimates of the second-order parameter ρ as function of k (*right*)

In Fig. 2, we present the value of the automobile claim amounts by year (left) and the sample path of the estimates $\hat{\rho}_n^{FH}(k)$, $\hat{\rho}_n^{FAGH(\tau)}(k)$, and $\hat{\rho}_n^{CG(\tau)}(k)$ in (4), (5), and (7), respectively. We have chosen $\tau = 0$ and $\tau = -1$ for $\hat{\rho}_n^{FAGH(\tau)}$ and $\hat{\rho}_n^{CG(\tau)}$, respectively, based on the stability of the sample paths, as function of k. We conclude that, for large values of k, the estimates given by $\hat{\rho}_n^{FAGH(0)}$ and $\hat{\rho}_n^{CG(-1)}$ are very close and are both much stable than the estimates given by $\hat{\rho}_n^{FH}$.

4.2   Simulated Data

We have implemented a multi-sample Monte Carlo simulation experiment, with 200 samples of size 2,000, to obtain the distributional behavior of the estimators $\hat{\rho}_n^{FH}$, $\hat{\rho}_n^{FAGH(\tau)}$, and $\hat{\rho}_n^{CG(\tau)}$ in (4), (5), and (7), respectively, for the Fréchet model with $\gamma = 0.5$ and d.f. given in Remark 3.

In Fig. 3, we present the simulated mean values (E) and root mean square errors (RMSE) patterns of the abovementioned ρ-estimators, as functions of k, for $n = 2,000$. Since we have $\rho = -1$, we have used $\tau = 0$ in $\hat{\rho}_n^{FAGH(\tau)}(k)$ (see Remark 1). Using again the stability of the sample paths of the mean values and the RMSE, for large k, we elect $\tau = -1$ in $\hat{\rho}_n^{CG(\tau)}(k)$.

Figure 3 evidences that, with the proper choice of τ, $\hat{\rho}_n^{CG(-1)}(k)$ has the best performance (not only in terms of bias but also in terms of RMSE), $\hat{\rho}_n^{FAGH(0)}(k)$ is the second best and finally $\hat{\rho}_n^{FH}(k)$ has the worst performance (although it is a maximum likelihood estimator). The adaptive choice of τ is outside the scope of this paper.

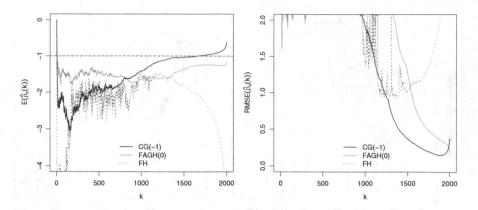

Fig. 3 Simulated mean values (*left*) and root mean squared errors (*right*) for the Fréchet model, with $\gamma = 0.5$ and $n = 2,000$

Acknowledgments Research partially supported by FCT—Fundação para a Ciência e a Tecnologia, projects PEst-OE/MAT/UI0006/2011 (CEAUL), PEst-OE/MAT/UI0297/2011 (CMA/UNL) and PTDC/FEDER, EXTREMA Project.

References

1. Beirlant, J., Caeiro, F., Gomes, M.I.: An overview and open research topics in statistics of univariate extremes. Revstat **10**, 1–31 (2012)
2. Beirlant, J., Goegebeur, Y., Segers, J., Teugels, J.: Statistics of Extremes. Theory and Applications. Wiley, Chichester (2004)
3. Caeiro, F., Gomes, M.I.: A new class of estimators of the "scale" second order parameter. Extremes **9**, 193–211 (2006)
4. Caeiro, F., Gomes, M.I.: Asymptotic comparison at optimal levels of reduced-bias extreme value index estimators. Stat. Neerl. **65**, 462–488 (2011)
5. Caeiro, F., Gomes, M.I., Henriques Rodrigues, L.: Reduced-bias tail index estimators under a third order framework. Commun. Stat. Theory Methods **38**(7), 1019–1040 (2009)
6. Ciuperca, G., Mercadier, C.: Semi-parametric estimation for heavy tailed distributions. Extremes **13**, 55–87 (2010)
7. Feuerverger, A., Hall, P.: Estimating a tail exponent by modelling departure from a Pareto distribution. Ann. Stat. **27**, 760–781 (1999)
8. Fraga Alves, M.I., Gomes, M.I., de Haan, L.: A new class of semi-parametric estimators of the second order parameter. Port. Math. **60**(2), 193–213 (2003)
9. Geluk, J., de Haan, L.: Regular Variation, Extensions and Tauberian Theorems. CWI Tract, vol. 40. Center for Mathematics and Computer Science, Amsterdam (1987)
10. Goegebeur, Y., Beirlant, J., de Wet, T.: Kernel estimators for the second order parameter in extreme value statistics. J. Stat. Plann. Inference **140**, 2632–2652 (2010)
11. Gomes, M.I., de Haan, L., Peng, L.: Semi-parametric estimation of the second order parameter in statistics of extremes. Extremes **5**, 387–414 (2002)
12. Hall, P., Welsh, A.H.: Adaptative estimates of parameters of regular variation. Ann. Stat. **13**, 331–341 (1985)

Peaks Over Random Threshold Asymptotically Best Linear Estimation of the Extreme Value Index

Lígia Henriques-Rodrigues and M. Ivette Gomes

Abstract

A new class of *location invariant* estimators of a positive extreme value index (EVI) is introduced. On the basis of second-order best linear unbiased estimators of the EVI, a class of PORT best linear EVI-estimators is considered, with PORT standing for peaks over random thresholds. A heuristic procedure for the adaptive choice of the tuning parameters under play is proposed and applied to a set of financial data.

1   Introduction and Scope of the Paper

The extreme value index (EVI) is the shape parameter in the extreme value (EV) distribution function (d.f.), with the functional form

$$EV_\gamma(x) = \begin{cases} \exp\left(-(1 + \gamma x)^{-1/\gamma}\right), & 1 + \gamma x > 0 \text{ if } \gamma \neq 0 \\ \exp(-\exp(-x)), & x \in \mathbb{R} \qquad\quad \text{if } \gamma = 0. \end{cases} \tag{1}$$

Let $X_{i:n}$ denote the i-th ascending order statistic (o.s.), $1 \leq i \leq n$, associated with the random sample $(X_1, \ldots, X_n)$ from a model F. The EV d.f. in (1) appears as the possible limiting distribution of the linearized sequence of maximum values $\{X_{n:n}\}_{n \geq 1}$, whenever such a nondegenerate limit exists. We then say that F is in the max-domain of attraction of EV_γ, and use the notation $F \in \mathcal{D}_\mathcal{M}(EV_\gamma)$. We shall work in a context of heavy-tailed models, i.e., we shall consider that $\gamma > 0$ in (1). This type of heavy-tailed models appears often in practice, in fields such as

L. Henriques-Rodrigues (✉)
CEAUL and Instituto Politécnico de Tomar, Tomar, Portugal
e-mail: Ligia.Rodrigues@aim.estt.ipt.pt

M.I. Gomes
CEAUL and FCUL, Campo Grande, 1749-016 Lisboa, Portugal
e-mail: ivette.gomes@fc.ul.pt

A. Pacheco et al. (eds.), *New Advances in Statistical Modeling and Applications*,
Studies in Theoretical and Applied Statistics, DOI 10.1007/978-3-319-05323-3_14,
© Springer International Publishing Switzerland 2014

finance, insurance, and ecology, among others (see [17]). Given the underlying d.f. F and with $F^{\leftarrow}(y) := \inf\{x : F(x) \geq y\}$ being the generalized inverse function of F, let us denote $U(t) = F^{\leftarrow}(1 - 1/t), t > 1$, the reciprocal quantile function. Let us also use the notation RV_τ for the class of regularly varying functions with index of regularly variation $\tau \in \mathbb{R}$, i.e. positive measurable functions g such that $\lim_{t\to\infty} g(tx)/g(t) = x^\tau$, for all $x > 0$ (see [3] for details on regular variation). Then, $F \in \mathscr{D}_M(EV_{\gamma>0}) \iff 1 - F \in RV_{-1/\gamma} \iff U \in RV_\gamma$, the so-called first-order condition.

The second-order parameter rules the rate of convergence in the first-order condition, and it is the nonpositive value ρ (≤ 0) which appears in

$$\lim_{t\to\infty} \left(\ln U(tx) - \ln U(t) - \gamma \ln x\right)/A(t) = \begin{cases} (x^\rho - 1)/\rho & \text{if } \rho < 0 \\ \ln x & \text{if } \rho = 0, \end{cases} \qquad (2)$$

which is often assumed to hold for every $x > 0$, and where $|A|$ must then be in RV_ρ (see [7]). For technical simplicity, we shall assume that $\rho < 0$ and we shall also slightly restrict the whole domain of attraction considering that [13],

$$U(t) = C\, t^\gamma \left(1 + A(t)/\rho + o(t^\rho)\right), \quad A(t) = O(t^\rho), \; \rho < 0, \; C > 0.$$

As usual in a semiparametric estimation of parameters of extreme events we shall consider intermediate k-sequences, i.e. sequences of integer values $k = k(n)$, between 1 and n, such that

$$k = k(n) \to \infty \quad \text{and} \quad k/n \to 0 \quad \text{as } n \to \infty. \qquad (3)$$

In this paper, to make location invariant the asymptotically best linear unbiased (ABLU) estimators considered in [9], we apply to them the peaks over random threshold (PORT) methodology, introduced in [2]. These PORT EVI-estimators, to be presented in Sect. 2, depend upon an extra *tuning parameter* $q, 0 \leq q < 1$, which makes them highly flexible. In Sect. 3, we describe the ABLU estimators, studied in [9], and introduce the PORT asymptotically best linear (ABL) EVI-estimators. In Sect. 4, we provide an algorithm for the adaptive choice of the tuning parameters k and q under play. Section 5 is dedicated to an application of the algorithm to a set of financial data.

2   PORT EVI-Estimation

The classical EVI-estimators are Hill estimators [15], consistent for $\gamma > 0$. They are the averages of the k log-excesses over a high random threshold $X_{n-k:n}$, i.e.

$$H(k) \equiv H(k; \underline{X}_n) := \frac{1}{k} \sum_{i=1}^{k} \left\{ \ln \frac{X_{n-i+1:n}}{X_{n-k:n}} \right\}. \qquad (4)$$

The Hill estimators are scale-invariant, but not location-invariant, as often desired, contrarily to the PORT-Hill estimators, introduced in [2] and further studied in [10]. The class of PORT-Hill estimators is based on a sample of excesses over a random level $X_{[nq]+1:n}$, with $[x]$ denoting, as usual, the integer part of x,

$$\mathbf{X}_n^{(q)} := \left(X_{n:n} - X_{[nq]+1:n}, \ldots, X_{n-k:n} - X_{[nq]+1:n} \right), \tag{5}$$

with $1 \leq k < n_q$, $n_q := n - [nq] - 1$ and where
- $0 < q < 1$, for any $F \in \mathcal{D}_{\mathcal{M}}(EV_{\gamma>0})$ (the random level is an empirical quantile);
- $q = 0$, for d.f.'s with a finite left endpoint $x_F := \inf\{x : F(x) > 0\}$ (the random threshold is the minimum).

The PORT-Hill estimators have the same functional form of the Hill estimators, given in (4), but with the original sample $\mathbf{X}_n = (X_1, \ldots, X_n)$ replaced by the sample of excesses $\mathbf{X}_n^{(q)}$. For $0 \leq q < 1$ and $k < n_q$, they are thus given by

$$H^{(q)}(k) := H(k; \mathbf{X}_n^{(q)}) = \frac{1}{k} \sum_{i=1}^{k} \left\{ \ln \frac{X_{n-i+1:n} - X_{[nq]+1:n}}{X_{n-k:n} - X_{[nq]+1:n}} \right\}, \quad 0 \leq q < 1.$$

These estimators are now invariant for both changes of scale and location in the data, and depend on the tuning parameter q, that provide a highly flexible class of EVI-estimators. In what follows, we use the notation χ_q for the q-quantile of the d.f. F. Then (see [17], among others),

$$X_{[nq]+1:n} \xrightarrow[n\to\infty]{p} \chi_q := F^{\leftarrow}(q), \quad \text{for} \quad 0 \leq q < 1 \quad \left(F^{\leftarrow}(0) = x_F \right). \tag{6}$$

2.1 Second-Order Framework for Heavy-Tailed Models Under a Non-Null Shift

If we induce any arbitrary shift, s, in the model X underlying our data, with quantile function $U_X(t)$, the transformed r.v. $Y = X + s$ has an associated quantile function given by $U_s(t) \equiv U_Y(t) = U_X(t) + s$. When applying the PORT-methodology, we are working with the sample of excesses in (5), or equivalently, we are inducing a random shift, strictly related to χ_q, in (6). We shall thus use the subscript q instead of the subscript s, whenever we think of a shift χ_q. Consequently, the parameter ρ, as well as the A-function, in (2), depends on such a shift χ_q, i.e. $\rho = \rho_q$, $A = A_q$, and

$$\left(A_q(t), \rho_q \right) := \begin{cases} \left(\gamma \chi_q / U_0(t), -\gamma \right), & \text{if } \gamma + \rho_0 < 0 \wedge \chi_q \neq 0 \\ \left(A_0(t) + \gamma \chi_q / U_0(t), \rho_0 \right), & \text{if } \gamma + \rho_0 = 0 \\ \left(A_0(t), \rho_0 \right), & \text{otherwise,} \end{cases} \tag{7}$$

where ρ_0, U_0, and A_0 are, respectively, the second-order parameter, the quantile function, and the A-function associated with an unshifted model. For any intermediate sequence k as in (3), under the validity of the second-order condition in (2), for any real $q, 0 \leq q < 1$, and with Z_k^H an asymptotic standard normal distribution, the PORT-Hill estimator has an asymptotic distributional representation (see [2]),

$$H^{(q)}(k) \stackrel{d}{=} \gamma + \frac{\gamma^2 Z_k^H}{\sqrt{k}} + \left(\frac{1}{1-\rho_0} A_0(n/k) + \frac{\gamma}{1+\gamma} \frac{\chi_q}{U_0(n/k)} \right) (1 + o_p(1)).$$

(8)

3 Asymptotically Best Linear Unbiased Estimation of the EVI

In the general theory of Statistics, whenever we ask the question whether the combination of information can improve the performance of an estimator, we are led to think on BLU estimators, i.e., on unbiased linear combinations of an adequate set of statistics, with minimum variance among the class of such linear combinations.

Given a vector of m statistics directly related to the EVI, γ, say $\mathbf{T} \equiv (T_{ik}, i = k - m + 1, \ldots, k)$, $1 \leq m \leq k$, let us assume that, asymptotically, the covariance matrix of $\mathbf{T}$ is well approximated by $\gamma^2 \Sigma$, i.e., it is known up to the scale factor γ^2, and that its mean value is well approximated by $\gamma \mathbf{s} + \varphi(n, k) \mathbf{b}$, as in the main theorem underlying this theory (see [1] for further details). The linear combination of our set of statistics with minimum variance and unbiased, in an asymptotic sense, is called an ABLU estimator, and denoted by BL_T. The ABLU estimator is then given by $BL_T^{(\rho)}(k; m) := \mathbf{a}' \mathbf{T}$, where $\underline{\mathbf{a}}' = (a_1, a_2, \ldots, a_m)$ is such that $\mathbf{a}' \Sigma \mathbf{a}$ is minimum, subject to the conditions $\mathbf{a}' \mathbf{s} = 1$ and $\mathbf{a}' \mathbf{b} = 0$. The solution of such a problem is given by (see [9])

$$\mathbf{a} = -\frac{1}{\Delta} \mathbf{b}' \Sigma^{-1} \left(\mathbf{s} \, \mathbf{b}' - \mathbf{b} \, \mathbf{s}' \right) \Sigma^{-1},$$

with $\Delta = ||\mathbf{P}' \Sigma^{-1} \mathbf{P}||$. Provided the results were not asymptotic, we could derive that

$$\mathbb{V}ar \left(BL_T^{(\rho)}(k; m) \right) = \gamma^2 \frac{\mathbf{b}' \Sigma^{-1} \mathbf{b}}{\Delta}.$$

The ABLU estimators considered in [9] are ABLU-Hill EVI-estimators, i.e. asymptotically unbiased linear combinations of Hill's estimators computed at different intermediate levels $k - m + 1, k - m + 2, \ldots, k$, i.e., linear combinations based on the vector $\underline{\mathbf{H}} \equiv (H(k - m + 1), \ldots, H(k))$, with the functional form

$$BL(k) \equiv BL_H^{(\rho)}(k) = \sum_{i=1}^{k} a_i^H(\rho) H(i),$$

and where the weights $a_i^H = a_i^H(\rho)$, $i = 1, \ldots, k$, are given in the Proposition 2.4 of [9], whenever we consider $m = k > 2$ levels.

We now advance with PORT-ABL-Hill EVI-estimators, based on the vector of m statistics $\underline{\mathbf{H}}^{(q)} \equiv \left(H^{(q)}(k-m+1), \ldots, H^{(q)}(k)\right)$, with the functional form

$$BL^{(q)}(k) \equiv BL_{H^{(q)}}^{(\rho_q)}(k) = \sum_{i=1}^{k} a_i^{H^{(q)}}(\rho_q) H^{(q)}(i), \tag{9}$$

and where the weights $a_i^{H^{(q)}}(\rho_q) = a_i^H(\rho = \rho_q)$, with ρ_q given in (7).

Remark 1. The ABLU-Hill EVI-estimators introduced in [9] are *reduced bias* (RB) estimators. But for the class of PORT-ABL-Hill EVI-estimators introduced in this work we can no longer guarantee a null asymptotic dominant component of asymptotic bias. All depend on the value χ_q that appears in the distributional representation of $H^{(q)}(k)$, in (8).

---

4   Adaptive PORT-ABL-Hill Estimation

The estimates of the second-order shape parameter ρ_q, $\hat{\rho}_\tau^{(q)}(k)$, $k = 1, \ldots, n_q - 1$, are obtained using the functional form introduced in [14]. The choice of k and q in the PORT-ABL EVI-estimators is performed on the basis of the bootstrap methodology (for more details on the bootstrap methodology see [8], among others) and the adaptive PORT-ABL-Hill estimation was implemented according to the following algorithm:

1. Given the observed sample $(x_1, x_2, \ldots, x_n)$, consider for $q = 0(0.1)0.95$, the observed sample, $\underline{x}_n^{(q)}$, with $\underline{X}_n^{(q)}$ given in (5), and compute $\hat{\rho}_q \equiv \hat{\rho}_0^{(q)} \equiv \hat{\rho}_0^{(q)}(k_1; \underline{x}_n^{(q)})$, with $k_1 = [n_q^{0.999}]$.

2. Next compute, for $k = 1, 2, \ldots, n_q - 1$ the observed values of $BL^{(q)}(k)$, with $BL^{(q)}(k)$ given in (9).

3. Consider sub-samples' sizes $m_1 = o(n_q) = [n_q^{1-\varepsilon}]$ and $m_2 = [m_1^2/n_q] + 1$.

4. For l from 1 until $B = 250$, generate independently B bootstrap samples $(x_1^*, \ldots, x_{m_2}^*)$ and $(x_1^*, \ldots, x_{m_2}^*, x_{m_2+1}^*, \ldots, x_{m_1}^*)$, of sizes m_2 and m_1, respectively, from the empirical d.f., $F_{n_q}^*(x) = \frac{1}{n_q} \sum_{i=1}^{n_q} I_{\{X_i \leq x\}}$, associated with $\underline{x}_n^{(q)}$.

5. Denoting $T^*(k)$ the bootstrap counterpart of the auxiliary statistic $T(k) := BL^{(q)}([k/2]) - BL^{(q)}(k)$, obtain $(t_{m_1,l}^*(k), t_{m_2,l}^*(k))$, $1 \leq l \leq B$, observations of the statistic $T_{m_i}^*(k)$, $i = 1, 2$, and compute

$$MSE^*(m_i, k) = \frac{1}{B} \sum_{l=1}^{B} \left(t_{m_i,l}^*(k)\right)^2, \quad k = 1, 2, \ldots, m_i - 1, \ i = 1, 2.$$

6. Obtain

$$\hat{k}^*_{0|T}(m_i) := \arg \min_{1 \leq k < m_i} MSE^*(m_i, k), \quad i = 1, 2.$$

7. Compute the threshold estimate

$$\hat{k}^{*(q)}_0 \equiv \hat{k}^*_{0|BL^{(q)},T} := \min \left(n_q - 1, \left[(1 - 2^{\hat{\rho}_q})^{1/(1-2\hat{\rho}_q)} \frac{\left(\hat{k}^*_{0|T}(m_1) \right)^2}{\hat{k}^*_{0|T}(m_2)} \right] + 1 \right).$$

If $k^{*(q)}_0 \geq m_2$ return to Step 3 and decrease the value of ε.

8. Obtain $BL^{*(q)}_{m_1|T} := BL^{(q)}(\hat{k}^{*(q)}_0)$.

9. With $B^*(m_i, k) = \frac{1}{B} \sum_{l=1}^{B} t^*_{m_i,l}(k)$, $1 \leq k < m_i$, $i = 1, 2$, compute $\hat{q} :=$ $\arg\min_q \widehat{MSE}(\hat{k}^{*(q)}_0; q)$, where

$$\widehat{MSE}(k; q) := \frac{\left(BL^{*(q)}_{m_1|T} \right)^2}{k} + \left(\frac{\left(B^*(m_1, k) \right)^2}{\left(2^{\hat{\rho}_q} - 1 \right) B^*(m_2, k)} \right)^2,$$

with $\hat{\rho}_q$, $MSE^*(m_i, k)$, $i = 1, 2$ and $BL^{*(q)}_{m_1|T}$ given in Steps 1, 5, and 8, respectively.

10. With the notation $\hat{k}^{*(\hat{q})}_0 \equiv \hat{k}^*_{0|BL^{(\hat{q})},T}$, obtain the final adaptive PORT-ABL EVI-estimate,

$$BL^{**} \equiv BL^{*(\hat{q})}_{m_1|T} := BL^{*(\hat{q})}(\hat{k}^{*(\hat{q})}_0).$$

Remark 2. The value $\tau = 0$, considered in the description of the algorithm, has revealed to be the most adequate choice whenever we are in the region $|\rho| \leq 1$, a common region in the applications and the region where bias reduction is needed. If there are negative elements in the sample, the value n should be replaced by $n^+ = \sum_{i=1}^{n} I_{x_i > 0}$, the number of positive values in the sample. The Monte Carlo procedure in Steps 2 up to 10 of the algorithm can be replicated if we want to build bootstrap confidence intervals for the estimated parameters. The adaptive choice of the tuning parameters k and q can also be done through data-driven heuristic procedures, such as the ones introduced in [12] for the PORT-MVRB EVI-estimators, where MVRB stands for minimum variance reduced bias.

5   An Application to Financial Data

We shall now consider the performance of the algorithm provided in Sect. 4, when applied to the analysis of log-returns, X, and standardized log-returns associated with the daily closing values of the Microsoft Corp. (MSFT), collected from January

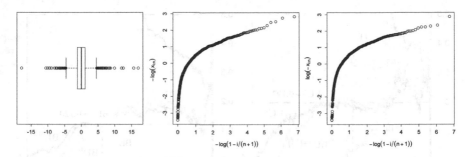

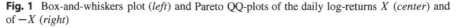

Fig. 1 Box-and-whiskers plot (*left*) and Pareto QQ-plots of the daily log-returns X (*center*) and of $-X$ (*right*)

4, 1999, through November 17, 2005, with a size equal to $n = 1{,}762$, and previously considered in [11, 12] under a weakly dependent and stationary setup.

From Fig. 1, with the box-and-whiskers plot of the data and the Pareto QQ-plots of X and $-X$, we see that the available data provides evidence on the heaviness of the right tail (gains), i.e. $\gamma > 0$, as well as the left tail, i.e. the left endpoint, x_F, is infinite. The value $q = 0$ is thus not admissible for this data set.

The possible presence of clustered volatility is a question of particular relevance in applied financial research, as extensively discussed in [16]. We have performed Engle's ARCH test for the presence of ARCH effects (see [6] and [4] for further details on the test), and the ARCH/GARCH model, a typical model for this type of empirical data, was not rejected for the analyzed log-returns. This test has also shown significant evidence on support of GARCH effects (i.e., heteroskedasticity). We have thus fitted the volatility model GARCH(1,1) in order to remove the observed stock returns heteroskedasticity, and have then applied the algorithm to the standardized log-returns of the MSFT data set, using an approach similar to the one in [16] together with the methodology in this paper for the estimation of the EVI of the standardized log-returns (for more details on the standardized log-returns of the MSFT data set see [12]).

The application of the algorithm for $m_1 = [n_q^{0.975}]$ led for the MSFT log-returns to $\hat{q} = 0.2$, $\hat{k}_0^{(0.2)} = 498$ and the adaptive EVI-estimate $BL^{**} = 0.197$, slightly below the value 0.243 obtained in [12]. For the MSFT standardized log-returns we were led to $\hat{q} = 0.1$, $\hat{k}_0^{(0.1)} = 534$ and the adaptive EVI-estimate $BL^{**} = 0.120$, also below the value 0.192 obtained in [12], as presented in Fig. 2, where we also present the MVRB EVI-estimator in [5].

5.1 Some Final Remarks

The PORT-ABL-Hill EVI-estimators do not always outperform the ABLU-Hill EVI-estimators. This fact had already happened with the PORT-Hill estimators when compared with the Hill estimator, and happens for all models with a

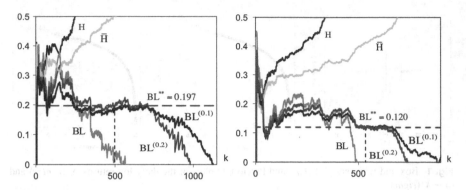

Fig. 2 Adaptive estimates of the PORT-ABL-Hill EVI-estimators for the daily log-returns (*left*) and for the standardized daily log-returns (*right*) of the MSFT data

left endpoint x_F greater than or equal to zero (see [2]). The double bootstrap methodology provided in this paper enables us to choose the value of q that provides what we think to be the best estimator of γ. Just as happens in [12], such an estimator is expected to compare favorably with second-order reduced bias estimators, such as the one introduced in [5], but a comparative study of the estimators is out of the scope of this paper. Note that the choice $q = 0$ is appealing in practice, but should be used with care, as it can induce a problem of sub-estimation and even inconsistency (see, for instance, [10]). As expected, the EVI-estimate of the standardized log-returns is smaller than the EVI-estimate of the original log-returns. The robustness of the algorithm to changes in the sub-sample size m_1 and the finite sample behavior of the PORT-ABL-Hill EVI-estimators are also two other relevant topics out of the scope of this paper.

Acknowledgments Research partially supported by National Funds through **FCT**—Fundação para a Ciência e a Tecnologia, project PEst-OE/MAT/UI0006/2011, PTDC/FEDER, EXTREMA project and grant SFRH/BPD/77319/2011.

References

1. Aitken, A.C.: On least squares and linear combinations of observations. Proc. R. Soc. Edinburgh **55**, 42–48 (1935)
2. Araújo Santos, P., Fraga Alves, M.I., Gomes, M.I.: Peaks over random threshold methodology for tail index and quantile estimation. Revstat **4**(3), 227–247 (2006)
3. Bingham, N.H., Goldie, C.M., Teugels, J.L.: Regular Variation. Cambridge University Press, New York (1989)
4. Box, G.E.P., Jenkins, G.M., Reinsel, G.C.: Time Series Analysis: Forecasting and Control, 3rd edn. Prentice-Hall, Englewood Cliffs (1994)
5. Caeiro, F., Gomes, M.I., Pestana, D.D.: Direct reduction of bias of the classical Hill estimator. Revstat **3**(2), 113–136 (2005)
6. Engle, R.F.: Autoregressive conditional heteroskedasticity with estimates of the variance of United Kingdom inflation. Econometrica **50**(4), 987–1007 (1982)

7. Geluk, J., de Haan, L.: Regular Variation, Extensions and Tauberian Theorems. CWI Tract, vol. 40. Center for Mathematics and Computer Science, Amsterdam (1987)
8. Gomes, M.I., Oliveira, O.: The bootstrap methodology in statistics of extremes—choice of the optimal sample fraction. Extremes **4**(4), 331–358 (2001)
9. Gomes, M.I., Figueiredo, F., Mendonça, S.: Asymptotically best linear unbiased tail estimators under a second order regular variation. J. Stat. Plan. Inference **134**(2), 409–433 (2005)
10. Gomes, M.I., Fraga Alves, M.I., Araújo Santos, P.: PORT Hill and moment estimators for heavy-tailed models. Commun. Stat. Simul. Comput. **37**(6), 1281–1306 (2008)
11. Gomes, M.I., Henriques-Rodrigues, L., Miranda, C.: Reduced-bias location-invariant extreme value index estimation: a simulation study. Commun. Stat. Simul. Comput **40**, 424–447 (2011)
12. Gomes, M.I., Henriques-Rodrigues, L., Fraga Alves, I., Manjunath, B.G.: Adaptive PORT-MVRB estimation: an empirical comparison of two heuristic algorithms. J. Stat. Comput. Simul. (2012). doi:10.1080/00949655.2011.652113
13. Hall, P., Welsh, A.H.: Adaptive estimates of parameters of regular variation. Ann. Stat. **13**, 331–341 (1985)
14. Henriques-Rodrigues, L., Gomes, M.I.: A note on the PORT methodology in the estimation of a shape second-order parameter. In: Oliveira, P.E., et al. (eds.) Recent Developments in Modeling and Applications in Statistics, pp. 127–137. Studies in Theoretical and Applied Statistics. Selected Papers of the Statistical Societies. Springer, Berlin (2013)
15. Hill, B.M.: A simple general approach to inference about the tail of a distribution. Ann. Stat. **3**(5), 1163–1174 (1975)
16. McNeil, A., Frey, R.: Estimation of tail-related risk measures for heteroscedastic financial times series: an extreme value approach. J. Empirical Finance **7**, 271–300 (2000)
17. Reiss, R.-D., Thomas, M.: Statistical Analysis of Extreme Values, with Application to Insurance, Finance, Hydrology and Other Fields, 2nd/3rd edn. Birkhäuser Verlag, Basel (2001/2007)

Extremal Quantiles, Value-at-Risk, Quasi-PORT and DPOT

P. Araújo Santos and M.I. Fraga Alves

Abstract

Under the context of high quantiles, Value-at-Risk (VaR) models based on the PORT Hill estimator, VaR models based on the DPOT method and other unconditional and conditional models are compared through a out-of-sample accuracy study. To obtain a reasonable number of violations for backtesting, the log returns have been used from the Down Jones Industrial Average index, which constitutes a financial time series with a very large data size.

1   Introduction

We are concerned with extraordinary events in financial markets—the so known as "Black Swans" events—such as the Black Thursday (stock market crash on 24 October 1929), the Black Monday (stock market crash on 19 October 1987), the turmoil in the bond market in February 1994 and the recent 2008 financial crisis. These crises are characterized by extreme price changes and a major concern for regulators and owners of financial institutions is the adequacy of capital to ensure that they can still be in business after such extreme price changes. Considering time-series of daily log returns $R_t = \log(P_t/P_{t-1})$, where P_t is the value of the portfolio at time t, the VaR(p) for time $t + 1$, $\mathrm{VaR}_{t+1}(p)$, is defined by

P.A. Santos (✉)
Departamento de Informática e Métodos Quantitativos, Escola Superior de Gestão e Tecnologia, Instituto Politécnico de Santarém, Complexo Andaluz, Apartado 295, 2001-904 Santarém, Portugal
e-mail: paulo.santos@esg.ipsantarem.pt

M.I.F. Alves
Faculdade de Ciências, Departamento de Estatística e Investigação Operacional, Universidade de Lisboa, Campo Grande, 1749-016 Lisboa, Portugal
e-mail: isabel.alves@fc.ul.pt

A. Pacheco et al. (eds.), *New Advances in Statistical Modeling and Applications*, Studies in Theoretical and Applied Statistics, DOI 10.1007/978-3-319-05323-3_15, © Springer International Publishing Switzerland 2014

$$P[R_{t+1} \leq \text{VaR}_{t+1}(p)] = p,$$

where p is the coverage rate or probability level. $\text{VaR}_{t+1}(p)$ is a quantile p of the return R_{t+1} distribution. Value-at-Risk (VaR) emerged as the primary tool for financial risk assessment. Here, we are dealing with rare events and thus with much lower probabilities than the usual $p = 0.01$ used for daily capital requirements calculations under the Basel II Accord. In this work it will be considered the probability of an adverse extreme price movement that is expected to occur approximately once every 4 years ($p = 0.001$) or once every 8 years ($p = 0.0005$); therefore, we fall in the context of extremal quantiles. This context may have interest in the development of stress tests (e.g., [19]), which are directly related to the occurrence of extremes in financial markets. Some authors (e.g., [5]) argued that when small probabilities come into play, an unconditional approach is better suited for VaR estimation, because extreme price changes do not appear to be related to a particular level of volatility, nor exhibit time dependence. In fact, it is demonstrated by de Haan et al. [6], that for certain dependent processes, such as ARCH, volatility clustering vanishes at the level of extremes. Moreover, Resnick and Starica [23] have shown the consistency of the Hill estimator under certain types of dependence, such as GARCH.

In this work, both unconditional and conditional VaR models are compared. In Sect. 2 the VaR methods used in the comparative study are summarized. In Sect. 3, the results of the comparative out-of-sample study are presented.

2   VaR Models

For the out-of-sample study, the following models were considered.

2.1   Quasi-PORT

The Hill estimator for the tail index γ [16] may exhibit a high asymptotic bias. Recent developments in EVT involve the reduction of bias (see Peng [21], Feuerverger and Hall [8], Gomes et al. [10, 12], among others). They achieved γ estimators with asymptotic variance equal or higher than $(\gamma(1-\rho)/\rho)^2 > \gamma^2$, where ρ is a second order parameter. More recently, Caeiro et al. [4], Gomes and Pestana [11], Gomes et al. [13] and Gomes et al. [14] have proposed minimum variance reduced bias (MVRB) estimators for γ. They reduce bias without increasing the asymptotic variance, which is kept at the value γ^2. A simple class of MVRB-estimators is the one introduced in Caeiro et al. [4] with the functional form,

$$\hat{\gamma}^{\overline{H}}_{n,k,\hat{\beta},\hat{\rho}} := \hat{\gamma}^H_n \left(1 - \frac{\hat{\beta}}{1 - \hat{\rho}} \left(\frac{n}{k}\right)^{\hat{\rho}}\right), \tag{1}$$

where $\hat{\gamma}_n^H$ is the Hill estimator and $\hat{\rho}$ and $\hat{\beta}$ are consistent estimators of the second order parameters ρ and β. See Fraga Alves et al. [9] for ρ estimation and Gomes et al. [14] for β estimation.

The MVRB tail index estimators are not location invariant, but they are much less sensitive to changes in location than the classical Hill estimator, thus, they are "approximately" location invariant. Gomes et al. [15] have proposed to use the PORT Hill estimator [3] instead of the Hill estimator [16] in the MVRB estimator (1). This estimator was named "quasi-PORT" tail index estimator. For the case of high quantiles, Gomes et al. [15] proposed to use the "quasi-PORT" tail index estimator instead of the PORT Hill estimator in the PORT-Weissman-Hill high quantile estimator [3]. This estimator was named "quasi-PORT" VaR_p estimator. The PORT estimators involve a tuning parameter q. With $q = 0.25$ and $q = 0.5$, two unconditional VaR models based on the "quasi-PORT" VaR_p estimator were chosen. The estimates of ρ and β were obtained using the algorithm suggested in Gomes and Pestana [11].

2.2 DPOT

Araújo Santos and Fraga Alves [2] proposed a Peaks Over a Threshold (POT) model with durations between excesses as covariates (DPOT) and the out-of-sample performance was compared with other models, for forecasting 1-day-ahead VaR(0.01) denoted by $VaR_{t+1|t}(0.01)$. This is the VaR used by financial institutions to compute daily capital requirements under the Basel II Accord. Here, for very small values of p, the DPOT model was used in the comparative study of Sect. 3 with the most simple specification ($v = 1, c = 1$) and with the specification with better out-of-sample results in the comparative studies in [2] ($v = 3, c = 0.75$). In this model v denotes the number of previous excesses considered in the model and c is a tuning parameter (details in [2]).

2.3 Other Models

We have chosen three more models from Extreme Value Theory (EVT). The unconditional POT model, the conditional EVT model with normal innovations (denoted here by CEVT-n) and the conditional EVT model with skewed t innovations (denoted here by CEVT-sst). For a review of these models, see McNeil and Frey [20] and Kuester et al. [17].

Finally, three parametric conditional models were used in the comparative study. Fully parametric models in the location-scale class assumes for the returns,

$$R_t = \mu_t + \varepsilon_t = \mu_t + Z_t \sigma_t,$$

where Z_t are a sequence of iid rv's with zero mean and unit variance (also known as innovations), μ_t the conditional mean and σ_t the conditional standard deviation. We choose the RiskMetrics model [25] and the AR-APARCH model. In this last model the μ_t is expressed as a first order autoregressive process based on returns and σ_t expressed as in the APARCH(1,1) model proposed by Ding et al. [7]. We denote by APARCH-n the AR-APARCH model with normal innovations and by APARCH-sst the AR-APARCH model with skewed-t innovations.

3   Out-of-Sample Study with the DJIA Index

Under the context of extremal quantiles, we set $p = 0.001$ and $p = 0.0005$. To achieve a reasonable number of violations for backtesting, it is important to have a very large data size and this leads us to use the log returns of the Down Jones Industrial Average index, one of the oldest stock indexes. From October 2, 1928, until March 25, 2011, we compute 20,713 returns and with a moving windows of size $n_w = 1,000$ days, we obtain 19,713 one-day-ahead VaR forecasts for each model. As in previous studies, for the EVT methods, we choose the number of top order statistics $k = 100$; see McNeil and Frey [20] for a simulation study that supports a similar choice. To test the unconditional coverage (UC) hypothesis we apply the Kupiec test [18] and to test the independence (IND) hypothesis we apply the maximum to median ratio test (Araújo Santos and Fraga Alves [1]) denoted by MM independence test. The programs were written in the R language [22] and with the fGarch [26] and POT [24] packages.

Tables 1 and 2 summarize the results, respectively, for $p = 0.001$ and $p = 0.0005$. The APARCH-sst based on Skewed-t errors performs well in terms of UC under $p = 0.001$. Empirical findings show that the Skewed-t is clearly preferable to the normal for the distribution of the errors. The performance of conditional parametric models based on the normal distribution (RiskMetrics and APARCH-n) is extremely poor with the number of violations exceeding more than five times the expected under UC.

Further to the tail, more poor are the results. With the smaller probability level $p = 0.0005$, RiskMetrics produced 101 violations which represent more than ten times the expected value equal to 9.8565, under UC. The APARCH-n model produced 74 violations, more than seven times the expected. These results confirm what is well known in the literature (see, for instance, Danielsson and Vries, 1997). On the other hand, the accuracy of the best performers quasi-PORT($q = 0.5$) and DPOT($v = 1$) is very good, with the number of violations very close to the expected under UC. These two models have also good results in terms of independence. In the group of EVT models they perform clearly better than the classic POT model and than the CEVT hybrid model. Finally, it is interesting to note that one of the best performers, quasi-PORT($q = 0.5$), is based on the iid assumption and this provides evidence that the iid assumption can work well when we are dealing with very small probability levels.

Table 1 Out-of-sample accuracy for VaR(0.001) applied to Down Jones Industrial Average index returns from October 2, 1928 until March 25, 2011, with a rolling window of size 1,000[a]

Model	Number of violations	Violation frequencies	Kupiec p-value	MM ratio p-value
Unconditional EVT models				
POT	36	0.001826	0.0000	0.1254
Quasi-PORT($q = 0.25$)	31	0.001573	0.0190	0.1048
Quasi-PORT($q = 0.5$)	20	0.001015	0.9486	0.1849
Conditional EVT models				
DPOT($v = 1$)	22	0.001116	0.6130	0.1966
DPOT($v = 3$)	32	0.001623	0.0112	0.5849
CEVT-n	31	0.001573	0.0190	0.8919
CEVT-sst	31	0.001573	0.0190	0.9631
Conditional parametric models				
RiskMetrics	128	0.006493	0.0000	0.0015
APARCH-n	101	0.005124	0.0000	0.0141
APARCH-sst	22	0.001116	0.6130	0.0564

[a]For each model, the number of 1-day-ahead VaR(0.001) forecasts is 19,713 and the expected number of violations under the UC hypothesis is 19.713

Table 2 Out-of-sample accuracy for VaR(0.0005) applied to Down Jones Industrial Average index returns from October 2, 1928 until March 25, 2011, with a rolling window of size 1,000[a]

Model	Number of violations	Violation frequencies	Kupiec p-value	MM ratio p-value
Unconditional EVT models				
POT	27	0.001370	0.0000	0.2102
quasi-PORT($q = 0.25$)	14	0.000710	0.2146	0.0981
quasi-PORT($q = 0.5$)	11	0.000558	0.7206	0.1562
Conditional EVT models				
DPOT($v = 1$)	10	0.000507	0.9636	0.2035
DPOT($v = 3$)	24	0.001217	0.0001	0.1048
CEVT-n	24	0.001217	0.0001	0.7834
CEVT-sst	25	0.001268	0.0001	0.9922
Conditional parametric models				
RiskMetrics	101	0.005124	0.0000	0.0288
APARCH- n	74	0.003754	0.0000	0.0105
APARCH- sst	16	0.000812	0.0729	0.0799

[a]For each model, the number of 1-day-ahead VaR(0.0005) forecasts is 19,713 and the expected number of violations under the UC hypothesis is 9.8565.

As future research, we plan to extend the out-of-sample study presented in this section, to other indexes and other types of large financial time series, such as individual stocks and foreign currencies.

Acknowledgments Research partially supported by National Funds through *FCT*—Fundação para a Ciência e a Tecnologia, FCT/PROTEC, project PEst-OE/MAT/UI0006/2011, and FCT/PTDC/MAT/101736/2008, EXTREMA project.

References

1. Araújo Santos, P., Fraga Alves, M.I.: A new class of independence tests for interval forecasts evaluation. Comput. Stat. Data Anal. (2010, in press). doi:10.1016/j.csda2010.10.002
2. Araújo Santos, P., Fraga Alves, M.I.: Forecasting Value-at-Risk with a duration based POT method. Notas e Comunicações CEAUL 17/2011 (2011)
3. Araújo Santos, P., Fraga Alves, M.I., Gomes, M.I.: Peaks over random threshold methodology for tail index and quantile estimation. Revstat Stat. J. **4**(3), 227–247 (2006)
4. Caeiro, F., Gomes, M.I., Pestana, D.: Direct reduction of bias of the classical Hill estimator. Revstat Stat. J. **3**(2), 111–136 (2005)
5. Danielsson, J., de Vries C.: Value at Risk and Extreme Returns, Discussion Paper 273. LSE Financial Markets Group, London School of Economics (1997)
6. de Haan, L., Resnick, S., Rootzén, H., de Vries, C.: Extremal behavior of solutions to a stochastic difference equation with application to ARCH processes. Stoch. Process. Appl. **32**, 213–224 (1989)
7. Ding, Z., Engle, R.F., Granger, C.W.J.: A long memory property of stock market return and a new model. J. Empir. Finance **1**, 83–106 (1993)
8. Feuerverger, A., Hall, P.: Estimating a tail exponent by modelling departure from a Pareto distribution. Ann. Stat. **27**(2), 760–781 (1999)
9. Fraga Alves, M.I., Gomes, M.I., de Haan, L.: A new class of semi-parametric estimators of the second order parameter. Port. Math. **60**(1), 193–213 (2003)
10. Gomes, M.I., Martins, M.J.: "Asymptotically unbiased" estimators of the tail index based on the external estimation of the second order parameter. Extremes **5**(1), 5–31 (2002)
11. Gomes, M.I., Pestana D.: A sturdy reduced bias extreme quantile (VaR) estimator. J. Am. Stat. Assoc. **102**(477), 280–292 (2007)
12. Gomes, M.I., Martins, M.J., Neves M.M.: Alternatives to a semi-parametric estimator of parameters of rare events - the Jackknife methodology. Extremes **3**(3), 207–229 (2000)
13. Gomes, M.I., Martins, M.J., Neves, M.M.: Improving second order reduced-bias tail index estimator. Revstat Stat. J. **5**(2), 177–207 (2007)
14. Gomes, M.I., de Haan, L., Henriques Rodrigues, L.: Tail Index estimation for heavy-tailed models: accommodation of bias in weighted log-excesses. J. R. Stat. Soc. Ser. B Stat. Methodol. B **70**(1), 31–52 (2008)
15. Gomes, M.I., Figueiredo, F., Henriques-Rodrigues, L., Miranda, M.C.: A quasi-PORT methodology for VaR based on second-order reduced-bias estimation. Notas e Comunicações CEAUL 05/2010 (2010)
16. Hill, B.M.: A simple general approach to inference about the tail of a distribution. Ann. Stat. **3**(5), 1163–1174 (1975)
17. Kuester, K., Mittik, S., Paolella, M.S.: Value-at-risk prediction: a comparison of alternative strategies. J. Finance Econ. **4**(1), 53–89 (2006)
18. Kupiec, P.: Techniques for verifying the accuracy of risk measurement models. J. Derivatives **3**, 73–84 (1995)
19. Longin, F.: From VaR to stress testing: the extreme value approach. J. Bank. Finance **24**, 1097–1130 (2001)
20. McNeil, A.J., Frey, R.: Estimation of tail-related risk measures for heteroscedastic financial time series: an extreme value approach. J. Empir. Finance **7**, 271–300 (2000)
21. Peng, L.: Asymptotically unbiased estimator for the extreme-value-index. Stat. Probab. Lett. **38**(2), 107–115 (1998)

22. R Development Core Team: R: A Language and Environment for Statistical Computing. R Foundation for Statistical Computing, Vienna. http://www.R-project.org (2008). ISBN: 3-900051-07-0

23. Resnick, S.I., Starica, C.: Tail Index Estimation for Dependent Data. Mimeo, School of ORIE, Cornell University (1996)

24. Ribatet, M.: POT: Generalized Pareto Distribution and Peaks Over Threshold. R Package Version 1.0-9. http://people.epf1.ch/mathieu.ribatet,http://r-forge.r-project.org/projects/pot/ (2009)

25. Riskmetrics: J.P. Morgan Technical Document, 4th edn. J.P. Morgan, New York (1996)

26. Wuertz, D., Chalabi, Y., Miklovic, M.: fGarch: Rmetrics - Autoregressive Conditional Heteroskedastic Modelling. R Package Version 290.76. http://www.rmetrics.org (2008)

The MOP EVI-Estimator Revisited

M. Fátima Brilhante, M. Ivette Gomes, and Dinis Pestana

Abstract

A simple generalisation of the classical Hill estimator of a positive extreme value index (EVI) has been recently introduced in the literature. Indeed, the Hill estimator can be regarded as the logarithm of the geometric mean, or equivalently the logarithm of the mean of order $p = 0$, of a set of adequate statistics. Instead of such a geometric mean, it is thus sensible to consider the mean of order p (MOP) of those statistics, with $p \geq 0$. In this paper, a small-scale simulation study and a closer look at the asymptotic behaviour at optimal levels of the class of MOP EVI-estimators enable us to better understand their properties and to suggest simple adaptive EVI-estimates.

1 Introduction and Preliminaries

Let $(X_1, \ldots, X_n)$ denote the available random sample of size n, from an underlying distribution function (d.f.) F. Let $X_{1:n} \leq \cdots \leq X_{n:n}$ denote the associated ascending order statistics (o.s.'s) and let us assume that there exist sequences of real constants $\{a_n > 0\}$ and $\{b_n \in \mathbb{R}\}$ such that the maximum, linearly normalised, converges in

M.F. Brilhante (✉)
CEAUL and Universidade dos Açores, DM, Campus de Ponta Delgada, Apartado 1422, 9501-801 Ponta Delgada, Portugal
e-mail: fbrilhante@uac.pt

M.I. Gomes • D. Pestana
CEAUL and FCUL, DEIO, Universidade de Lisboa, Campo Grande, 1749-016 Lisboa, Portugal
e-mail: ivette.gomes@fc.ul.pt; dinis.pestana@fc.ul.pt

A. Pacheco et al. (eds.), *New Advances in Statistical Modeling and Applications*, Studies in Theoretical and Applied Statistics, DOI 10.1007/978-3-319-05323-3__16, © Springer International Publishing Switzerland 2014

distribution to a non-degenerate random variable (r.v.). Then, the limit distribution is necessarily of the type of the general *extreme value* d.f.,

$$EV_\gamma(x) = \begin{cases} \exp(-(1+\gamma x)^{-1/\gamma}), & 1+\gamma x > 0 \text{ if } \gamma \neq 0 \\ \exp(-\exp(-x)), & x \in \mathbb{R} \qquad\quad \text{if } \gamma = 0. \end{cases} \tag{1}$$

The d.f. F is said to belong to the *max-domain of attraction* of EV_γ in (1) and we write $F \in \mathscr{D}_{\mathscr{M}}(EV_\gamma)$. The parameter γ is the *extreme value index* (EVI), the primary parameter of extreme events. This index measures the heaviness of the right *tail function* $\overline{F} := 1 - F$ and the heavier the right tail the larger γ is. In this paper we shall work with heavy-tailed distributions in $\mathscr{D}_{\mathscr{M}}^+ := \mathscr{D}_{\mathscr{M}}(EV_\gamma)_{\gamma>0}$. These heavy-tailed models are quite common in the most diversified areas of application, among which we mention insurance, finance and biostatistics.

For models in $\mathscr{D}_{\mathscr{M}}^+$, the classical EVI-estimators are the Hill estimators [12], which are averages of log-excesses. We have

$$H(k) := \frac{1}{k} \sum_{i=1}^{k} V_{ik}, \quad V_{ik} := \ln X_{n-i+1:n} - \ln X_{n-k:n}, \quad 1 \le i \le k < n, \tag{2}$$

and $H(k)$ is a consistent estimator of γ if $k = k_n$ is an *intermediate* sequence of integers, i.e. if

$$k = k_n \to \infty \quad \text{and} \quad k_n = o(n), \text{ as } n \to \infty. \tag{3}$$

Note now that we can write

$$H(k) = \sum_{i=1}^{k} \ln\left(\frac{X_{n-i+1:n}}{X_{n-k:n}}\right)^{1/k} = \ln\left(\prod_{i=1}^{k} \frac{X_{n-i+1:n}}{X_{n-k:n}}\right)^{1/k}, \quad 1 \le i \le k < n,$$

the logarithm of the *geometric mean* of the statistics

$$U_{ik} := X_{n-i+1:n}/X_{n-k:n}, \quad 1 \le i \le k. \tag{4}$$

More generally, and just as in [2], we shall now consider as basic statistics for the EVI estimation, the *mean of order* p (MOP) of U_{ik}, in (4), i.e. the class of statistics

$$A_p(k) = \begin{cases} \left(\frac{1}{k} \sum_{i=1}^{k} U_{ik}^p\right)^{1/p} & \text{if } p > 0 \\[2mm] \left(\prod_{i=1}^{k} U_{ik}\right)^{1/k} & \text{if } p = 0. \end{cases} \tag{5}$$

We shall provide in Sect. 2 an indication on how to generalise the Hill estimator, defined in (2), on the basis of the statistics in (5). In Sect. 3 we illustrate the finite sample properties of such a class of EVI-estimators, through a small-scale simulation study related to a model that depends directly not only on the EVI, γ, but also on a second-order parameter $\rho \leq 0$, to be defined in Sect. 2. In Sect. 4, the asymptotic behaviour of the class of MOP EVI-estimators at optimal levels, in the sense of minimal mean square error (MSE) leads us to suggest a simple estimation of ρ. Finally, in Sect. 5 we provide simple adaptive choices of the tuning parameters p and k, which are simple and nice alternatives to the computer-intensive bootstrap algorithm in [2].

2   The Class of MOP EVI-Estimators

Note that with $F^{\leftarrow}(x) := \inf\{y : F(y) \geq x\}$ denoting the generalised inverse function of F, and $U(t) := F^{\leftarrow}(1 - 1/t)$, $t \geq 1$, the reciprocal quantile function, we can write the distributional identity $X \overset{d}{=} U(Y)$, with Y a unit Pareto r.v., i.e. a r.v. with d.f. $F_Y(y) = 1 - 1/y$, $y \geq 1$. For the o.s.'s associated with a random Pareto sample $(Y_1, \ldots, Y_n)$, the distributional identity $Y_{n-i+1:n}/Y_{n-k:n} = Y_{k-i+1:k}$, $1 \leq i \leq k < n$, holds. Moreover, $k\, Y_{n-k:n}/n \xrightarrow[n\to\infty]{P} 1$, i.e. $Y_{n-k:n} \overset{P}{\sim} n/k$. Consequently, and provided that (3) holds, we get

$$U_{ik} := \frac{X_{n-i+1:n}}{X_{n-k:n}} = \frac{U(Y_{n-i+1:n})}{U(Y_{n-k:n})} = \frac{U(Y_{n-k:n} Y_{k-i+1:k})}{U(Y_{n-k:n})} = Y_{k-i+1:k}^{\gamma}(1 + o_p(1)),$$

i.e. $U_{ik} \overset{P}{\sim} Y_{k-i+1:k}^{\gamma}$. Hence, we have the approximation $\ln U_{ik} \approx \gamma \ln Y_{k-i+1:k} = \gamma E_{k-i+1:k}$, $1 \leq i \leq k$, with E denoting a standard exponential r.v. The log-excesses $V_{ik} = \ln U_{ik}$, $1 \leq i \leq k$, are thus approximately the k top o.s.'s of a sample of size k from an exponential parent with mean value γ. This justifies the Hill EVI-estimator, in (2). We can further write $U_{ik}^p = Y_{k-i+1:k}^{\gamma p}(1 + o_p(1))$. Since $\mathbb{E}(Y^a) = 1/(1 - a)$ if $a < 1$, the law of large numbers enables us to say that if $p < 1/\gamma$,

$$A_p(k) \xrightarrow[n\to\infty]{p} \left(1/(1 - \gamma p)\right)^{1/p}, \quad \text{i.e.} \quad \left(1 - A_p^{-p}(k)\right)/p \xrightarrow[n\to\infty]{p} \gamma,$$

with $A_p(k)$ given in (5). Hence the reason for the class of MOP EVI-estimators, introduced in [2], dependent on a *tuning* parameter $p \geq 0$, and defined by

$$H_p(k) := \begin{cases} \dfrac{1 - A_p^{-p}(k)}{p} & \text{if } p > 0 \\[2mm] \ln A_0(k) = H(k) \text{ if } p = 0. \end{cases} \tag{6}$$

Let us denote by RV_a the class of regularly varying functions at infinity, with an index of regular variation equal to $a \in \mathbb{R}$, i.e. positive measurable functions g such that for all $x > 0$, $g(tx)/g(t) \to x^a$, as $t \to \infty$ (see [1], for details on regular variation). We can guarantee that

$$F \in \mathscr{D}_{\mathscr{M}}^+ \quad \Longleftrightarrow \quad \overline{F} \in RV_{-1/\gamma} \quad \Longleftrightarrow \quad U \in RV_\gamma. \tag{7}$$

The first characterisation in (7) was proved in [7] and the second one in [3].

The *second-order parameter* ρ (≤ 0) rules the rate of convergence in the first-order condition, in (7), i.e. the rate of convergence of $\{\ln U(tx) - \ln U(t) - \gamma \ln x\}$ to zero, and it is the non-positive parameter appearing in the limiting relation

$$\lim_{t \to \infty} \frac{\ln U(tx) - \ln U(t) - \gamma \ln x}{A(t)} = \begin{cases} \frac{x^\rho - 1}{\rho} & \text{if } \rho < 0 \\ \ln x & \text{if } \rho = 0, \end{cases} \tag{8}$$

which is assumed to hold for every $x > 0$, and where $|A|$ must then be of regular variation with index ρ (see [6]).

Regarding the asymptotic properties of the Hill and more generally the MOP EVI-estimators, we now state the following theorem, proved in [2], a generalisation to $p \geq 0$ of Theorem 1, in [4], related to $p = 0$, i.e. related to the Hill estimator.

Theorem 1 ([2], Theorem 2). *Under the validity of the first-order condition, in (7), and for intermediate sequences $k = k_n$, i.e. if (3) holds, the class of estimators $H_p(k)$, in (6), is consistent for the estimation of γ, provided that $p < 1/\gamma$.*

If we moreover assume the validity of the second-order condition, in (8), the asymptotic distributional representation

$$H_p(k) \overset{d}{=} \gamma + \frac{\gamma(1 - p\gamma)Z_k^{(p)}}{\sqrt{k}\sqrt{1 - 2p\gamma}} + \frac{1 - p\gamma}{1 - p\gamma - \rho} A(n/k) + o_p(A(n/k)), \tag{9}$$

holds for $p < 1/(2\gamma)$, with $Z_k^{(p)}$ asymptotically standard normal.

Remark 1. Note that for all $\gamma > 0$, the asymptotic standard deviation $\sigma_p(\gamma) = \gamma(1 - p\gamma)/\sqrt{1 - 2p\gamma}$, appearing in (9), increases with $p \geq 0$. But for all $\gamma > 0$, $\rho < 0$ and $p \neq (1 - \rho)/\gamma$, the asymptotic bias ruler, $b_p(\gamma|\rho) = (1 - p\gamma)/(1 - p\gamma - \rho)$, also in (9), decreases with p.

The facts in Remark 1 claim for a comparison of the class of estimators in (6), already performed in [2], and revisited in Sects. 3 and 4, respectively, through a small-scale Monte-Carlo simulation and a closer look at the asymptotic behaviour, at optimal levels, of the class of MOP EVI-estimators.

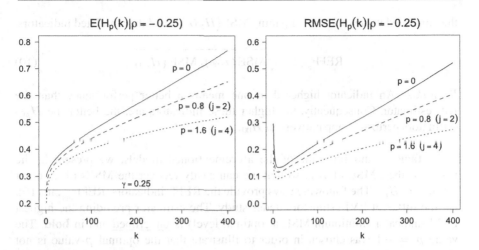

Fig. 1 Mean values (*left*) and root mean square errors (*right*) of $H_p(k)$ for a *Burr*(γ, ρ) d.f. with (γ, ρ) = (0.25, −0.25), and a sample size $n = 1,000$

3   Finite Sample Properties of the MOP Class of EVI-Estimators

We shall now present some results associated with multi-sample Monte-Carlo simulation experiments of size $5,000 \times 20$ for the class of MOP EVI-estimators, in (6), and for sample sizes $n = 100, 200, 500, 1,000, 2,000$ and $5,000$, from underlying *Burr*(γ, ρ) parents, with d.f. $F(x) = 1 - (1 + x^{-\rho/\gamma})^{1/\rho}, x \geq 0, \gamma > 0,$ $\rho < 0$, with γ varying from 0.1 until 1 and ρ varying from −1 until −0.1, with step 0.05. Other parents have been considered in [2]. Details on multi-sample simulation can be found in [9].

For each value of n and for each of the models, we have first simulated the mean value (E) and the root mean square error (RMSE) of the estimators $H_p(k)$, in (6), as functions of the number of top order statistics k involved in the estimation and for $p = j/(10\gamma), j = 0, 1, 2, 3, 4$. As an illustration, some of those p-values, based on the first replicate with a size 5,000, are pictured in Fig. 1, related to an underlying *Burr*(γ, ρ) parent, with (γ, ρ) = (0.25, −0.25) and a sample size $n = 1,000$.

The simulation of the mean values at optimal levels (levels where MSEs are minima as functions of k) of the EVI-estimators $H_p(k)$, in (6), again for $p = j/(10\gamma), j = 0, 1, 2, 3, 4$, i.e. the simulation of the mean values of $H_{p0} := H_p(k_{p0}), k_{p0} := \arg\min_k \text{MSE}(H_p(k))$, on the basis of the 20 replicates with 5,000 runs each, has shown that, as intuitively expected, H_{p0} are decreasing in p until a value smaller than $4/(10\gamma)$, and approaching the true value of γ. Regarding bias, we can safely take $p = 4/(10\gamma)$. But we have to pay attention to variance, which increases with p.

For the EVI-estimators $H_p(k)$, we have considered H_{p0}, the estimator H_p computed at its simulated optimal level, again in the sense of minimum MSE, i.e. at the simulated value of $k_{p0} := \arg\min_k \text{MSE}\big(H_p(k)\big)$, and the simulated indicators

$$\text{REFF}_{p|0} := \sqrt{\text{MSE}\,(H_{00})/\text{MSE}\,(H_{p0})}. \tag{10}$$

Remark 2. An indicator higher than one means a better performance than the Hill estimator. Consequently, the higher these indicators are, the better the H_{p0}-estimators perform comparatively to H_{00}.

In Table 1, and for some of the aforementioned models, we present in the first row, the RMSE of H_{00}, so that we can easily recover the MSEs of all other estimators H_{p0}. The following rows provide the REFF-indicators, $\text{REFF}_{p|0}$, in (10), for the different EVI-estimators under study. The estimator providing the highest REFF-indicator (minimum MSE at optimal level) is <u>underlined</u> and in **bold**. The value $\rho = -1$ was chosen in order to illustrate that the optimal p-value is not always $p = 4/(10\gamma)$.

---

4   A Brief Note on the Asymptotic Comparison of MOP EVI-Estimators at Optimal Levels

With $\sigma_p = \sigma_p(\gamma)$ and $b_p = b_p(\gamma|\rho)$ given in Remark 1, the so-called *asymptotic mean square error* (AMSE) is then given by $\text{AMSE}\,(H_p(k)) := \sigma_p^2/k + b_p^2\,A^2(n/k)$. Regular variation theory enables us to assert that, whenever $b_p \neq 0$, there exists a function $\varphi(n) = \varphi(n, \gamma, \rho)$, such that

$$\lim_{n\to\infty} \varphi(n)\,\text{AMSE}\,(H_{p0}) = \left(\sigma_p^2\right)^{-\frac{2\rho}{1-2\rho}} \left(b_p^2\right)^{\frac{1}{1-2\rho}} =: \text{LMSE}\,(H_{p0}).$$

Moreover, if we slightly restrict the second-order condition in (8), assuming that $A(t) = \gamma\beta t^\rho$, $\rho < 0$, we can write $k_{p0} \equiv k_{p0}(n) = \arg\min_k \text{MSE}\,(H_p(k)) = \left(\sigma_p^2\,n^{-2\rho}/(b_p^2\gamma^2\beta^2(-2\rho))\right)^{1/(1-2\rho)}(1+o(1))$. We again consider the usual asymptotic relative efficiency,

$$\text{AREFF}_{p|0} \equiv \text{AREFF}_{H_{p0}|H_{00}} := \sqrt{\text{LMSE}(H_{00})/\text{LMSE}(H_{p0})}.$$

Consequently, as derived in [2],

$$\text{AREFF}_{p|0} = \left(\left(\frac{\sqrt{1-2p\gamma}}{1-p\gamma}\right)^{-2\rho}\left|\frac{1-p\gamma-\rho}{(1-\rho)(1-p\gamma)}\right|\right)^{\frac{1}{1-2\rho}}. \tag{11}$$

For every (γ, ρ) there is always a positive p-value p_0, such that $\text{AREFF}_{p|0} \geq 1$, for any $p \in (0, p_0)$. Let p_M denote the value of p in $(0, p_0)$ where $\text{AREFF}_{p|0}$ is

Table 1 Simulated RMSEs of H_{00}, denoted by RMSE$_{00}$ (*first row*) and REFF-indicators of $H_p(k)$, $p = j/(10\gamma)$, $j = 1, 2, 3, 4$, for Burr(γ, ρ) underlying parents, together with 95 % confidence intervals

n	100	200	500	1,000	2,000	5,000
$(\gamma, \rho) = (0.25, -0.1)$						
RMSE$_{00}$	0.648 ± 0.0055	0.533 ± 0.0043	0.431 ± 0.0044	0.376 ± 0.0040	0.332 ± 0.0034	0.283 ± 0.0024
$p = 0.4$	1.188 ± 0.0013	1.163 ± 0.0009	1.142 ± 0.0011	1.128 ± 0.0015	1.113 ± 0.0030	1.092 ± 0.0022
$p = 0.8$	1.475 ± 0.0026	1.408 ± 0.0016	1.353 ± 0.0022	1.322 ± 0.0025	1.291 ± 0.0042	1.249 ± 0.0037
$p = 1.2$	1.905 ± 0.0043	1.779 ± 0.0025	1.672 ± 0.0035	1.614 ± 0.0039	1.560 ± 0.0058	1.494 ± 0.0044
$p = 1.6$	$\underline{\mathbf{2.535}} \pm 0.0066$	$\underline{\mathbf{2.323}} \pm 0.0039$	$\underline{\mathbf{2.144}} \pm 0.0053$	$\underline{\mathbf{2.046}} \pm 0.0058$	$\underline{\mathbf{1.960}} \pm 0.0080$	$\underline{\mathbf{1.855}} \pm 0.0057$
$(\gamma, \rho) = (0.25, -0.25)$						
RMSE$_{00}$	0.237 ± 0.0023	0.196 ± 0.0016	0.155 ± 0.0017	0.131 ± 0.0012	0.112 ± 0.0011	0.092 ± 0.0008
$p = 0.4$	1.119 ± 0.0011	1.097 ± 0.0018	1.074 ± 0.0014	1.062 ± 0.0010	1.053 ± 0.0010	1.043 ± 0.0008
$p = 0.8$	1.282 ± 0.0021	1.233 ± 0.0039	1.175 ± 0.0029	1.144 ± 0.0022	1.121 ± 0.0022	1.096 ± 0.0019
$p = 1.2$	1.515 ± 0.0032	1.432 ± 0.0056	1.326 ± 0.0048	1.265 ± 0.0034	1.219 ± 0.0035	1.170 ± 0.0033
$p = 1.6$	$\underline{\mathbf{1.850}} \pm 0.0046$	$\underline{\mathbf{1.718}} \pm 0.0065$	$\underline{\mathbf{1.554}} \pm 0.0062$	$\underline{\mathbf{1.455}} \pm 0.0046$	$\underline{\mathbf{1.378}} \pm 0.0046$	$\underline{\mathbf{1.293}} \pm 0.0045$
$(\gamma, \rho) = (0.25, -1)$						
RMSE$_{00}$	0.066 ± 0.0066	0.051 ± 0.0052	0.037 ± 0.0034	0.029 ± 0.0026	0.023 ± 0.0020	0.016 ± 0.0015
$p = 0.4$	1.047 ± 0.0013	1.038 ± 0.0011	1.029 ± 0.0010	1.026 ± 0.0012	1.023 ± 0.0009	1.020 ± 0.0009
$p = 0.8$	1.097 ± 0.0029	1.075 ± 0.0026	1.054 ± 0.0024	1.046 ± 0.0026	1.038 ± 0.0022	$\underline{\mathbf{1.031}} \pm 0.0022$
$p = 1.2$	1.161 ± 0.0040	1.117 ± 0.0043	1.075 ± 0.0044	1.056 ± 0.0041	$\underline{\mathbf{1.039}} \pm 0.0045$	1.022 ± 0.0043
$p = 1.6$	$\underline{\mathbf{1.257}} \pm 0.0048$	$\underline{\mathbf{1.178}} \pm 0.0056$	$\underline{\mathbf{1.100}} \pm 0.0063$	$\underline{\mathbf{1.060}} \pm 0.0058$	1.025 ± 0.0071	0.985 ± 0.0074

Table 2 Values of p_0 (*first entry*), $p_M = \arg\sup_p \text{AREFF}_{p|0}$ (*second entry*) and $\text{AREFF} \equiv \text{AREFF}_{p_M|0}$ (*third entry*), as a function of γ (*first column*) and ρ (*first row*)

γ	ρ	−0.1	−0.2	−0.3	−0.4	−0.5	−0.6	−0.7	−0.8	−0.9	−1.0
0.2	p_0	2.001	1.923	1.851	1.784	1.722	1.663	1.609	1.558	1.510	1.465
	p_M	1.369	1.287	1.215	1.152	1.096	1.046	1.000	0.958	0.921	0.886
	AREFF	1.015	1.022	1.025	1.025	1.025	1.023	1.022	1.021	1.020	1.018
0.4	p_0	1.000	0.962	0.926	0.892	0.861	0.832	0.804	0.779	0.755	0.732
	p_M	0.684	0.643	0.608	0.576	0.548	0.523	0.500	0.479	0.460	0.443
	AREFF	1.015	1.022	1.025	1.025	1.025	1.023	1.022	1.021	1.020	1.018
0.6	p_0	0.667	0.641	0.617	0.595	0.574	0.554	0.536	0.519	0.503	0.488
	p_M	0.456	0.429	0.405	0.384	0.365	0.349	0.333	0.319	0.307	0.295
	AREFF	1.015	1.022	1.025	1.025	1.025	1.023	1.022	1.021	1.020	1.018
0.8	p_0	0.500	0.481	0.463	0.446	0.430	0.416	0.402	0.389	0.377	0.366
	p_M	0.342	0.322	0.304	0.288	0.274	0.261	0.250	0.240	0.230	0.221
	AREFF	1.015	1.022	1.025	1.025	1.025	1.023	1.022	1.021	1.020	1.018
1.0	p_0	0.400	0.385	0.370	0.357	0.344	0.333	0.322	0.312	0.302	0.293
	p_M	0.274	0.257	0.243	0.230	0.219	0.209	0.200	0.192	0.184	0.177
	AREFF	1.015	1.022	1.025	1.025	1.025	1.023	1.022	1.021	1.020	1.018

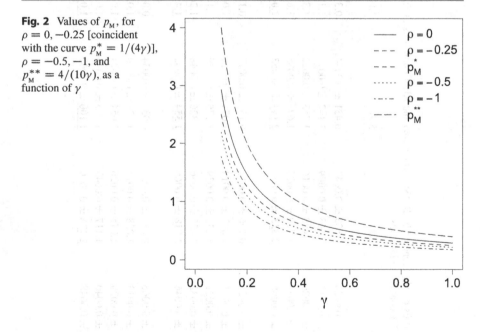

Fig. 2 Values of p_M, for $\rho = 0, -0.25$ [coincident with the curve $p_M^* = 1/(4\gamma)$], $\rho = -0.5, -1$, and $p_M^{**} = 4/(10\gamma)$, as a function of γ

maximised, i.e., $p_M := \arg\sup_p \text{AREFF}_{p|0}$. The values of p_0, p_M and the AREFF-indicator, $\text{AREFF}_{p_M|0}$, have been obtained analytically and are presented in Table 2 for a few values of (γ, ρ).

Note that $\text{AREFF}_{p|0}$, in (11), depends on (p, γ) through $p\gamma$. There thus exists a function $\varphi(\rho)$ such that $p_M = \varphi(\rho)/\gamma$. If we consider the equation $d \ln \text{AREFF}_{p|0}/dp = 0$, with $\text{AREFF}_{p|0}$ given in (11), we are led to the equation

$2\rho/(1-2a) - 1/(1-a-\rho) + (1-2\rho)/(1-a) = 0$, with $a = p\gamma$, and finally to the second-order equation $2a^2 - 2a(2-\rho) + 1 = 0$. We thus get

$$\varphi(\rho) = 1 - \rho/2 - \sqrt{\rho^2 - 4\rho + 2}\Big/2. \tag{12}$$

Note further that for $\rho = -0.25$, a quite common value in most practical situations, $\varphi(\rho) = \varphi(-1/4) = 1/4$. Consequently, the choice $p_M^* = 1/(4\gamma)$ can also work as a simple nice approximation to p_M, as can be seen in Fig. 2.

5   Simple Adaptive Selections of the Tuning Parameters

The maximisation of the AREFF indicator, in (11), led us to the value $p_M = \varphi(\rho)/\gamma$, with $\varphi(\rho)$ given in (12). As we have nowadays reliable techniques for the estimation of second-order parameters, we can easily provide optimal simple choices of p and k, to be used in the building of an adaptive MOP estimate of γ. Alternatively, we could also rely on the heuristic choice $p_M^* = 1/(4\gamma)$. The first two steps in the algorithm were adapted from [10] and are suitable for models with $\rho \geq -1$, the most relevant in practice, and where the Hill EVI-estimator does not perform usually quite well. The estimators of β and ρ used are the ones in [8] and [5], respectively.

1. Given an observed sample $(x_1, \ldots, x_n)$, compute the observed values of the most simple class of estimators in [5], given by

$$\hat{\rho}(k) := \min\left(0, \ \frac{3(W_{k,n} - 1)}{W_{k,n} - 3}\right), \quad W_{k,n} := \frac{\ln\left(M_{k,n}^{(1)}\right) - \frac{1}{2}\ln\left(M_{k,n}^{(2)}/2\right)}{\frac{1}{2}\ln\left(M_{k,n}^{(2)}/2\right) - \frac{1}{3}\ln\left(M_{k,n}^{(3)}/6\right)}$$

where $M_{k,n}^{(j)} := \sum_{i=1}^{k} V_{ik}^j/k$, $j = 1, 2, 3$, with V_{ik} given in (2).

2. Using the notation $\lfloor x \rfloor$ for the integer part of x, work with $\hat{\rho} \equiv \hat{\rho}(k_1)$ and $\hat{\beta} \equiv \hat{\beta}_{\hat{\rho}}(k_1)$, with $k_1 = \lfloor n^{0.999} \rfloor$, being $\hat{\beta}_{\hat{\rho}}(k)$ the estimator in [8], given by

$$\hat{\beta}_{\hat{\rho}}(k) := \left(\frac{k}{n}\right)^{\hat{\rho}} \frac{d_k(\hat{\rho}) D_k(0) - D_k(\hat{\rho})}{d_k(\hat{\rho}) D_k(\hat{\rho}) - D_k(2\hat{\rho})}, \tag{13}$$

dependent on the estimator $\hat{\rho} = \hat{\rho}(k_1)$, and where, for any $\alpha \leq 0$,

$$d_k(\alpha) := \frac{1}{k}\sum_{i=1}^{k}\left(\frac{i}{k}\right)^{-\alpha}, \quad D_k(\alpha) := \frac{1}{k}\sum_{i=1}^{k}\left(\frac{i}{k}\right)^{-\alpha} U_i, \quad U_i = i\left(\ln\frac{X_{n-i+1:n}}{X_{n-i:n}}\right).$$

3. Estimate the optimal level for the estimation through the Hill estimator (see [11]), i.e. estimate $k_{00} := \arg\min_k \mathrm{MSE}(H_0(k))$ through

Table 3 Simulated Bias $\hat{b}_{00}$, $\hat{b}_{\hat{p}_M 0}$ and $\hat{b}_{\hat{p}_M^* 0}$ of $\hat{H}_{00}$, $\hat{H}_{\hat{p}_M 0}$ and $\hat{H}_{\hat{p}_M^* 0}$, for $Burr(\gamma, \rho)$ underlying parents, together with 95 % confidence intervals

n	100	200	500	1,000	2,000	5,000
$(\gamma, \rho) = (0.25, -0.1)$						
$\hat{b}_{00}$	0.954 ± 0.0013	0.829 ± 0.0011	0.705 ± 0.0010	0.633 ± 0.0006	0.571 ± 0.0004	0.505 ± 0.0003
$\hat{b}_{\hat{p}_M 0}$	0.948 ± 0.0012	0.823 ± 0.0011	0.699 ± 0.0008	0.627 ± 0.0005	0.565 ± 0.0004	0.500 ± 0.0003
$\hat{b}_{\hat{p}_M^* 0}$	**0.932** ± 0.0012	**0.809** ± 0.0010	**0.688** ± 0.0008	**0.617** ± 0.0005	**0.557** ± 0.0004	**0.493** ± 0.0003
$(\gamma, \rho) = (0.25, -0.25)$						
$\hat{b}_{00}$	0.320 ± 0.0007	0.271 ± 0.0005	0.223 ± 0.0005	0.195 ± 0.0003	0.172 ± 0.0002	0.147 ± 0.0002
$\hat{b}_{\hat{p}_M 0}$	0.317 ± 0.0006	0.268 ± 0.0005	0.220 ± 0.0004	0.192 ± 0.0003	0.169 ± 0.0002	0.144 ± 0.0002
$\hat{b}_{\hat{p}_M^* 0}$	**0.310** ± 0.0006	**0.262** ± 0.0005	**0.215** ± 0.0004	**0.188** ± 0.0003	**0.166** ± 0.0002	**0.142** ± 0.0002
$(\gamma, \rho) = (0.25, -1)$						
$\hat{b}_{00}$	0.033 ± 0.0004	0.025 ± 0.0003	0.017 ± 0.0002	0.013 ± 0.0002	0.010 ± 0.0001	0.007 ± 0.0001
$\hat{b}_{\hat{p}_M 0}$	0.031 ± 0.0004	0.023 ± 0.0003	0.016 ± 0.0002	0.012 ± 0.0001	0.009 ± 0.0001	0.007 ± 0.0001
$\hat{b}_{\hat{p}_M^* 0}$	**0.029** ± 0.0004	**0.021** ± 0.0003	**0.015** ± 0.0002	**0.011** ± 0.0001	**0.009** ± 0.0001	**0.006** ± 0.0001

Table 4 Simulated RMSEs of $\hat{H}_{00}$, denoted by $\hat{r}_{00}$ (*first row*), and REFF-indicators of $\hat{H}_{\hat{p}_M 0}$ and $\hat{H}_{\hat{p}_M^* 0}$, for $Burr(\gamma, \rho)$ underlying parents, together with 95% confidence intervals

n	100	200	500	1,000	2,000	5,000
$(\gamma, \rho) = (0.25, -0.1)$						
$\hat{r}_{00}$	0.984 ± 0.0015	0.847 ± 0.0011	0.715 ± 0.0010	0.639 ± 0.0006	0.575 ± 0.0004	0.508 ± 0.0003
$\hat{p}_M$	1.011 ± 0.0006	1.011 ± 0.0005	1.011 ± 0.0005	1.011 ± 0.0003	1.012 ± 0.0004	1.012 ± 0.0002
$\hat{p}_M^*$	$\mathbf{1.029} \pm 0.0006$	$\mathbf{1.028} \pm 0.0005$	$\mathbf{1.027} \pm 0.0005$	$\mathbf{1.027} \pm 0.0003$	$\mathbf{1.026} \pm 0.0004$	$\mathbf{1.026} \pm 0.0002$
$(\gamma, \rho) = (0.25, -0.25)$						
$\hat{r}_{00}$	0.340 ± 0.0008	0.284 ± 0.0005	0.231 ± 0.0005	0.200 ± 0.0003	0.175 ± 0.0002	0.149 ± 0.0002
$\hat{p}_M$	1.019 ± 0.0009	1.019 ± 0.0008	1.019 ± 0.0007	1.018 ± 0.0005	1.018 ± 0.0005	1.018 ± 0.0003
$\hat{p}_M^*$	$\mathbf{1.042} \pm 0.0009$	$\mathbf{1.041} \pm 0.0008$	$\mathbf{1.040} \pm 0.0007$	$\mathbf{1.039} \pm 0.0005$	$\mathbf{1.038} \pm 0.0006$	$\mathbf{1.037} \pm 0.0003$
$(\gamma, \rho) = (0.25, -1)$						
$\hat{r}_{00}$	0.067 ± 0.0003	0.052 ± 0.0002	0.037 ± 0.0002	0.029 ± 0.0001	0.023 ± 0.0001	0.017 ± 0.0001
$\hat{p}_M$	1.056 ± 0.0023	1.051 ± 0.0022	1.043 ± 0.0014	$\mathbf{1.041} \pm 0.0020$	$\mathbf{1.034} \pm 0.0019$	$\mathbf{1.031} \pm 0.0023$
$\hat{p}_M^*$	$\mathbf{1.074} \pm 0.0027$	$\mathbf{1.063} \pm 0.0028$	$\mathbf{1.047} \pm 0.0018$	1.039 ± 0.0023	1.029 ± 0.0023	1.021 ± 0.0030

$$\hat{k}_{00} = \left\lfloor \left((1 - \hat{\rho})^2 n^{-2\hat{\rho}} / \left(\hat{\beta}^2(-2\hat{\rho}) \right) \right)^{1/(1-2\hat{\rho})} \right\rfloor.$$

4. Consider the EVI-estimate, $\hat{H}_{00} := H_0(\hat{k}_{00})$.
5. Consider $\hat{p}_{\mathrm{M}} = \varphi(\hat{\rho})/\hat{H}_{00}$, with $\varphi(\rho)$ given in (12), and $p_{\mathrm{M}}^* = 1/(4\hat{H}_{00})$.
6. Compute $\hat{H}_{\hat{p}_{\mathrm{M}}0} \equiv H_{\hat{p}_{\mathrm{M}}}(\hat{k}_{M0}(\hat{p}_{\mathrm{M}}))$ and $\hat{H}_{\hat{p}_{\mathrm{M}}^*0} \equiv H_{\hat{p}_{\mathrm{M}}^*}(\hat{k}_{M0}(\hat{p}_{\mathrm{M}}^*))$, with

$$\hat{k}_{M0}(p) := \left\lfloor \left((1 - p\hat{H}_{00} - \hat{\rho})^2 n^{-2\hat{\rho}} / \left(\hat{\beta}^2(-2\hat{\rho})(1 - 2p\hat{H}_{00}) \right) \right)^{1/(1-2\hat{\rho})} \right\rfloor.$$

The application of the algorithm to several randomly simulated samples led us to the conclusion that, for values of ρ close to zero, $\hat{k}_{00}$ and $\hat{k}_{M0}$ provide a clear over-estimation of the corresponding optimal levels. This leads to a high bias of the EVI-estimators under consideration for values of ρ close to zero. But, in general, the bias of $\hat{H}_{\hat{p}_{\mathrm{M}}0}$ is smaller than the bias of $\hat{H}_{00}$, as can be seen in Table 3, where we present the simulated bias of $\hat{H}_{00}$, $\hat{H}_{\hat{p}_{\mathrm{M}}0}$ and $\hat{H}_{\hat{p}_{\mathrm{M}}^*0}$, respectively, denoted by $\hat{b}_{00}$, $\hat{b}_{\hat{p}_{M}0}$ and $\hat{b}_{\hat{p}_{M0}^*}$, as well as associated 95 % confidence intervals. The estimator providing the smallest bias is underlined and in **bold**.

We further present Table 4, similar to Table 1, where we show in the first row, the RMSE of $\hat{H}_{00}$, so that we can easily recover the MSEs of the other two final EVI-estimators, $\hat{H}_{\hat{p}_{M}0}$ and $\hat{H}_{\hat{p}_{M}^*0}$. The following rows provide the REFF-indicators for the aforementioned EVI-estimators. The estimator providing the highest REFF indicator (minimum MSE at optimal level) is again underlined and in **bold**.

We also provide the following comments:
- The comparison of Tables 1 and 4 leads us to point out the loss of efficiency in the final estimates, when ρ is close to zero.
- However, regarding MSE the adaptive MOP EVI-estimates overpass the adaptive Hill estimate for all values of (γ, ρ).
- The heuristic estimate $\hat{H}_{p_{M}^*0}$ outperforms, in general, $\hat{H}_{p_{M}0}$.
- The obtained results claim for a further simulation study, out of the scope of this paper.
- Also corrected-bias MOP EVI-estimators are welcome.

Acknowledgements Research partially supported by National Funds through **FCT**—Fundação para a Ciência e a Tecnologia, project PEst-OE/MAT/UI0006/2011, and EXTREMA, PTDC/FEDER.

References

1. Bingham, N., Goldie, C.M., Teugels, J.L.: Regular Variation. Cambridge University Press, Cambridge (1987)
2. Brilhante, M.F., Gomes, M.I., Pestana, D.: A simple generalisation of the Hill estimator. Comput. Stat. Data Anal. (2012). doi:10.1016/j.csda.2012.07.019
3. de Haan, L.: Slow variation and characterization of domains of attraction. In: Tiago de Oliveira, J. (ed.) Statistical Extremes and Applications, pp. 31–48. D. Reidel, Dordrecht (1984)

4. de Haan, L., Peng, L.: Comparison of extreme value index estimators. Stat. Neerl. **52**, 60–70 (1998)
5. Fraga Alves, M.I., Gomes, M.I., de Haan, L.: A new class of semi-parametric estimators of the second order parameter. Port. Math. **60**(2), 194–213 (2003)
6. Geluk, J., de Haan, L.: Regular Variation, Extensions and Tauberian Theorems. CWI Tract, vol. 40. Center for Mathematics and Computer Science, Amsterdam (1987)
7. Gnedenko, B.V.: Sur la distribution limite du terme maximum d'une série aléatoire. Ann. Math. **44**, 423–453 (1943)
8. Gomes, M.I., Martins, M.J.: "Asymptotically unbiased" estimators of the tail index based on external estimation of the second order parameter. Extremes **5**(1), 5–31 (2002)
9. Gomes, M.I., Oliveira, O.: The bootstrap methodology in statistical extremes—choice of the optimal sample fraction. Extremes **4**(4), 331–358 (2001)
10. Gomes, M.I., Pestana, D.: A sturdy reduced-bias extreme quantile (VaR) estimator. J. Am. Stat. Assoc. **102**(477), 280–292 (2007)
11. Hall, P.: On some simple estimates of an exponent of regular variation. J. R. Stat. Soc. B **44**, 37–42 (1982)
12. Hill, B.M.: A simple general approach to inference about the tail of a distribution. Ann. Stat. **3**, 1163–1174 (1975)

Tail Dependence of a Pareto Process

Marta Ferreira

Abstract

Heavy-tailed autoregressive processes defined with minimum or maximum operator are good alternatives to classic linear ARMA with heavy tail noises, in what concerns extreme values modeling. In this paper we present a full characterization of the tail dependence of the autoregressive minima process, Yeh–Arnold–Robertson Pareto(III).

1   Introduction

Extreme value theory (EVT) provides tools that enable to estimate the probability of events that are more extreme than any that have already been observed. The classical result in EVT states that if the maximum of an independent and identically distributed (i.i.d.) sequence of random variables (r.v.'s) converges to some nondegenerate function G_γ, then it must be the generalized extreme value (*GEV*) function,

$$G_\gamma(x) = \exp(-(1 + \gamma x)^{-1/\gamma}), \ 1 + \gamma x > 0, \ \gamma \in \mathbf{R},$$

with the usual continuity correction $G_0(x) = exp(-e^{-x})$. The shape parameter γ, known as the tail index, determines the tail behavior: if $\gamma > 0$ we have a heavy tail (Fréchet max-domain of attraction), $\gamma = 0$ means an exponential tail (Gumbel max-domain of attraction) and $\gamma < 0$ indicates a short tail (Weibull max-domain of attraction).

M. Ferreira (✉)
University of Minho, DMA/CMAT, Braga, Portugal
e-mail: msferreira@math.uminho.pt

A. Pacheco et al. (eds.), *New Advances in Statistical Modeling and Applications*,
Studies in Theoretical and Applied Statistics, DOI 10.1007/978-3-319-05323-3__17,
© Springer International Publishing Switzerland 2014

The first results in EVT were developed under independence but, more recently, models for extreme values have been constructed under more realistic assumption of temporal dependence.

MARMA processes (*maximum autoregressive moving average*) with Fréchet marginals, in particular ARMAX [or MARMA(1,0)], given by,

$$X_i = \max(c\, X_{i-1}, W_i),$$

with $0 < c < 1$ and $\{W_i\}_{i \geq 1}$ i.i.d., have been successfully applied to time series modeling in alternative to classical linear heavy-tailed ARMA (see [2] and references therein). Generalizations of MARMA processes and respective applications to financial time series can be seen in, e.g., [13] and [4]. Here we shall focus on autoregressive Pareto processes, i.e., an autoregressive process whose marginal distributions are of the Pareto or generalized Pareto form. As Pareto observed [10], many economic variables have heavy-tailed distributions not well modeled by the normal curve. Instead, he proposed a model, subsequently called, in his honor, the Pareto distribution, whose tail function decreases at a negative power of x as $x \to \infty$, i.e., $1 - F(x) \sim cx^{-\alpha}$, as $x \to \infty$. Generalizations of Pareto's distribution have been proposed for modeling economic variables (a survey can be seen in [1]).

We consider autoregressive Pareto(III) processes, more precisely, the Yeh–Arnold–Robertson Pareto(III) [12], in short YARP(III)(1), given by

$$X_n = \min\left(p^{-1/\alpha} X_{n-1}, \frac{1}{1-U_n}\varepsilon_n\right),$$

where innovations $\{\varepsilon_n\}_{n \geq 1}$ are i.i.d. r.v.'s with distribution Pareto(III)$(0, \sigma, \alpha)$, i.e., a generalized type III Pareto, such that

$$1 - F_\varepsilon(x) = \left[1 + \left(\frac{x}{\sigma}\right)^\alpha\right]^{-1}, \quad x > 0.$$

with $\sigma, \alpha > 0$. The sequence $\{U_n\}_{n \geq 1}$ has i.i.d. r.v.'s with a Bernoulli(p) distribution (independent of the innovations). We interpret $1/0$ as $+\infty$. By conditioning on U_n, it is readily verified that the YARP(III)(1) process has a Pareto(III)$(0, \sigma, \alpha)$ stationary distribution and will be completely stationary if the distribution of the starting r.v. X_0 is also Pareto(III)$(0, \sigma, \alpha)$.

In this paper we analyze the dependence behavior of the YARP(III)(1) process in the right tail (the most used for applications). This process is almost unknown in literature but has large potential as it presents a quite similar tail behavior to ARMAX and more robust parameters estimation [3]. We characterize the lag-m tail dependence ($m = 1, 2, \ldots$) by computing several coefficients considered in [5, 6], defined under a temporal approach. The lag-m tail dependence allows a characterization of the process in time, analogous to the role of the ACF of a linear time series. In addition, these measures are also important in applications, such as risk assessment in financial time series or in engineering, to investigate how the best performer in a system is attracted by the worst one.

2   Measures of Tail Dependence

The *tail-dependence coefficient* (TDC), usually denoted λ, was the first tail
dependence concept appearing in literature in a paper by Sibuya, who has shown
that, no matter how high we choose the correlation of normal random pairs, if we
go far enough into the tail, extreme events tend to occur independently in each
margin [11]. It measures the probability of occurring extreme values for one r.v.
given that another assumes an extreme value too. More precisely,

$$\lambda = \lim_{t \downarrow 0} P(F_1(X_1) > 1 - t | F_2(X_2) > 1 - t),$$

where F_1 and F_2 are the distribution functions (d.f.'s) of r.v.'s X_1 and X_2,
respectively. It characterizes the dependence in the tail of a random pair (X_1, X_2), in
the sense that, $\lambda > 0$ corresponds to tail dependence whose degree is measured by
the value of λ, whereas $\lambda = 0$ means tail independence. Modern risk management
is highly interested in assessing the amount of tail dependence. As an example, the
Value-at-Risk at probability level $1 - t$ (VaR_{1-t}) of a random asset Z is given by
the quantile function evaluated at $1 - t$, $F_Z^{-1}(1 - t) = \inf\{x : F_Z(x) \geq 1 - t\}$, and
estimation is highly sensitive towards the tail behavior and the tail dependence of
the portfolio's asset-return distribution. Observe that the TDC can be formulated as

$$\lambda = \lim_{t \downarrow 0} P(X_1 > VaR_{1-t}(X_1) | X_2 > VaR_{1-t}(X_2)).$$

Generalizations of the TDC have been considered with several practical applications. In [6], for integers s and k such that $1 \leq s < d - k + 1 \leq d$, it was considered
the *upper s, k-extremal coefficient* of random vector $\mathbf{X} = (X_1, \ldots, X_d)$, defined by

$$\lambda_U(X_{s:d} | X_{d-k+1:d}) \equiv \lambda_U(U_{s:d} | U_{d-k+1:d})$$
$$= \lim_{t \downarrow 0} P(U_{s:d} > 1 - t | U_{d-k+1:d} > 1 - t),$$

where $U_{1:d} \leq \ldots \leq U_{d:d}$ are the order statistics of $(F_1(X_1), \ldots, F_d(X_d))$ and $X_{i:d}$
the inverse probability integral transform of $U_{i:d}$. In engineering, the coefficient
$\lambda_U(X_{s:d} | X_{d-k+1:d})$ can be interpreted as the limiting probability that the sth worst
performer in a system is attracted by the kth best one, provided the latter has an
extremely good performance. In mathematical finance, $\lambda_U(X_{s:d} | X_{d-k+1:d})$ can be
viewed as the limiting conditional probability that $X_{s:d}$ violates its value-at-risk at
level $1 - t$, given that $X_{d-k+1:d}$ has done so. If $s = k = 1$, we obtain the *upper
extremal dependence coefficient*, ϵ^U, considered in [7].

The study of systemic stability is also an important issue within the context of
extreme risk dependence. The fragility of a system has been addressed through
the *Fragility Index* (FI) introduced in [8]. More precisely, consider a random
vector $\mathbf{X} = (X_1, \ldots, X_d)$ with d.f. F and $N_x := \sum_{i=1}^{d} 1_{\{X_i > x\}}$ the number of
exceedances among $X_1, \ldots, X_d$ above a threshold x. The FI corresponding to $\mathbf{X}$

is the asymptotic conditional expected number of exceedances, given that there is at least one exceedance, i.e., $FI = \lim_{x\to\infty} E(N_x|N_x > 0)$. The stochastic system $\{X_1,\ldots,X_d\}$ is called fragile whenever $FI > 1$. In [5] it can be seen as a generalization of the FI that measures the stability of a stochastic system divided into blocks. More precisely, the block-FI of a random vector $\mathbf{X} = (X_1,\ldots,X_d)$ relative to a partition $\mathscr{D} = \{I_1,\ldots,I_s\}$ of $D = \{1,\ldots,d\}$ is

$$FI(\mathbf{X},\mathscr{D}) = \lim_{x\to\infty} E(N_\mathbf{x}|N_\mathbf{x} > 0),$$

where $N_\mathbf{x}$ is the number of blocks where it occurs at least one exceedance of $\mathbf{x}$, i.e.,

$$N_\mathbf{x} = \sum_{j=1}^{s} \mathbf{1}_{\{\mathbf{X}_{I_j} \nleq \mathbf{x}_{I_j}\}},$$

and where $\mathbf{X}_{I_j}$ is a sub-vector of $\mathbf{X}$ whose components have indexes in I_j, with $j = 1,\ldots,s$ (i.e., $\mathbf{X}_{I_j}$ is the jth block of random vector $\mathbf{X}$) and $\mathbf{x}_{I_j}$ is a vector of length $|I_j|$ with components equal to $x \in \mathbf{R}$. Observe that if we consider a partition $\mathscr{D}^* = \{I_j = \{j\} : j = 1,\ldots d\}$, then the coefficient $FI(\mathbf{X},\mathscr{D}^*)$ is the FI introduced in [8]. All operations and inequalities on vectors are meant componentwise.

Here we shall consider the abovementioned tail dependence coefficients defined in a time series perspective. More precisely, consider a stationary process $\{X_i\}_{i\geq 1}$ with marginal d.f. F_X. The lag-m TDC $(m = 1,2,\ldots)$ is given by

$$\lambda_m = \lim_{t\downarrow 0} P(F_X(X_{1+m}) > 1 - t | F_X(X_1) > 1 - t),$$

measuring the probability of occurring one extreme value observation given that another assumes an extreme value too, whenever separated in time by a lag-m. Analogously, we define the *lag-m upper s,k-extremal coefficient*,

$$\lambda_U(X_{s:m}|X_{m-k+1:m}) \equiv \lambda_U(U_{s:m}|U_{m-k+1:m})$$

$$= \lim_{t\downarrow 0} P(U_{s:m} > 1 - t | U_{m-k+1:m} > 1 - t),$$

a measure of the probability that, for a horizon of m successive time instants, the sth worst performer is attracted by the kth best one, provided the latter has an extremely good performance. If $s = k = 1$, we obtain the *lag-m upper extremal dependence coefficient*, ϵ_m^U. Finally, the lag-m block-FI relative to a partition $\mathscr{D}_m$ of $D_m = \{1,\ldots,m\}$ is

$$FI(\mathbf{X},\mathscr{D}_m) = \lim_{x\to\infty} E(N_\mathbf{x}|N_\mathbf{x} > 0),$$

where, for a horizon of m successive time instants, $N_\mathbf{x}$ is the number of blocks where it occurs at least one exceedance of $\mathbf{x}$. Hence it measures the stability within

m successive time instants of a stochastic process divided into blocks. Analogously we define the $FI(\mathbf{X}, \mathscr{D}_m^*)$ for a partition $\mathscr{D}^* = \{I_j = \{j\} : j = 1, \ldots m\}$ as the lag-m FI version of [8].

3   Tail Dependence of YARP(III)(1)

In this section we shall present a characterization of the dependence structure and tail behavior of the YARP(III)(1) process. We start with the reference to some existing results and then we compute the above mentioned measures.

In order to determine the distribution of the maximum, $M_n = \max_{0 \le i \le n} X_i$, it is convenient to consider a family of level crossing processes $\{Z_n(x)\}$ indexed by $x > 0$, defined by

$$Z_n(x) = \begin{cases} 1 & \text{if } X_n > x \\ 0 & \text{if } X_n \le x. \end{cases}$$

These two processes are themselves Markov chains with corresponding transition matrices given by

$$P = \left(1 + \left(\tfrac{x}{\sigma}\right)^\alpha\right)^{-1} \begin{bmatrix} p + \left(\tfrac{x}{\sigma}\right)^\alpha & 1 - p \\ (1-p)\left(\tfrac{x}{\sigma}\right)^\alpha & 1 + p\left(\tfrac{x}{\sigma}\right)^\alpha \end{bmatrix}.$$

Hence, we have

$$F_{M_n}(x) = P(M_n \le x) = P(Z_0(x) = 0, Z_1(x) = 0, \ldots, Z_n(x) = 0)$$
$$= P(X_0 \le x)P(Z_i(x) = 0 | Z_{i-1}(x) = 0)^n = \frac{\left(\tfrac{x}{\sigma}\right)^\alpha}{1 + \left(\tfrac{x}{\sigma}\right)^\alpha} \left(\frac{p + \left(\tfrac{x}{\sigma}\right)^\alpha}{1 + \left(\tfrac{x}{\sigma}\right)^\alpha}\right)^n$$

and $\frac{n^{-1/\alpha}}{\sigma} M_n \xrightarrow{d} Fréchet(0, (1-p)^{-1}, \alpha)$.

In [3] it was proved that the YARP(III)(1) process presents a β-mixing dependence structure. Hence, it satisfies the local dependence condition $D(u_n)$ of Leadbetter [9] for any real sequence $\{u_n\}_{n \ge 1}$ and so, for each $\tau > 0$ such that $n(1 - F_X(u_n)) \to \tau$, as $n \to \infty$, we have $P(M_n \le u_n) \to e^{-\theta \tau}$ as $n \to \infty$, with $\theta = 1 - p$ (Proposition 2.2 of [3]). The parameter θ, known in literature as *extremal index*, is associated with the tendency of clustering of high levels: in case $\theta < 1$ large values tend to occur in clusters, i.e., near each other and tail dependence takes place. Indeed, the YARP(III)(1) process presents tail dependence with lag-m TDC, $\lambda_m = p^m$ (see Proposition 2.8 of [3]).

The one-step transition probability function (tpf) of the YARP(III)(1) process is given by:

$$Q(x,]0, y]) = P(X_n \leq y|X_{n-1} = x) = P(\min(p^{-1/\alpha}x, \tfrac{\varepsilon_n}{1-U_n}) \leq y)$$
$$= \begin{cases} 1 - P(\tfrac{\varepsilon_n}{1-U_n} > y) \ , \ x > yp^{1/\alpha} \\ 1 \qquad\qquad\qquad , \ x \leq yp^{1/\alpha} \end{cases} = \begin{cases} (1-p)F_\varepsilon(y) \ , \ x > yp^{1/\alpha} \\ 1 \qquad\qquad , \ x \leq yp^{1/\alpha}. \end{cases}$$

Similarly, we derive the m-step tpf:

$$Q^m(x,]0, y]) = \begin{cases} 1 - \prod_{j=0}^{m-1}[\overline{F}_\varepsilon(p^{j/\alpha}y)(1-p) + p] \ , \ x > yp^{m/\alpha} \\ 1 \qquad\qquad\qquad\qquad\qquad\qquad\quad , \ x \leq yp^{m/\alpha}. \end{cases} \tag{1}$$

In the sequel we shall denote a_t the quantile function at $1 - t$, i.e.,

$$a_t \equiv F_X^{-1}(1 - t) = \sigma(t^{-1} - 1)^{1/\alpha} \tag{2}$$

and, for a set A, $\alpha(A)$ and $\zeta(A)$ denote the maximum and the minimum of A, respectively.

Proposition 1. *The YARP(III)(1) process has lag-m upper s, k-extremal coefficient,*

$$\lambda_U(X_{s:m}|X_{m-k+1:m})$$
$$= \frac{\displaystyle\sum_{i=0}^{s-1}\sum_{I\in\mathscr{F}_i}\sum_{J\subset I}(-1)^{|J|}p^{\alpha(\overline{I}\cup J)-\zeta(\overline{I}\cup J)}}{\displaystyle -\sum_{\emptyset\neq J\subset D_m}(-1)^{|J|}p^{\alpha(J)-\zeta(J)} - \sum_{i=1}^{k-1}\sum_{I\in\mathscr{F}_i}\sum_{J\subset\overline{I}}(-1)^{|J|}p^{\alpha(I\cup J)-\zeta(I\cup J)}},$$

where $\mathscr{F}_i$ denotes the family of all subsets of $D_m = \{1,\ldots,m\}$ with cardinal equal to i and $\overline{I}$ the complement set of $I \in \mathscr{F}_i$ in D_m.

Proof. Consider notation $P_A(t) = P(\bigcap_{a\in A}\{F_X(X_a) > 1 - t\})$, for any set A. From Propositions 2.1 and 2.9 in [6], we have

$$\lambda_U(X_{s:m}|X_{m-k+1:m}) = \lim_{t\downarrow 0} \frac{\displaystyle\sum_{i=0}^{s-1}\sum_{I\in\mathscr{F}_i}\sum_{J\subset I}(-1)^{|J|}P_{\overline{I}\cup J}(t)/t}{\displaystyle -\sum_{\emptyset\neq J\subset\{1,\ldots,m\}}(-1)^{|J|}P_J(t)/t - \sum_{i=1}^{k-1}\sum_{I\in\mathscr{F}_i}\sum_{J\subset\overline{I}}(-1)^{|J|}P_{I\cup J}(t)/t}.$$

Now just observe that, for $i_1 < i_2 < i_3$, we have successively

$$P_{\{i_1,i_2,i_3\}}(t) = \int_{a_t}^{\infty} P(X_{i_3} > a_t, X_{i_2} > a_t | X_{i_1} = u_1) dF_X(u_1)$$

$$= \int_{a_t}^{\infty} \int_{a_t}^{\infty} P(X_{i_3} > a_t | X_{i_2} = u_2) Q(u_1, du_2) dF_X(u_1)$$

$$= \int_{a_t}^{\infty} \int_{a_t}^{\infty} [1 - Q^{i_3-i_2}(u_2,]0, a_t])] Q^{i_2-i_1}(u_1, du_2) dF_X(u_1),$$

where a_t is given in (2). Applying (1), we obtain

$$P_{\{i_1,i_2,i_3\}}(t) = t[t + p^{i_3-i_2}(1-t)] \int_{a_t}^{\infty} \int_{a_t}^{\infty} Q^{i_2-i_1}(u_1, du_2) dF_X(u_1)$$

$$= t[t + p^{i_3-i_2}(1-t)] \int_{a_t}^{\infty} [1 - Q^{i_2-i_1}(u_1,]0, a_t])] dF_X(u_1)$$

$$= [t + p^{i_3-i_2}(1-t)][t + p^{i_2-i_1}(1-t)]t.$$

A similar reasoning leads us to, for $i_1 < i_2 < \ldots < i_k$,

$$P_{\{i_1,\ldots,i_k\}}(t)$$

$$= \int_{a_t}^{\infty} \cdots \int_{a_t}^{\infty} (1 - Q^{i_k-i_{k-1}}(u_{i_{k-1}},]0, a_t])) \prod_{j=2}^{k-1} Q^{i_{k-j}-i_{k-j+1}+1}(u_{i_{k-j}}, du_{i_{k-j+1}}) dF_X(u_{i_1})$$

$$= \prod_{j=2}^{k} (t + p^{i_j-i_{j-1}}(1-t))t,$$

and hence

$$\lim_{t \downarrow 0} P_{\{i_1,\ldots,i_k\}}(t)/t = \lim_{t \downarrow 0} \prod_{j=2}^{k} (t + p^{i_j-i_{j-1}}(1-t)) = p^{i_k-i_1}. \qquad (3)$$

$\square$

Corollary 1. *The YARP(III)(1) process has lag-m upper extremal dependence coefficient,*

$$\epsilon_m^U = \frac{p^{m-1}}{m - (m-1)p}.$$

A positive ϵ_m^U means the existence of extremal dependence on a time horizon of m time instants.

Proposition 2. *The YARP(III)(1) process has lag-m block-FI, relative to a partition $\mathscr{D}_m$ of $D_m = \{1,\ldots,m\}$, given by*

$$FI(\mathbf{X}, \mathscr{D}_m) = \frac{\sum_{j=1}^{s} \sum_{k \in I_j} (-1)^{k-1} \sum_{J \subset I_j; |J|=k} p^{\alpha(J)-\zeta(J)}}{m - (m-1)p}.$$

Proof. Based on Propositions 3.1 and 5.2 in [5], we have

$$
\begin{aligned}
FI(\mathbf{X}, \mathscr{D}_m) &= \lim_{t \downarrow 0} \frac{\sum_{j=1}^{s} P(\bigcup_{i \in I_j} \{F_X(X_i) > 1-t\})}{1 - P(\bigcap_{i \in \{1,\dots m\}} \{F_X(X_i) < 1-t\})} \\
&= \lim_{t \downarrow 0} \frac{\sum_{j=1}^{s} \sum_{k \in I_j} (-1)^{k-1} \sum_{J \subset I_j; |J|=k} P(\bigcap_{i \in J} \{F_X(X_i) > 1-t\})}{1 - F_{M_{m-1}}(a_t)}.
\end{aligned}
$$

Now observe that, from (3), we have

$$\lim_{t \downarrow 0} P(\cap_{i \in J} \{F_X(X_i) > 1-t\})/t = p^{\alpha(J)-\zeta(J)}$$

and from (1) and (2), we have

$$\lim_{t \downarrow 0}(1 - F_{M_{m-1}}(a_t))/t = \lim_{t \downarrow 0} \frac{1}{t}\left(1 - \frac{t^{-1}-1}{t^{-1}}\left(\frac{p+t^{-1}-1}{t^{-1}}\right)^{m-1}\right) = m-(m-1)p.$$

$\square$

Corollary 2. *The YARP(III)(1) process has lag-m FI,*

$$FI(\mathbf{X}, \mathscr{D}_m^*) = \frac{m}{m - (m-1)p}.$$

Therefore, on a time horizon of m $(m > 1)$ time instants the process is strongly fragile since $FI > 1$.

We remark that the tail measures given above only depend on the parameter p of the YARP(III)(1) process and thus can be estimated through this latter. For a survey on the estimation of p, see [3].

References

1. Arnold, B.C.: Pareto Distributions. International Cooperative Publishing House, Fairland (1983)
2. Davis, R., Resnick, S.: Basic properties and prediction of max-ARMA processes. Adv. Appl. Probab. **21**, 781–803 (1989)
3. Ferreira, M.: On the extremal behavior of a pareto process: an alternative for armax modeling. Kybernetika **48**(1), 31–49 (2012)
4. Ferreira, M., Canto e Castro, L.: Modeling rare events through a pRARMAX process. J. Stat. Plann. Inference **140**(11), 3552–3566 (2010)
5. Ferreira, M., Ferreira, H.: Fragility index block tailed vectors. J. Stat. Plann. Inference **142**(7), 1837–1848 (2012)

6. Ferreira, M., Ferreira, H.: Tail dependence between order statistics. J. Multivariate Anal. **105**(1), 176–192 (2012)
7. Frahm G.: On the extremal dependence coefficient of multivariate distributions. Stat. Probab. Lett. **76**, 1470–1481 (2006)
8. Geluk, J.L., De Haan, L., De Vries, C.G.: Weak and strong financial fragility. Tinbergen Institute Discussion Paper, TI 2007-023/2 (2007)
9. Leadbetter, M.R.: On extreme values in stationary sequences. Z. Wahrsch. verw. Gebiete **28**, 289–303 (1974)
10. Pareto, V.: Cours d'economie Politique. Rouge, Lausanne (1897)
11. Sibuya, M.: Bivariate extreme statistics. Ann. Inst. Stat. Math. **11** 195–210 (1960)
12. Yeh, H.C., Arnold, B.C., Robertson, C.A.: Pareto processes. J. Appl. Probab. **25**, 291–301 (1988)
13. Zhang, Z., Smith, R.L.: On the estimation and application of max-stable processes. J. Stat. Plann. Inference **140**(5), 1135–1153 (2010)

Application of the Theory of Extremes to the Study of Precipitation in Madeira Island: Statistical Choice of Extreme Domains of Attraction

Délia Gouveia, Luiz Guerreiro Lopes, and Sandra Mendonça

Abstract

In the past and nowadays, hydrology is one of the most natural fields of application for the theory of extremes. This work presents an application of univariate extreme value theory to the study of precipitation in Madeira Island. The method for testing extreme value conditions investigated by Dietrich et al. (Extremes 5:71–85, 2002) was applied to the monthly 1-day maxima precipitation data for the rainy season from seven pluviometric stations maintained by the Portuguese Meteorological Institute. The statistical procedures for the problem of statistical choice of extreme domains of attraction analysed by Neves and Fraga Alves (TEST 16:297–313, 2007) were also applied to each station data set. The results of this analysis indicate the possible k upper extremes to be used for each local sample and the sign of each extreme value index γ.

1 Introduction

In the first book on statistics of extremes, Emil Gumbel [8] wrote that the oldest problems connected with extreme values arise from the study of floods. As stated by Katz et al. [11], early work in hydrology usually assumed an exponential distribution for the excess over a high threshold, which is equivalent to a Gumbel distribution

D. Gouveia (✉)
CEAUL, CIMO/IPB and University of Madeira, Funchal, Portugal
e-mail: delia@uma.pt

L.G. Lopes
CIMO/IPB, ICAAM/UE and University of Madeira, Funchal, Portugal
e-mail: lopes@uma.pt

S. Mendonça
CEAUL and University of Madeira, Funchal, Portugal
e-mail: smendonca@uma.pt

A. Pacheco et al. (eds.), *New Advances in Statistical Modeling and Applications*,
Studies in Theoretical and Applied Statistics, DOI 10.1007/978-3-319-05323-3_18,
© Springer International Publishing Switzerland 2014

for the maximum. Nowadays, statistics of extremes for independent and identical random variables are applied following two different methodologies: parametric and semi-parametric. Unlike the parametric methodology, the only assumption in the semi-parametric approach is that the distribution function F is in the domain of attraction of an extreme value distribution. The statistical methodologies associated with a semi-parametric set-up for peaks over random threshold (PORT) approach are a result of the research of Laurens de Haan and collaborators [4].

In this work, data from seven pluviometric stations in Madeira Island provided by the Portuguese Meteorological Institute (IM) were used. The assumption that the distribution function F belongs to the domain of attraction of an extreme value distribution for monthly 1-day maxima precipitation data was tested for the rainy season, and the available data from each station was then analysed in order to find the most suitable domain of attraction for the sampled distribution.

The volcanic island of Madeira, located in the Atlantic Ocean off the coast of Northwest Africa has a near E–W-oriented orographic barrier, approximately perpendicular to the prevailing NE wind direction, which induces a remarkable variation of precipitation between the northern and southern slopes [2].

In the last two centuries, extreme precipitation events triggered at least thirty significant flash floods (in terms of damages and loss of lives) in Madeira Island [19]. Since 2001, at least nine events of this nature, with different intensities, have occurred in the island, and the last and most significant one occurred on the 20th of February 2010 [2].

The structure of this paper is as follows. Section 2 presents the methods of univariate extreme value theory applied in this study and the available precipitation data used. This is followed by Sect. 3, where the results of the analysis are presented. Finally, Sect. 4 contains a summary and some final comments.

2   Methods and Data

Let $X_1, X_2, \ldots, X_n$ be independent random variables with common distribution function F. The assumption that F is in the domain of attraction of an extreme value distribution, $F \in \mathcal{D}(G_\gamma)$, means that there are normalising constants $a_n > 0$ and $b_n \in \mathbb{R}$ such that, for all x,

$$\lim_{n \to \infty} P\left(\max_{1 \le i \le n} \frac{X_i - b_n}{a_n} \le x\right) = G_\gamma(x) \tag{1}$$

with

$$G_\gamma(x) = \begin{cases} \exp(-(1 + \gamma x)^{-1/\gamma}), \text{ if } 1 + \gamma x > 0, \gamma \neq 0; \\ \exp(-\exp(-x)), \text{ if } x \in \mathbb{R}, \gamma = 0. \end{cases} \tag{2}$$

We shall use here the approach of Dietrich et al. [1] to test $H_0 : F \in D(G_\gamma)$, for some γ in $\mathbb{R}$. Our choice is reinforced by Hüsler and Li's work [10] and the study they made comparing the mentioned approach to two others. The test statistic is given (with the usual notations) by:

$$E_n(k) = k \int_0^1 \left(\frac{\log X_{n-[kt]:n} - \log X_{n-k:n}}{\hat{\gamma}_+} - \frac{t^{-\hat{\gamma}_-} - 1}{\hat{\gamma}_-}(1 - \hat{\gamma}_-) \right)^2 t^\eta dt \qquad (3)$$

where $\eta > 0$ and the estimates $\hat{\gamma}_+$ and $\hat{\gamma}_-$ for $\gamma_+ = \max\{\gamma, 0\}$ and $\gamma_- = \min\{\gamma, 0\}$, respectively, are the moment estimators. The integer k satisfies $k \to \infty$, $k/n \to 0$ and $k^{1/2}A(n/k) \to 0$, as $n \to \infty$, with A related to the second order condition.

The recommended procedure (see [10, 21]) for application of the test is:
1. Estimate $\hat{\gamma}_+$ and $\hat{\gamma}_-$ by the moment estimator and calculate the value of the test statistic $E_n(k)$;
2. Determine the corresponding quantile $Q_{1-\alpha,\hat{\gamma}}$ using the Table 1 in [10];
3. If $E_n(k) > Q_{1-\alpha,\hat{\gamma}}$, then reject H_0 with nominal type I error α.

When the hypothesis that F belongs to the domain of attraction of an extreme value distribution is not rejected, it may be useful for applications to know what is the most suitable domain of attraction for the sampled distribution. We shall use the normalized versions of the Hasofer and Wang's test statistic [9],

$$W_n^*(k) = \sqrt{k/4} \left(k \left(\frac{1}{k} \frac{(k^{-1} \sum_{i=1}^k Z_i)^2}{k^{-1} \sum_{i=1}^k Z_i^2 - (k^{-1} \sum_{i=1}^k Z_i)^2} \right) - 1 \right) \qquad (4)$$

and of the Greenwood's test statistic [7],

$$R_n^*(k) = \sqrt{k/4} \left(\frac{k^{-1} \sum_{i=1}^k Z_i^2}{(k^{-1} \sum_{i=1}^k Z_i)^2} - 2 \right) \qquad (5)$$

both given by Neves and Fraga Alves [17], where $Z_i = \{X_{n-i+1:n} - X_{n-k:n}\}_{i=1,...,k}$ and k is the number of observations above the random threshold $X_{n-k:n}$. Under the null hypothesis of the Gumbel domain of attraction and some additional (second order) conditions, the test statistics $W_n^*(k)$ and $R_n^*(k)$ are asymptotically normal, as $n \to \infty$.

The critical region for the two-sided test of a nominal size α, $H_0 : \{F \in \mathcal{D}(G_0)\}$ vs. $H_1 : \{F \in \mathcal{D}(G_\gamma)_{\gamma \neq 0}\}$, is given by $|W_n^*(k)| > z_{1-\frac{\alpha}{2}}$ and by $|R_n^*(k)| > z_{1-\frac{\alpha}{2}}$, where $z_{1-\frac{\alpha}{2}}$ denotes the $(1 - \alpha/2)$—quantile of the standard normal distribution. The one-sided testing problem of testing the Gumbel domain against the Weibull domain, $H_0 : \{F \in \mathcal{D}(G_0)\}$ vs. $H_1 : \{F \in \mathcal{D}(G_\gamma)_{\gamma < 0}\}$, has the critical region given by $W_n^*(k) > z_{1-\alpha}$ for the Hasofer and Wang's test. To test H_0 against the Fréchet domain ($H_1 : \{F \in \mathcal{D}(G_\gamma)_{\gamma > 0}\}$), the rejection criterion to use is $W_n^*(k) < -z_{1-\alpha}$. Using the test statistic $R_n^*(k)$, the rejection criterion is $R_n^*(k) < -z_{1-\alpha}$ ($R_n^*(k) > z_{1-\alpha}$) when we have $H_1 : \{F \in \mathcal{D}(G_\gamma)_{\gamma < 0}\}$ ($H_1 : \{F \in \mathcal{D}(G_\gamma)_{\gamma > 0}\}$).

In this paper we present the results of the application of these testing procedures to the monthly 1-day maxima precipitation data for the rainy season from seven

Table 1 Details of the pluviometric stations used in this study

Number	Name	Latitude	Longitude	Altitude (m)	Period	n
365	Santana	32°48'N	16°53'W	380	1942–2007	382
370	Bica da Cana	32°45'N	17°03'W	1,560	1961–2009	286
373	Areeiro	32°43'N	16°55'W	1,610	1961–1993	196
375	Santo da Serra	32°43'N	16°49'W	660	1970–2009	240
385	Lugar de Baixo	32°40'N	17°05'W	15	1961–2004	264
521	Santa Catarina	32°41'N	16°46'W	49	1961–2009	293
522	Funchal	32°38'N	16°53'W	58	1949–2009	366

weather stations in Madeira Island maintained by the IM. Each station is identified by the name of the place where it is located. Besides the name, Table 1 provides other information about each station, namely its number, geographical location and altitude, the period considered and the sample size.

3 Results and Discussion

Our first step was to check if we can consider that, for each data set, the corresponding distribution function F belongs to the domain of attraction of an extreme value distribution. To apply the test statistic $E_n(k)$ to the data from each station we used the R program code provided by Li at www.imsv.unibe.ch/~deyuan/research.html, taking $\eta = 2$, as suggested by Hüsler and Li [10]. The values of the test statistic $E_n(k)$ and its corresponding 0.95 quantile for varying k are shown in Figs. 1 and 2. Table 2 shows possible values of k for which $F \in \mathscr{D}(G_\gamma)$ is not rejected.

This preliminary data analysis assumes that there exists an underlying distribution for the data in the attraction domain of some classical extreme value distribution, either Gumbel, Fréchet, or Weibull. Classical extreme value distributions arise assuming stability of the limiting distribution of suitably normalized independent and identical random variables. The restriction of the analysis to the 6 months period of the rainy season minimizes the heterogeneity of the data, but does not guarantee the homogeneity for Santana, Santa Catarina and Santo da Serra data sets according to the Anderson–Darling test [22]. Nevertheless, we applied the same statistical procedure to all the data sets for the reasons exposed in Sect. 4.

Our next step was to find the most suitable domain of attraction for the sampled distribution for each station data set. The test statistics $W_n^*(k)$ and $R_n^*(k)$ were implemented in the R software language [20].

The application of test statistics $W_n^*(k)$ and $R_n^*(k)$ to the seven available data sets yielded the plots in Figs. 3 and 4. For the values of k in Table 2 the choice of the domains of attraction suggested by the test statistics $W_n^*(k)$ and $R_n^*(k)$ is presented in Tables 3 and 4 by choices A and B, respectively. We observe that when the Weibull domain is suggested by the test statistic $W_n^*(k)$, the choice by the test

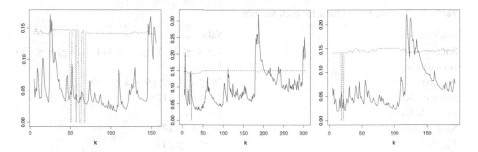

Fig. 1 Values of the test statistic $E_n(k)$ (*solid*) and the 0.95 quantile (*dotted*) applied to the data sets from Areeiro (*left*), Santana (*centre*) and Santo da Serra (*right*) stations

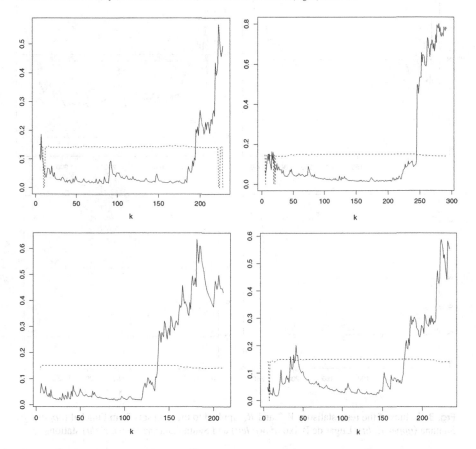

Fig. 2 Values of the test statistic $E_n(k)$ (*solid*) and the 0.95 quantile (*dotted*) applied to the data sets from Bica da Cana (*upper left*), Funchal (*upper right*), Lugar de Baixo (*down left*) and Santa Catarina (*down right*) stations

Table 2 Possible values for k for each location

Station name	Values of k
Santana	$8 \leq k \leq 21; 23 \leq k \leq 111 ; 113 \leq k \leq 179$
Bica da Cana	$12 \leq k \leq 194$
Areeiro	$28 \leq k \leq 49; 69 \leq k \leq 145$
Santo da Serra	$23 \leq k \leq 117$
Lugar de Baixo	$k \leq 137$
Santa Catarina	$44 \leq k \leq 177$
Funchal	$22 \leq k \leq 245$

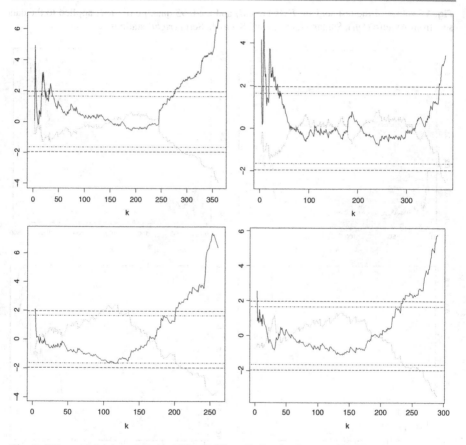

Fig. 3 Values of the tests statistics W_n^* and R_n^* applied to the data sets from Funchal (*upper left*), Santana (*upper right*), Lugar de Baixo (*down left*) and Santa Catarina (*down right*) stations

statistic $R_n^*(k)$ can be the Gumbel domain or also the Weibull domain, but for fewer values of k. The difference between the test statistics was pointed out by Neves and Fraga Alves [17], who refer that the Greenwood-type test barely detects small negative values of γ, and that the Hasofer and Wang's test is the most powerful test when analysing alternatives in the Weibull domain of attraction.

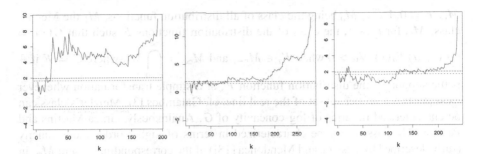

Fig. 4 Values of the tests statistics W_n^* and R_n^* applied to the data sets from Areeiro (*left*), Bica da Cana (*centre*) and Santo da Serra (*right*) stations

Table 3 Statistical choice of domain of attraction for the four locations with lower altitude

Choice	Funchal	Santana	Lugar de Baixo	Santa Catarina
A	Weibull	Weibull	Fréchet	Gumbel
	(some values of k)	(some values of k)	(some values of k)	
B	Gumbel	Gumbel	Gumbel	Gumbel

Table 4 Statistical choice of domain of attraction for the three locations with higher altitude

Choice	Areeiro	Bica da Cana	Santo da Serra
A	Weibull	Weibull (some values of k)	Weibull (some values of k)
B	Weibull	Weibull (some values of k)	Gumbel

4   Final Remarks

The aim of this work was to present a preliminary application of univariate extreme value theory to the study of precipitation in Madeira Island.

The extreme value index γ is of primary interest in extreme value analysis and it is the only parameter estimated under a semi-parametric approach. The estimation of γ is based on the k top order statistics in the sample and our analysis provides information about the region of k values to use for each location. We notice that for almost all locations there is an evidence for non-positive values of the shape parameter for some values of k. However, the sign of the estimates of γ, when applying maximum likelihood and probability-weighted moments estimators, was non-negative in all the cases.

For three locations the identical distribution hypothesis was rejected but that does not invalidate our analysis because much progress has been achieved in relaxing the independence and identicality assumptions. The identical distribution hypothesis has been relaxed by Mejzler [12–16] who described a class of limit laws which is the simile in extreme value theory to the Lévy–Khinchine's L class of self-decomposable laws. Observing that any univariate distribution is max-infinite divisible, Graça Martins and Pestana [5, 6] defined classes of distribution functions

M_r, $r = 0, 1, \ldots, M_0$ being the class of all distribution functions, M_1 the Mejzler class, M_r, for $r > 1$, the class of the distribution functions F such that $G(x) = G(x + a) F_a(x)$, $\forall a > 0$ when $F_a \in M_{r-1}$ and $M_\infty = \bigcap_{r=0}^{\infty} M_r$, where $G = F$ if $\mathbb{R}$ is the support of the distribution function F, or a simple transformation whenever its support is a proper subset of the real line, cf. Galambos [3]. Mejzler's class can be characterized in terms of log-concavity of G. Analogously, Graça Martins and Pestana M_r classes may be characterized in terms of higher order monotonicity (fully described by Pestana and Mendonça [18]) of the corresponding G, and M_∞ in terms of complete monotonicity: $F \in M_\infty$ if and only the corresponding G satisfies $G(x) = \exp[-K(x)]$, with K completely monotone. Hence, M_∞ is a non-trivial extension of both the superclass of stable extreme value distributions for maxima, and a subclass of Mejzler's laws, that can provide a proper framework to analyse maxima of linearly transformed data arising from various parent distributions. This is postponed for future work, namely on kernel estimation of a fitting K.

Acknowledgements To the Portuguese Foundation for Science and Technology (FCT), for the financial support through the PhD grant SFRH/BD/39226/2007, financed by national funds of MCTES. To the Center of Statistics and Applications of the University of Lisbon (CEAUL) for the bibliographical support. To the Portuguese Meteorological Institute (IM), namely to Dr. Victor Prior, for providing the precipitation data, and to the University of Madeira for the logistic support. The authors would also like to thank Prof. Dinis Pestana and the anonymous reviewers for their valuable comments and suggestions to improve this paper.

References

1. Dietrich, D., de Haan, L., Hüsler, J.: Testing extreme value conditions. Extremes **5**, 71–85 (2002)
2. Fragoso, M., Trigo, R.M., Pinto, J.G., Lopes, S., Lopes, A., Ulbrich, S., Magro, C.: The 20 February 2010 Madeira flash-floods: synoptic analysis and extreme rainfall assessment. Nat. Hazards Earth Syst. Sci. **12**, 715–730 (2012)
3. Galambos, J.: The Asymptotic Theory of Extreme Order Statistics, 2nd edn. Krieger, Malabar (1987)
4. Gomes, M.I., Canto e Castro, L., Fraga Alves, M.I., Pestana, D.: Statistics of extremes for IID data and breakthroughs in the estimation of the extreme value index: Laurens de Haan leading contributions. Extremes **11**, 3–34 (2008)
5. Graça Martins, M.E., Pestana, D.D.: The extremal limit theorem – extensions. In: Grossman, W., Mogyoródi, J., Vincze, I., Wertz, W. (eds.) Probability Theory and Mathematical Statistics with Application, pp. 143–153. D. Reidel, Boston (1985)
6. Graça Martins, M.E., Pestana, D.D.: Nonstable limit laws in extreme value theory. In: Puri, M.L., Vilaplana, J.P., Wertz, W. (eds.) New Perspectives in Theoretical and Applied Statistics, pp. 449–457. Wiley, New York (1987)
7. Greenwood, M.: The statistical study of infectious diseases. J. R. Stat. Soc. Ser. A **109**, 85–109 (1946)
8. Gumbel, E.J.: Statistics of Extremes. Columbia University Press, New York (1958)
9. Hasofer, A.M., Wang, Z.: A test for extreme value domain of attraction. J. Am. Stat. Assoc. **87**, 171–177 (1992)
10. Hüsler, J., Li, D.: On testing extreme value conditions. Extremes **9**, 69–86 (2006)

11. Katz, R.W., Parlange, M.B., Naveau, P.: Statistics of extremes in hydrology. Adv. Water Resour. **25**, 1287–1304 (2002)
12. Mejzler, D.: On the problem of the limit distributions for the maximal term of a variational series. L'vov. Politechn. Inst. Naučn. Zapiski Ser. Fiz.-Mat. **38**, 90–109 (1956)
13. Mejzler, D.: On a certain class of limit distributions and their domain of attraction. Trans. Am. Math. Soc. **117**, 205–236 (1965)
14. Mejzler, D.: Limit distributions for the extreme order statistics. Can. Math. Bull. **21**, 447–459 (1978)
15. Mejzler, D.: Asymptotic behaviour of the extreme order statistics in the non identically distributed case. In: Tiago de Oliveira, J. (ed.) Statistical Extremes and Applications, pp. 535–547. D. Reidel, Boston (1984)
16. Mejzler, D.: Extreme value limit laws in the nonidentically distributed case. Isr. J. Math. **57**, 1–27 (1987)
17. Neves, C., Fraga Alves, M.I.: Semi-parametric approach to the Hasofer-Wang and Greenwood statistics in extremes. TEST **16**, 297–313 (2007)
18. Pestana, D.D., Mendonça, S.: Higher-order monotone functions and probability theory. In: Hadjisavvas, N., Martínez-Legaz, J.E., Penot, J.P. (eds.) Generalized Convexity and Generalized Monotonicity, pp. 317–331. Springer, New York (2001)
19. Quintal, R.: Aluviões da Madeira; Séculos XIX e XX. Territorium **6**, 31–48 (1999)
20. R Development Core Team: R: A Language and Environment for Statistical Computing. R Foundation for Statistical Computing, Vienna (2011)
21. Reiss, R.D., Thomas, M.: Statistical Analysis of Extreme Values with Applications to Insurance, Finance, Hydrology and Other Fields, 3rd edn.Birkhäuser, Basel (2007)
22. Scholz, F.W., Stephens, M.A.: K-sample Anderson-Darling tests. J. Am. Stat. Assoc. **82**, 918–924 (1987)

The Traveling Salesman Problem and the Gnedenko Theorem

Tiago Salvador and Manuel Cabral Morais

Abstract

The traveling salesman problem (TSP) has three characteristics common to most problems, which have attracted and intrigued mathematicians: the simplicity of its definition, the wealth of its applications, and the inability to find its optimal solution in polynomial-time.

In this paper, we provide point and interval estimates for the optimal cost of several instances of the TSP, by using the solutions obtained by running four approximate algorithms—the 2-optimal and 3-optimal algorithms and their greedy versions—and considering the three-parameter Weibull model, whose location parameter represents the (unknown) optimal cost of the TSP.

1 Traveling Salesman Problem: Definition and a Few Milestones

Consider a salesperson seeking to visit each city on a given list of N ($N > 3$) cities exactly once and to return to his(her) city of origin, and assume that (s)he knows the cost of traveling between any two cities i and j, c_{ij} ($i, j = 1, \ldots, N$). The

T. Salvador
Instituto Superior Técnico, Technical University of Lisbon, Av. Rovisco Pais 1, 1049-001 Lisboa, Portugal
e-mail: tiago.salvador@ist.utl.pt

M.C. Morais (✉)
CEMAT and Mathematics Department, Instituto Superior Técnico, Technical University of Lisbon, Av. Rovisco Pais 1, 1049-001 Lisboa, Portugal
e-mail: maj@math.ist.utl.pt

A. Pacheco et al. (eds.), *New Advances in Statistical Modeling and Applications*, Studies in Theoretical and Applied Statistics, DOI 10.1007/978-3-319-05323-3_19, © Springer International Publishing Switzerland 2014

traveling salesman problem (TSP) consists in finding a sequence of cities such that the associated total traveling cost is minimal—the optimal tour. In mathematical terms, the TSP corresponds to the identification of the cyclic permutation π of the integers from 1 to N that minimizes $\sum_{i=1}^{N} c_{i\pi(i)}$.

It is hard to determine the origins of the TSP. It is believed that the term "Traveling Salesman Problem" was introduced in mathematical circles in 1931–1932 by H. Whitney [9, p. 5]; but, according to Hoffman and Wolfe [9, p. 5], the TSP was first mentioned in a handbook for traveling salesmen from 1832. This handbook includes example tours through Germany and Switzerland, however, no mathematical treatment had been given to the problem [18]. Curiously enough, mathematical problems related to the TSP problem were treated in the 1800s by the mathematicians W.R. Hamilton and T.P. Kirkman [9, p. 3].

[9, pp. 6–9] also refers that the seminal paper [1] expressed the TSP as an integer linear program problem, developed the cutting plane method for its solution and solved an instance with 49 (American) cities to optimality, by constructing a tour and proving that no other tour could be shorter. Since then the TSP is used as a benchmark for many optimization methods, has several applications (such as in scheduling, logistics, manufacture of microchips), and, unsurprisingly, is one of the most intensively studied problems in computational mathematics.

2   Complexity, Approximate Algorithms, and Statistical Approach

The TSP is an NP-hard problem [9, p. 7]; hence, we do not expect to find a polynomial-time algorithm to solve it. However, several tour construction algorithms and iterative improvement algorithms have been proposed; they do not necessarily yield the optimal solution but can provide an approximate solution in a reasonable amount of time.

In this paper, we will focus on the results yielded by the λ-optimal algorithm ($\lambda = 2, 3$), an iterative improvement algorithm whose description requires two preliminary definitions taken from [11] and [4].

Definition 1. A tour is said to be λ-optimal (or simply λ-opt) if it is impossible to obtain a tour with smaller cost by replacing any λ of its edges by any other set of λ edges.

Definition 2. The λ-neighborhood of a tour T, $N_\lambda(T)$, is the set of all tours we can obtain of T by replacing λ of its edges.

The λ-optimal algorithm can be described as follows:
Step 1   Randomly choose a tour T.
Step 2   If T is λ-optimal, stop. Otherwise, go to Step 3.

Step 3   Compute $N_\lambda(T)$ and the associated costs of its tours.

Step 4   Choose T^* as the tour in $N_\lambda(T)$ with smallest cost and return to Step 2 with $T = T^*$.

If instead of choosing the tour with the smallest cost at Step 4, we choose the first one we find that has a smaller cost than T, we end up dealing with a variant (and well-known first improvement strategy) of the original algorithm rightly called λ-optimal greedy algorithm and also addressed in this paper.

The computational complexity of the TSP and the inability to find the optimal tour in polynomial-time justify a statistical approach to this optimization problem [6]. This approach naturally requires the collection of data and the use of a probabilistic model in order to make inferences on the cost of the optimal tour.

As for the data, let A be an algorithm for solving the TSP that will be run n times, X_{ij} be the cost of the tour at the iteration j of run i with $n_i(N)$ iterations for N cities, $X_{i(1)} = \min_{j=1,\ldots,n_i(N)} X_{ij}$ be the cost of the approximate solution yielded by algorithm A at run i, $i = 1,\ldots,n$. Then $(X_{1(1)},\ldots,X_{n(1)})$ can be thought as a random sample of n minimum costs (one random sample per run).

As far as the probabilistic model is concerned, the (Fisher–Tippett–)Gnedenko theorem [3, 5], a fundamental result in extreme-value theory, provides a suitable model. This theorem can be informally stated as follows: under certain conditions, the standardized minimum of a random sample converges to one of the following distributions—Fréchet, Weibull, or Gumbel. If we bear in mind that the Weibull distribution is the only of those three limiting distributions with a range limited from below and that the costs of the admissible tours are limited to the left by the cost of the optimal tour, the Weibull model is the obvious choice for the distribution of $X_{i(1)}$, $F_{X_{i(1)}}(.)$, as we increase the number of cities N (thus, $n_i(N)$). It is important to note that the fact that the range of the costs X_{ij} is limited to the left is a necessary but not a sufficient condition for $F_{X_{i(1)}}(.)$ to belong to the domain of attraction of the Weibull distribution. Moreover, as [6] noticed, there is a clearly interdependence among the intermediate solutions, hence assuming independence and applying the Gnedenko theorem is disputable.

Although debatable, the idea of using the Weibull model is not new: [6] refers that [13] was the first author to use this model, while dealing with combinatorially explosive plant-layout problems. [6, 7, 12, 14, 16] have proceeded in a similar way, with remarkable results. The Weibull model has been used by further authors in the statistical approach to other combinatorial optimization problems, such as the covering problem [17].

A random variable X has a three-parameter Weibull distribution if its probability density function is given by $f_X(x) = \frac{c}{b} \times \left(\frac{x-a}{b}\right)^{c-1} e^{-\left(\frac{x-a}{b}\right)^c} \times I_{[a,+\infty)}(x)$, where $a > 0$ (because costs c_{ij} are assumed nonnegative), $b > 0$, and $c > 0$ represent the location, scale, and shape parameters, respectively.

Assuming that the three-parameter Weibull model is suitable to characterize the costs of the approximate solutions of the TSP, we are supposed to estimate its location parameter a, i.e. the cost of the optimal tour, by making use of the sample $(x_{1(1)},\ldots,x_{n(1)})$ with the results of the n runs of an approximate algorithm A.

The obvious choice is to consider the sample minimum, $x_{(1)} = \min_{i=1,\ldots,n} x_{i(1)}$. A more reasonable choice is the first component of the maximum likelihood (ML) estimate of (a, b, c), $(\hat{a}, \hat{b}, \hat{c})$, which is not as trivial to obtain as one might think [15], thus, the need to resource to alternative estimates, such as the ones proposed by Zanakis [20] and Wyckoff et al. [19], $(\bar{a}, \bar{b}, \bar{c})$ and $(\tilde{a}, \tilde{b}, \tilde{c})$, respectively, given by

$$
\left(\frac{x_{(1)} \times x_{(n)} - x_{(2)}^2}{x_{(1)} + x_{(n)} - 2 \times x_{(2)}}, -\bar{a} + x_{(\lceil 0.63n \rceil)}, \{\log[-\log(1 - p_k)] \right.
$$

$$
- \log[-\log(1 - p_j)]\} / \log \left(\frac{x_{(\lceil np_k \rceil)} - \bar{a}}{x_{(\lceil np_j \rceil)} - \bar{a}} \right) \Big)
$$

$$
\left(\frac{x_{(1)} - \frac{\bar{x}}{n^{1/\tilde{c}_0}}}{1 - \frac{1}{n^{1/\tilde{c}_0}}}, \exp\left\{ \frac{\gamma}{\tilde{c}_0} + \frac{1}{n} \sum_{i=1}^{n} \log(x_{(i)} - \tilde{a}) \right\}, \right.
$$

$$
\left. \frac{n \times k_n}{-\sum_{i=1}^{s} \log(x_{(i)} - \tilde{a}) + \frac{s}{n-s} \sum_{i=s+1}^{n} \log(x_{(i)} - \tilde{a})} \right),
$$

where $x_{(i)}$ represents the ith order statistic of the sample, $\lceil y \rceil$ denotes the ceiling of the real number y, $p_j = 0.16731$, $p_k = 0.97366$, $\bar{x}$ is the arithmetic mean, $s = \lceil 0.84n \rceil$, $\tilde{c}_0 = \{\log[-\log(1 - p_k)] - \log[-\log(1 - p_j)]\} / \{\log \hat{E}[x_{(\lceil np_k \rceil)} - x_{(1)}] - \log[x_{(\lceil np_j \rceil)} - x_{(1)}]\}$, $\gamma \simeq 0.577215665$ is the Euler constant, and k_n is a constant whose value depends on the dimension of the sample (for exact values of k_n see Table 6 in [2]).

As for confidence intervals (CI) for a, due to min-stability, $X_{(1)} \sim$ Weibull $(a, b/n^{1/c}, c)$, and we can use this result to derive $P[X_{(1)} - b < a < X_{(1)}] = 1 - e^{-n}$ and $P[X_{(1)} - b/[-n/\log(\alpha)]^{1/c} < a < X_{(1)}] = 1 - \alpha$. Thus, two CI for the cost of the optimal tour immediately follow: $CI_{(1-e^{-n}) \times 100\%}(a) = [x_{(1)} - b; x_{(1)}]$ (see [7]); $CI_{(1-\alpha) \times 100\%}(a) = [x_{(1)} - b/[-n/\log(\alpha)]^{1/c}; x_{(1)}]$ (see [12]). Finally, let us remind the reader that the parameters b and c are unknown, hence, by plugging-in the point estimates of b and c, we can obtain the following approximate confidence intervals:

Golden–Alt [7]	$CI_{GA}(a) = [x_{(1)} - \hat{b}; x_{(1)}]$
Golden–Alt–Zanakis [14]	$CI_{GAZ}(a) = [x_{(1)} - \bar{b}; x_{(1)}]$
Golden–Alt–Wyckoff–Bain–Engelhardt [14]	$CI_{GAWBE}(a) = [x_{(1)} - \tilde{b}; x_{(1)}]$
Los–Lardinois [12]	$CI_{LL}(a) = [x_{(1)} - \hat{b}/[-n/\log(\alpha)]^{1/\hat{c}}; x_{(1)}]$
Los–Lardinois–Zanakis [14]	$CI_{LLZ}(a) = [x_{(1)} - \bar{b}/[-n/\log(\alpha)]^{1/\bar{c}}; x_{(1)}]$
Los-Lardinois–Wyckoff–Bain–Engelhardt [14]	$CI_{LLWBE}(a) = [x_{(1)} - \tilde{b}/[-n/\log(\alpha)]^{1/\tilde{c}}; x_{(1)}]$

3 Statistical Analysis of the Results of the λ-Optimal and λ-Optimal Greedy Algorithms; Concluding Remarks

Now, we focus on the statistical analysis of the approximate solutions obtained by applying the λ-optimal and λ-optimal greedy algorithms ($\lambda = 2, 3$) to four different instances of the TSP—with known optimal cost: Dantzig42 (49 cities [1]), Krolak (100 cities [10]), Random (100 cities [16]), Gr120 (120 cities [16]).

These approximate solutions were obtained by existing programs, written in *Mathematica* (demonstrations.wolfram.com/ComparingAlgorithmsForThe TravelingSalesmanProblem) but modified to control the initial tour. The input of every program is a random tour; and, to allow the comparison of results (for a given problem), any run with the same number starts with the same randomly generated tour. The output of each run is the minimum cost and the associated tour. Then after 100 runs of each algorithm, we get a data set of minimum costs and compute the ML estimates of (a, b, c), using the *NMaximize* routine in *Mathematica*, as well as the Zanakis and Wyckoff–Bain–Engelhardt (WBE) estimates. We also obtain approximate confidence intervals for the optimal cost, as described in the previous section, and used the program *Concorde TSP Solver* (www.tsp.gatech.edu/concorde/index.html) in order to obtain the optimal cost and to confront it with its point and interval estimates.

Finally, we computed the observed values of the Kolmogorov–Smirnov (K–S) goodness-of-fit test statistic and the associated p-values, obtained by using the *ks.test* routine of R; the three conjectured distributions are three-parameter Weibull distributions with (a, b, c) equal to ML, Zanakis and WBE estimates.

It is apparent from Fig. 1 that different algorithms lead to samples with different sample location, scale, and skewness. For instance, the approximate solutions of the 2-optimal algorithms have a wider range than the ones obtained with the 3-optimal algorithms; in addition, the 3-optimal algorithms are associated with minimum costs closer to the optimal one, as expected.

The point estimates in Table 1 certainly deserve some comments. For example, the more accurate the approximate algorithm is, the smaller are the estimates of the location and scale parameters, confirming what was already apparent in the histograms and suggesting that the 3-optimal algorithms tend to perform remarkably better than the 2-optimal ones. As far as the WBE estimate of a is concerned, it is, in general, the smallest one, hence one might think that it overly underestimates the optimal cost. However, when we compare it to the optimal solution, we see that this is not the case for the 2-optimal and 2-optimal greedy algorithms; in fact, for these algorithms it is the most reasonable estimate of the location parameter.

We ought to add that, for the Dantzig42 instance of the TSP: we could not compute the Zanakis estimate of the shape parameter for the 3-optimal greedy algorithm because, when $x_{(\lceil np_j \rceil)} = x_{(17)} = x_{(1)}$ and $\bar{a} = x_{(1)}$, the denominator of the third component of $(\bar{a}, \bar{b}, \bar{c})$ involves a fraction with a null denominator; we were unable to obtain the WBE estimates of the scale and shape parameters because we had a similar problem while computing $\tilde{c}_0$ and dealing with repeated observations.

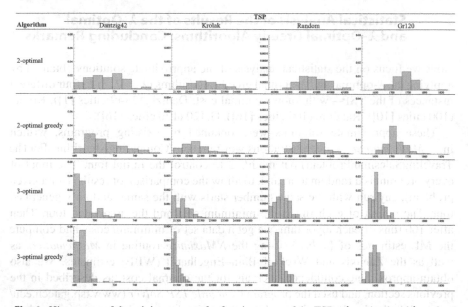

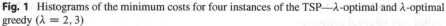

Fig. 1 Histograms of the minimum costs for four instances of the TSP—λ-optimal and λ-optimal greedy ($\lambda = 2, 3$)

Needless to say that the absence of these point estimates compromised the obtention of the corresponding interval estimates for a, as reported in Table 2.

Table 2 shows that the approximate confidence intervals are very accurate, and include the optimal cost in most cases. In fact, the three Golden–Alt approximate confidence intervals always contain the optimal cost a, as opposed to the three Los–Lardinois approximate 95 % confidence intervals. This can be explained by the approximate confidence level, $(1 - e^{-100}) \times 100\,\%$, which is very close to 100 % for the Golden-Alt confidence intervals; a possible way out is to choose a larger confidence level for the three Los–Lardinois confidence intervals. Table 2 also confirms the results obtained by Morais [14] and predicted by Golden and Alt [7]: when we replace the 2-optimal algorithms by the 3-optimal algorithms (respectively) there is a clear reduction of the relative ranges of the approximate confidence intervals for the optimal cost. This is essentially due to the fact that the range is an increasing function of the estimate of the scale parameter, which is smaller when we use the 3-optimal algorithms.

The results of the K–S goodness-of-fit test condensed in Table 3 confirm that the Weibull distribution is in fact reasonable for modelling the results of most instances of the TSP and algorithms: most p-values are larger than the usual significance levels, with the ML estimates being, in general, the ones that yield the best results.

This paper focuses on the statistical approach to the TSP, as described, for instance, by Golden [6] and other authors. As far as additions to previous statistical analyses of the TSP, we have: used greedy versions of the λ-optimal algorithm ($\lambda = 2, 3$) to obtain approximate costs; compared the approximate costs with the

Table 1 Estimates of the location, scale, and shape parameters for four instances of the TSP—listed in order corresponding to the 2-optimal, 2-optimal greedy, 3-optimal, and 3-optimal greedy algorithms

Estimates	TSP Dantzig42			Krolak			Random			Gr120		
	Location	Scale	Shape	Location	Scale	Shape	Location	Scale	Shape	Location	Scale	Shape
ML	671.545	44.419	2.472	21,385.322	1,553.030	2.245	40,294.739	2,781.148	2.598	1,636.970	111.754	3.470
	677.457	36.472	1.852	21,282.262	1,092.606	2.003	40,490.722	2,267.791	2.392	1,642.635	72.165	2.334
	679.202	0.332	0.159	21,282.565	450.916	1.176	39,299.312	1,379.657	2.851	1,627.257	34.150	1.842
	679.202	0.371	0.236	21,285.443	163.565	0.643	39,594.528	841.479	2.025	1,600.576	52.243	3.601
Zanakis	679.202	39.465	1.810	21,479.070	1,472.723	2.100	40,520.816	2,478.455	2.279	1,647.588	102.870	3.685
	679.202	37.012	1.755	21,349.383	980.603	2.005	40,694.053	2,077.385	2.156	1,652.319	61.819	2.626
	679.202	9.108	0.098	21,285.443	492.232	1.175	39,589.006	1,060.648	2.272	1,629.036	30.019	1.728
	679.202	9.108	–	21,285.443	176.490	0.908	39,632.591	803.010	2.092	1,610.719	40.982	3.064
WBE	676.496	37.621	2.122	21,327.217	1,625.231	2.363	40,214.590	2,869.768	2.731	1,615.914	132.984	4.864
	676.806	37.504	1.882	21,248.588	1,131.692	2.149	40,460.898	2,291.161	2.478	1,640.381	81.413	2.745
	679.202	–	–	21,276.805	438.391	1.282	39,446.252	1,206.764	2.435	1,626.910	34.966	1.811
	–	–	–	21,284.343	166.909	0.956	39,547.566	903.106	2.162	1,600.161	52.498	3.848
Minimum	679.202			21,485.534			40,545.272			1,649.850		
	679.202			21,349.445			40,700.098			1,652.669		
	679.202			21,285.443			39,589.006			1,629.047		
	679.202			21,285.443			39,634.711			1,610.750		
Optimal cost	679.202			21,285.443			39,486.840			1,596.347		

Table 2 Relative range of the six different confidence intervals and whether or not the optimal solution belongs to the confidence interval for four instances of the TSP—listed in order corresponding to the 2-optimal, 2-optimal greedy, 3-optimal, and 3-optimal greedy algorithms

Confidence intervals	TSP							
	Dantzig42		Krolak		Random		Gr120	
Golden–Alt	0.065399	Yes	0.072962	Yes	0.070432	Yes	0.070006	Yes
	0.053698	Yes	0.051331	Yes	0.057432	Yes	0.045206	Yes
	0.000489	Yes	0.021184	Yes	0.034940	Yes	0.021393	Yes
	0.000546	Yes	0.007684	Yes	0.021310	Yes	0.032727	Yes
Golden–Alt Zanakis	0.058104	Yes	0.069189	Yes	0.062767	Yes	0.064441	Yes
	0.054493	Yes	0.046069	Yes	0.052610	Yes	0.038725	Yes
	0.013410	Yes	0.023125	Yes	0.026861	Yes	0.018805	Yes
	0.013410	Yes	0.008292	Yes	0.020336	Yes	0.025672	Yes
Golden–Alt WBE	0.055390	Yes	0.076354	Yes	0.072677	Yes	0.083305	Yes
	0.055217	Yes	0.053167	Yes	0.058023	Yes	0.050999	Yes
	–	–	0.020596	Yes	0.030561	Yes	0.021904	Yes
	–	–	0.007841	Yes	0.022871	Yes	0.032887	Yes
Los–Lardinois	0.015824	Yes	0.015297	Yes	0.018258	No	0.025475	No
	0.008082	Yes	0.008907	Yes	0.013248	No	0.010057	No
	0.000000	Yes	0.001072	Yes	0.010210	Yes	0.003186	Yes
	0.000000	Yes	0.000033	Yes	0.003769	Yes	0.012354	Yes
Los–Lardinois Zanakis	0.008369	Yes	0.013022	Yes	0.013462	No	0.024874	No
	0.007386	Yes	0.008006	Yes	0.010335	No	0.010181	No
	–	–	0.001169	Yes	0.005737	Yes	0.002471	No
	–	–	0.000174	Yes	0.003803	Yes	0.008171	No
Los–Lardinois WBE	0.010605	Yes	0.017299	Yes	0.020117	No	0.040499	Yes
	0.008561	Yes	0.010391	Yes	0.014087	No	0.014211	No
	–	–	0.001335	Yes	0.007237	Yes	0.003156	No
	–	–	0.000200	Yes	0.004514	Yes	0.013215	Yes

optimal cost obtained by the Concorde TSP Solver; also considered more instances of the TSP with slightly more cities or with randomly generated coordinates. (Please note that we got similar results to the ones described in [14], for the 2- and 3-optimal algorithms applied to the Krolak instance of the TSP.)

A possibility of further work, that certainly deserves some consideration, is to confront the point estimates we obtained with the ones suggested by Hall and Wang [8]. Another one is to investigate the benefits of using the 4-optimal algorithms now that the computational power has largely increased and to apply the statistical approach to very large instances of the TSP, which cannot be solved by the Concorde TSP Solver in a reasonable amount of time.

We strongly hope that this paper gives a stimulus and contributes to fill the gap between the performance analysis of approximate algorithms and Statistics.

Table 3 Results of the K–S goodness-of-fit test for four instances of the TSP—listed in order corresponding to the 2-optimal, 2-optimal greedy, 3-optimal, and 3-optimal greedy algorithms.

	TSP						
	Dantzig42		Krolak		Random		Gr120	
Conjectured distribution	Obs. value	p-value	Obs. value	p-value	Obs. value	p-value	Obs. value	p-value
Weibull–ML	0.0814	0.5222	0.0496	0.9663	0.0396	0.9976	0.0784	0.5698
	0.0623	0.8318	0.0765	0.6020	0.0394	0.9978	0.1533	0.0182
	0.4375	0.0000	0.1263	0.0824	0.0508	0.9584	0.0750	0.6276
	0.5306	0.0000	0.2596	0.0000	0.0797	0.5487	0.0639	0.8091
Weibull–Zanakis	0.1021	0.2477	0.0565	0.9068	0.0558	0.915	0.0685	0.7357
	0.0958	0.3178	0.0971	0.3023	0.0447	0.9884	0.0935	0.3469
	0.3700	0.0000	0.1092	0.1837	0.0660	0.7760	0.1023	0.2463
	–	–	0.1831	0.0024	0.0951	0.3267	0.0691	0.7267
Weibull–WBE	0.1179	0.1242	0.0581	0.8889	0.0450	0.9875	0.0645	0.7997
	0.0655	0.7850	0.0678	0.7473	0.0400	0.9972	0.1014	0.2556
	–	–	0.1404	0.0388	0.0720	0.6785	0.0779	0.5789
	–	–	0.1931	0.0012	0.0917	0.3693	0.0531	0.9404

Acknowledgments The first author would like to thank Fundação Calouste Gulbenkian for the opportunity to study this topic within the scope of the program "Novos Talentos em Matemática." This work received financial support from Portuguese National Funds through FCT (*Fundação para a Ciência e a Tecnologia*) within the scope of project PEst-OE/MAT/UI0822/2011. The authors are grateful to the referees for their valuable suggestions and comments.

References

1. Dantzig, G., Fulkerson, D., Johnson, S.: Solution of a large-scale traveling salesman problem. J. Oper. Res. Am. **2**, 393–410 (1954)
2. Engelhardt, M., Bain, L.: Simplified statistical procedures for the Weibull or extreme-value distributions. Technometrics **19**, 323–331 (1977)
3. Fisher, R., Tippett, L.: Limiting forms of the frequency distribution of the largest or smallest member of a sample. Math. Proc. Camb. Philos. Soc. **24**, 180–190 (1928)
4. Frieze, A., Galbioti, G., Maffioli, F.: On the worst-case performance of some algorithms for the asymmetric traveling salesman problem. Netw. **12**, 23–39 (1982)
5. Gnedenko, B.: Sur la distribution limite du terme maximum d'une série aléatoire. Ann. Math. **44**, 607–620 (1943)
6. Golden, B.L.: A statistical approach to the TSP. Netw. **7**, 209–225 (1977)
7. Golden, B.L., Alt, F.: Interval estimation of a global optimum for large combinatorial problems. Nav. Res. Logist. Q. **26**, 69–77 (1979)
8. Hall, P., Wang, J.Z.: Estimating the end-point of a probability distribution using minimum-distance methods. Bernoulli **5**, 177–189 (1987)
9. Hoffman, A.J., Wolfe, P.: History. In: Lawler, E., Lenstra, J., Rinnooy Kan, A., Shmoys, D. (eds.) The Traveling Salesman Problem: A Guide Tour of Combinatorial Optimization, pp. 1–15. Wiley, Chichester (1985)
10. Krolak, P., Felts, W., Marble, G.: A man-machine approach toward solving the traveling salesman problem. Commun. ACM **14**, 327–334 (1971)

11. Lin, S.: Computer solutions of the traveling salesman problem. Bell Syst. Tech. J. **44**, 2245–2269 (1965)
12. Los, M., Lardinois, C.: Combinatorial programming, statistical optimization and the optimal transportation problem. Transp. Res. B **16**, 89–124 (1982)
13. McRoberts, K.: Optimization of facility layout. Ph.D. thesis, Iowa State, University of Science and Technology, Ames (1966)
14. Morais, M.: Problema do Caixeiro Viajante: Uma Abordagem Estatística. Report within the scope of the Masters in Applied Mathematics, Instituto Superior Técnico, Technical University of Lisbon, Portugal (1995)
15. Rockette, H., Antle, C., Klimko, L.: Maximum likelihood estimation with the Weibull model. J. Am. Stat. Assoc. **69**, 246–249 (1974)
16. Salvador, T.: The traveling salesman problem: a statistical approach. Report within the scope of the program "Novos Talentos em Matemática" — Fundação Calouste Gulbenkian, Portugal (2010)
17. Vasko, F.J., Wilson, G.R.: An efficient heuristic for large set covering problems. Nav. Res. Logist. Q. **31**, 163–171 (1984)
18. Wikipedia: http://en.wikipedia.org/wiki/Travelling_salesman_problem. Cited 1999.
19. Wyckoff, J., Bain, L., Engelhardt, M.: Some complete and censored sampling results for the three-parameter Weibull distribution. J. Stat. Comput. Sim. **11**, 139–151 (1980)
20. Zanakis, S.: A simulation study of some simple estimators for the three parameter Weibull distribution. J. Stat. Comput. Sim. **9**, 101–116 (1979)

Brugada Syndrome Diagnosis: Three Approaches to Combining Diagnostic Markers

Carla Henriques, Ana Cristina Matos, and Luís Ferreira dos Santos

Abstract

Brugada syndrome (BS) is an inherited cardiopathy that predisposes individuals without structural heart disease to sudden cardiac death. The diagnosis is performed by detecting a typical pattern in the electrocardiogram (ECG), called Type 1 Brugada pattern, but this is not always visible, so the diagnosis is not straightforward. In this study, we investigated other ECG markers, independent of the typical pattern, which exhibited a good ability to differentiate the carriers and the non-carriers of the genetic mutation responsible for this disease. The combination of these markers through linear models has led to enhancing the ability of each marker to discriminate between the two groups. We found linear combinations of these markers for which the area under the ROC curve (AUC) was greater than 0.9, which suggests an excellent ability to discriminate between the two groups. This study points towards good alternatives for diagnosing BS which may prevent searching for the Type 1 Brugada pattern in an ECG, but these alternatives should be investigated with a larger database in order to produce a good effective predictive model.

C. Henriques (✉)
CMUC and Escola Sup. Tecnologia e Gestão, Inst. Polit. de Viseu, Campus Politécnico de Repeses, 3504-510 Viseu, Portugal
e-mail: carlahenriq@estv.ipv.pt

A.C. Matos
Escola Sup. Tecnologia e Gestão, Inst. Polit. de Viseu, Campus Politécnico de Repeses, 3504-510 Viseu, Portugal
e-mail: amatos@estv.ipv.pt

L.F. dos Santos
Serviço de Cardiologia, Tondela-Viseu Hospital Center, Viseu, Portugal
e-mail: luisferreirasantos@gmail.com

A. Pacheco et al. (eds.), *New Advances in Statistical Modeling and Applications*, Studies in Theoretical and Applied Statistics, DOI 10.1007/978-3-319-05323-3_20, © Springer International Publishing Switzerland 2014

1   Introduction

Brugada syndrome (BS) is a disease characterized by a dysfunction of cardiac sodium channels that results from a genetic mutation, in most cases inherited from one parent, and which predisposes to malign cardiac arrhythmias and sudden cardiac death (SCD). It is estimated that this disease is responsible for at least 20 % of the cases of sudden cardiac death in individuals with normal hearts and at least 4 % of all cases of SCD [5], which clearly explains the growing scientific interest around it. In fact, it is a very recent clinical entity, since it was first documented in 1992 by Brugada and Brugada [1], but the number of publications covering this disorder has grown considerably in recent years. Individuals carrying the genetic mutation may never have any symptoms, however, SCD can be the first symptom of the disease and, therefore, it is a kind of "threat," sometimes "a silent threat," in relatives of carriers of the disease. The only way to eliminate the threat of sudden death is the implantation of an implantable cardioverter defibrillator, but this is very expensive, very uncomfortable for the patient, and can lead to complications, so it is recommended only in high risk patients. The diagnosis is usually performed detecting a specific pattern in an electrocardiogram (ECG), called Type 1 pattern or Brugada ECG, in combination with other easily identifiable clinical criteria. A major difficulty associated with the diagnosis is to detect the Type 1 pattern in an ECG, as this is often intermittent (affected individuals have intermittently normal ECGs and Brugada ECGs). Genetic tests are not an adequate solution, because they are very expensive and it is sometimes difficult to detect the genetic mutation associated with the disease. This work falls within this context. Based on the records of 113 members of two Portuguese families, with 42 carriers of the genetic mutation (identified by genetic tests), the electrocardiographic markers were investigated to find some that could discriminate between mutation carriers and non-carriers, in order to make a diagnosis without having to resort to Brugada pattern detection. Our records were obtained from a proband with a Type 1 Brugada pattern ECG, making the investigation of all family members mandatory (for more details see Santos et al. [5]). Through the use of receiver operating characteristic (ROC) curves and other statistical techniques, we have identified five ECG markers with a good ability to discriminate between mutation carriers (C) and non-carriers (NC). The natural question that followed was how to combine these markers in order to increase the discriminative ability of each one individually. Three approaches were explored: discriminant analysis, a distribution-free approach proposed by Pepe and Thompson [4], and logistic regression. As a result of this effort, in this study we envisage ways of combining these markers, which, in this data set, proved to be more efficient than the detection of a Brugada pattern. The SPSS, version 19, and R packages were used for the statistical analysis.

Table 1 Area under the ROC curve (AUC) and 95 % confidence intervals for AUC, for each of the five markers

	PR	QRSf	LAS	RMS40	dQT
AUC ($p < 0.001$)	0.86	0.83	0.77	0.77	0.72
95 % IC	(0.79, 0.93)	(0.73, 0.93)	(0.66, 0.88)	(0.65, 0.88)	(0.62, 0.82)

2 ECG Markers to Identify Mutation Carriers

All the known ECG parameters of depolarization and repolarization were measured and five were identified as being good discriminators between carriers and non-carriers of the genetic mutation: P-wave duration (PR), transmural dispersion of repolarization (dQT) between V1 and V3, filtered QRS duration (fQRS), where QRS stands for the combination of three of the graphical deflections seen on a typical ECG (typically an ECG has five deflections, arbitrarily named P, Q, R, S, and T waves), low-amplitude signal duration (LAS), and root-mean-square of the voltage in the last 40 ms of the fQRS (RMS40), the last three taken in a signal average ECG. To achieve these ECG markers the experience of the physician was conjugated with statistical tests to access the association of markers with the presence of the mutation.

Unfortunately, not all of the 113 records had information about those five markers, so, for the multivariate analysis, only 64 were considered, from which 37 were carriers. Some of the missing data relates to family members under 16 years old, and we know that the ECG manifestation of the disease is rare in pediatric age. We are involved in the construction of a national Brugada registry and we hope in a near future to have more data to present, but we have to realize that Brugada is a very rare diagnostic.

Univariate analysis for the five markers revealed that PR and QRSf were the most strongly associated with the genetic mutation. In fact, significant differences between C and NC were found in the five markers ($p < 0.0005$ both on T-test and Mann–Whitney test), but the highest values of eta-squared were obtained for PR and QRSf (0.4 and 0.3, respectively, while for the other three the values were approximately 0.2). ROC curves indicate good ability of each one of the five markers to identify mutation carriers, but again PR and QRSf appear to be the best ones (cf. Table 1 and Fig. 1).

Pearson correlation coefficients among the five markers are displayed in Table 2. The relatively high correlations between LAS and RMS40 and between these two and QRSf suggest that some of these markers may be dispensable in a multivariate model for the identification of mutation carriers.

Fig. 1 ROC curves for the five markers

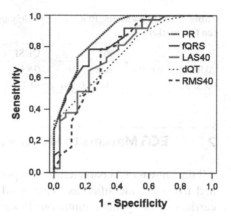

Table 2 Pearson correlation coefficients among the five markers

	QRSf	LAS	RMS40	dQT
PR	0.36 (n=64)	0.41 (n=64)	−0.3 (n=64)	0.2 (n=102)
QRSf		0.76 (n=71)	−0.71 (n=71)	0.37 (n=64)
LAS			−0.78 (n=71)	0.25 (n=64)
RMS40				0.25 (n=64)

3   Combining the Markers: Multivariate Analysis

The question which naturally followed the univariate and bivariate analysis was to find a simple way of combining the markers, that could be implemented easily as a diagnostic tool, aiming to enhance the ability to discriminate between carriers and non-carriers. In this work, several linear models were constructed and compared in terms of the AUC, since this is a very valuable tool to measure accuracy of a diagnostic test and is widely used in medical research. The simplest solution was to add PR and QRSf, the two markers that seemed to best separate the two groups. In fact, this simple solution was quite effective, as it yielded an ROC curve with AUC $= 0.927$ ($p < 0.001$). Later, we will present the ROC curve for this sum comparing it with others. To investigate other solutions of combining these markers, we resorted to three approaches: discriminant analysis, whose results are illustrated in the next subsection, a distribution-free methodology proposed by Pepe and Thompson [4], described in Sect. 3.2, and logistic regression analysis, whose results are discussed in Sect. 3.3. In each subsection, we compare some of the solutions given by these approaches, seeking effective guidelines toward a combination of this markers serving as a means of diagnosis.

Fig. 2 ROC curves for Y_{DA1} and $PR + QRSf$

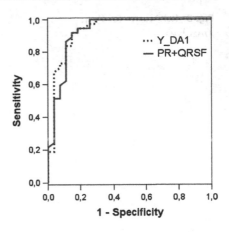

3.1   Discriminant Analysis

It is known that if the markers follow a multivariate normal distribution in both groups, C and NC, then the linear combination that optimizes AUC is the linear discriminant function [6]. However, the normality assumption is not supported by our data (for example, the Shapiro–Wilk test will give a p-value of 0.008 for the NC group when testing the normality of PR); consequently, while discriminant analysis will give good solutions, they will not be the best solutions when compared with other methodologies, as we will see in the following sections.

Applying discriminant analysis, first using solely PR and QRSf (which have emerged as more discriminative), then using a stepwise procedure and finally all five markers, the linear combinations obtained were, respectively:

$$Y_{DA1} = -11.62 + 0.031PR + 0.063\,QRSf\,,$$

$$Y_{DA2} = -10.877 + 0.029PR + 0.051QRSf + 0.021dQT\,,$$

$$Y_{DA3} = -7.664 + 0.031PR + 0.041QRSf - 0.043LAS - 0.04RMS + 0.022dQT\,.$$

The AUCs for the ROC curves associated with the above combinations are 0.935, 0.956, and 0.969, respectively ($p < 0.001$ for all three) each one higher than the AUC obtained just adding PR with QRSf. Figure 2 displays the ROC curve obtained for Y_{DA1} comparing it with the one yielded for the sum of PR with QRSf. On the right, Fig. 3 exhibits the ROC curves for the above three solutions of the discriminant analysis. All ROC curves exhibit an excellent ability of the underlying combinations to discriminate between C and NC, but Y_{DA2} and Y_{DA3} seem to perform better.

Fig. 3 ROC curves for Y_{DA1},
Y_{DA2}, Y_{DA3}

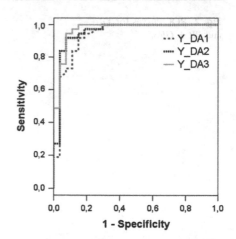

3.2   Distribution-Free Approach

Pepe and Thompson [4] proposed a distribution-free approach to find a linear combination of several diagnostic markers that optimizes the AUC. For two diagnostic markers, Y_1 and Y_2, given the fact that the ROC curve is invariant for monotone transformations, this problem is reduced to looking for the value α that maximizes the AUC of $Y_1 + \alpha Y_2$. To implement this on our data, we followed the strategy proposed in paper [4], which consists of evaluating the AUC for each combination $Y_1 + \eta Y_2$, with η ranging from -1 to 1 with increments of 0.01, and then do the same for $\eta Y_1 + Y_2$, which as the same AUC as $Y_1 + \frac{1}{\eta} Y_2$.

The best linear combination of PR and QRSf was $Y_{DF1} = 0.46 PR + QRSf$ and yielded an AUC=0.937 ($p < 0.01$), which seems better than the corresponding one of the discriminant analysis.

For more than $p > 2$ markers, this last strategy requires a search in the $p - 1$ dimensional space, which is computationally demanding. Pepe and Thompson [4] suggest a stepwise approach, searching first for the best linear combination of two markers (implementing the previous strategy), then searching for the marker that yields the best AUC when put in optimal linear combination with the linear combination found in the previous step, and so on, until all markers are investigated. Of course this strategy does not necessarily yield the optimal solution.

Implementing this approach on our data, we obtained precisely the last linear combination Y_{DF1} in the first step. So, as we have envisaged after the univariate statistical analysis, PR and QRSf seem to distinguish themselves from the other markers as having a greater potential to identify mutation carriers. In the second step, dQT entered in the combination yielding $Y_{DF2} = 0.46 PR + QRSf + 0.57 dQT$, with AUC=0.961, a greater AUC than the corresponding solution from discriminant analysis Y_{DA2}. In the third and final step, LAS enters into the linear combination giving $Y_{DF2} = 0.46 PR + QRSf + 0.57 dQT + 0.11 LAS$, with AUC=0.963. To compare the ROC curves of this three solutions we have plotted them in Fig. 4. As we can see,

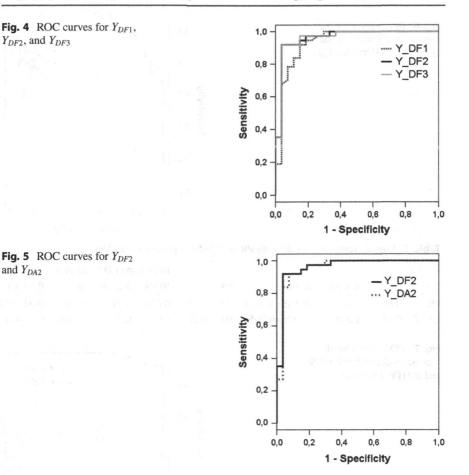

Fig. 4 ROC curves for Y_{DF1}, Y_{DF2}, and Y_{DF3}

Fig. 5 ROC curves for Y_{DF2} and Y_{DA2}

the difference between the solutions Y_{DF2} and Y_{DF3} is residual, so it seems that the inclusion of the LAS in the last step did not augment considerably the discriminative value of the combination. We have already noted that the AUC for Y_{DF2} is greater than the corresponding one obtained with the discriminant analysis. Comparing the ROC curves, now, we come to the same conclusion (cf. Fig. 5).

3.3   Logistic Regression

We first estimated a logistic regression model with covariates PR and QRSf, obtaining $Ln(odds) = -23.642 + 0.066PR + 0.13QRSF$, with AUC=0.936 ($p < 0.001$). Comparing this model with the Y_{DF1} (AUC=0.937) from the last section, we conclude that they are very similar (cf. Fig. 6). We also note that applying Bootstrap, we obtain similar p-values for the Wald test and similar confidence intervals for model coefficients, which supports the stability of the model against variation in the data set (see Table 3).

Fig. 6 ROC curves for the logistic model with PR and QRSf (Y_LR1) and for Y_{DF1}

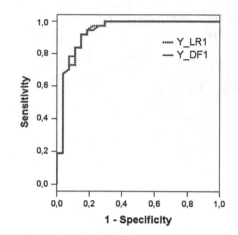

Table 3 Logistic regression model with PR and QRSf as predictor variables

	Coeff.	Std. err.	Wald test	95 % CI	Bootstrap (1,000 samples)			
					BIAS	Std. err.	p	95 % CI
PR	0.066	0.02	10.8 (p=0.001)	0.03–0.11	0.034	0.27	0.001	0.04–0.24
QRSf	0.13	0.046	7.9 (p=0.005)	0.04–0.22	0.06	0.63	0.006	0.05–0.41

Fig. 7 ROC curves for the logistic model with PR, QRSf and dQT(Y_LR2) and Y_{DF2}

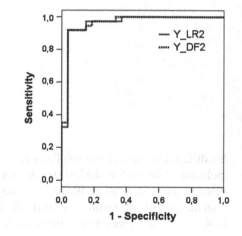

Applying a stepwise procedure for selecting or deleting variables, we end with a logistic model that includes PR, QRSf, and dQT. This is in accordance with the results of the previous subsections, where the third marker to come into the linear combination was dQT. The model obtained, $Ln(odds) = -26.459 + 0.069PR + 0.13QRSF + 0.068dQT$, is again very much similar to the one found in the distribution-free approach of the last subsection, in terms of discriminative ability, since the estimated AUC is 0.96 against the 0.961 of Y_{DF2}. Also, the ROC curves are almost indistinguishable (see Fig. 7). Still, our sample is too small to estimate a logistic model with more than two covariates with reasonable precision (Peduzzi et al. [3] suggest that the number of events per variable should be at least ten).

In fact, for the models with more than two covariates, bootstrap standard errors for the regression coefficients are quite big and bootstrap confidence intervals are very large. Nevertheless, we estimated the model with the five covariates and obtained

$$Ln(odds) = -22.1 + 0.127PR + .172QRSf - 0.373LAS - 0.252RMS40 + 0.109dQT,$$

with AUC=0.973, the best one obtained among all linear combinations studied.

4   Conclusions

All linear combinations studied lead to ROC curves with AUC> 0.9, which can be interpreted as a remarkable ability to discriminate between mutation carriers and non-carriers [2]. It seems clear that combining PR, QRSf, and dQT we obtain a quite effective model in discriminating between the two groups, but the gain of including also the information of LAS and RMS40 was not established. We emphasize that each of the solutions is more efficient in identifying mutation carriers than detecting a Brugada pattern in an ECG. As an example, we refer to Santos et al. [5], where 122 individuals were screened with several ECGs. For a single ECG, the authors report a sensitivity of 12.5 % and a specificity of 100 % and with several ECGs, performed during the follow-up, the sensitivity raised to 30 % maintaining the specificity (see Table 1 of Santos el al. [5]). From the ROC curves of the linear combinations studied in this paper, we can identify cutoff values with better sensitivity and/or specificity with only one ECG. For example, combining only PR with QRSf, we can obtain a sensitivity of 67.7 % and a specificity of 96.3 %. But combining PR, QRSf, and dQT, we can raise the sensitivity to 92 % maintaining the specificity. Combining the information of all five markers, cuttoff values may be chosen to have 100 % specificity and 60 % sensitivity. To sum up, the combination of these markers has a great potential for the diagnosis of Brugada Syndrome, but to be able to produce a good prediction model it is necessary to work with more data.

References

1. Brugada, P., Brugada, J.: Right bundle branch block, persistent ST segment elevation and sudden cardiac death: a distinct clinical and electrocardiographic syndrome. A multicenter report. J. Am. Coll. Cardiol. **20**, 1391–1396 (1992)
2. Hosmer, D.W., Lemeshow, S.: Applied Logistic Regression, 2nd edn. Wiley, New York (2000)
3. Peduzzi, P, Concato, J., Kemper E., Holford, T.R., Feinstein, A.R.: A simulation study of the number of events per variable in logistic regression analysis. J. Clin. Epidemiol. **49**, 1372–1379 (1996)
4. Pepe, M.S., Thompson, M.L.: Combining diagnostic test results to increase accuracy. Biostatistics **1**(2), 123–140 (2000)
5. Santos, L.F., Rodrigues, B., Moreira, D., Correia, E., Nunes, L., Costa, A., Elvas, L., Pereira, T., Machado, J.C., Castedo, S., Henriques, C., Matos, A., Santos, O.: Criteria to predict carriers of a novel SCN5A mutation in a large Portuguese family affected by the Brugada syndrome. Europace **14**(6), 882–888 (2012)
6. Su, J.Q., Liu, J.S.: Linear combinations of multiple diagnostic markers. J. Am. Stat. Assoc. **88**, 1350–1355 (1993)

Hierarchical Normal Mixture Model to Analyse HIV/AIDS LOS

Sara Simões Dias, Valeska Andreozzi, and Maria Oliveira Martins

Abstract

Inpatient length of stay (LOS) is an important measure of hospital activity and is often considered as a proxy of hospital resource consumption. In Portugal, hospitalizations related to HIV infection are some of the most expensive and the second major diagnosis category with greatest average LOS. This paper investigates factors associated with HIV/AIDS LOS. A hierarchical finite normal mixture model was fitted to the logarithm of LOS, to account for the inherent correlation of patients clustered in hospitals. We found that the model with two components had the best fit. In addition associated risk factors were identified for each component and the random effects make possible a comparison of relative efficiencies among hospitals.

1 Introduction

Comprehensive and accurate information about length of stay (LOS) is crucial to health planners and administrators, in order to develop strategic planning and to effectively deploy financial, human, and physical hospital resources [9]. LOS is

S.S. Dias (✉)
Departamento Universitário de Saúde Pública, FCM-UNL, Campo dos Mártires da Pátria, no. 130, Lisboa, Portugal
e-mail: sara.dias@fcm.unl.pt

V. Andreozzi
Centro de Estatística e Aplicações da Universidade de Lisboa, FCUL, Bloco C6 - Piso 4, Campo Grande, Lisboa, Portugal
e-mail: valeska.andreozzi@fc.ul.pt

M.O. Martins
Unidade de Parasitologia e Microbiologia Médicas, IHMT-UNL, Rua da Junqueira no. 100, Lisboa, Portugal
e-mail: mrfom@ihmt.unl.pt

A. Pacheco et al. (eds.), *New Advances in Statistical Modeling and Applications*, Studies in Theoretical and Applied Statistics, DOI 10.1007/978-3-319-05323-3_21, © Springer International Publishing Switzerland 2014

an indirect estimator of resources consumption and also an indicator of efficiency in the use of beds [1]. In Portuguese public hospitals the case-mix scheme is the one proposed by Fetter et al. [6]. The Diagnosis Related Group (DRG) provides a classification system of hospitalized patients with clinically recognized definitions. The DRG determines the payment allocated to the hospital and is based on the characteristics of patients consuming similar quantities of resources, as a result of a process of a similar care. The main assumption in funding is that patients with very long inpatient LOS have different resource consumption patterns from those assumed to have an usual LOS. For these reasons, it is crucial to understand and model the distribution of LOS.

Skewness and heterogeneity of LOS represents a challenge for statistical analysis [9]. In recent literature, several approaches have been adopted to analyse LOS, such as survival models [3, 14], frequentist and Bayesian frameworks [15], and latent class models [16, 18]. However, none of these approaches recognized that hospitalizations from the same hospital are more likely to be related. Neglecting the dependence of clustered (multilevel) data may result in spurious associations and misleading inferences. Some authors have attempted to accommodate this risk [11, 12], although not addressing the issues of skewness or the heterogeneity of LOS. Recognizing that there may be subgroups of patients regarding LOS and the multilevel structure of the DRG data, we propose a hierarchical modelling approach to overcome the challenge derived from these two features. The model includes variables at both levels (patients and hospitals) which allows us to estimate differences in outcome that are not fully explained by observed patient or other specific and known conditions.

2   Hierarchical Finite Mixture Model

It is very common that LOS data are skewed and contain atypical observations. For this reason a mixture distribution analysis will be performed in order to determine if the distribution is composed of several components or not. Data are initially investigated without restriction in number, size, and location of possible components to approximate the empirical distribution. This analysis results in models consisting of a finite number of components, which are then compared with each other. The suitable number of components is chosen via Akaike (AIC) and Bayesian (BIC) information criteria calculated for distributions with different numbers of components.

After the analysis of the LOS distribution, and taking into account two types of heterogeneity (within patients and among hospitals), a hierarchical finite mixture model will be fitted.

Consider y_{ij} $(i = 1, \ldots, m; j = 1, \ldots, n_i)$ the logarithm of LOS for the jth patient in the ith hospital, where m is the number of hospitals, n_i is the number of patients within hospital i resulting in $N = \sum_{i=1}^{m} n_i$ total patients. A finite mixture model [5, 13] for the probability density function of y_{ij} takes the form:

$$f(y_{ij}|x_{k,ij}, \Theta) = \sum_{k=1}^{K} \pi_k f_k(y_{ij}|x_{k,ij}, \beta_k, \phi_{ki}), \tag{1}$$

where π_k denotes the proportion of patients belonging to the kth component and $\sum_{k=1}^{K} \pi_k = 1$, K is the number of components in the mixture; f_k describes the normal distribution of the kth component with mean expressed by

$$\mu_{k,ij} = x_{k,ij}^T \beta_k + \phi_{ki} \tag{2}$$

where $x_{k,ij}$ is the vector of covariates for the kth component, β_k is the vector of linear effect regression parameters that can be different for the K components; ϕ_{ki} are the random effect parameters that capture the hospital heterogeneity in each component through the specification of a Gaussian density function with zero mean and variance–covariance equal to $\sigma_{\phi_k}^2 I_m$ (I_m denotes an $m \times m$ identity matrix).

The vector Θ, which contains all the unknown parameters in model 1, is estimated by the maximum likelihood approach through the application of EM algorithm [2]. Once this has been estimated, estimates of the posterior probabilities of population membership can be obtained by:

$$\hat{\tau}_k(y_{ij}; \hat{\Theta}) = \frac{\hat{\pi}_k f_k(y_{ij}|x_{ij}; \hat{\beta}_k, \hat{\phi}_{ki})}{\sum_{k=1}^{K} \hat{\pi}_k f_k(y_{ij}|x_{ij}; \hat{\beta}_k, \hat{\phi}_{ki})}. \tag{3}$$

In order to form no overlapping components (clusters), each y_{ij} will be assigned to the population that has the highest estimate posterior probability. All of the statistical analyses were performed using the statistical software R [17] and its package Flexmix [8, 10].

3   Application to HIV/AIDS LOS

This study is based on HIV/AIDS DRG 714 (infection with human immunodeficiency virus, with significant diagnosis related), and DRG 710 (infection with human immunodeficiency virus, with multiple diagnoses major or significant, without tuberculosis), occurred in 2008, which represent the DRG with more hospitalizations amongst the 17 created for HIV/AIDS patients.

3.1   Data

The DRG database for 2008 were provided by the Central Health System Administration (ACSS) and hospital characteristics were obtained from the Portuguese National Institute of Statistics.

All the hospitalizations meeting the following criteria were analysed: patients aged 18 years or older; geo-referenced episodes, i.e., hospitalization with patient

Table 1 Normal mixture distribution analysis for HIV/AIDS LOS DRG (without covariates)

DRG			One component		Two components		Three components	
714	μ_1	π_1	2.191	1	2.413	0.523	NC	NC
	μ_2	π_2			1.965	0.477	NC	NC
	μ_3	π_3					NC	NC
	AIC		2,852.132		2,820.130		NC	
	BIC		2,862.084		2,845.012		NC	
	Log-likelihood		−1,424.066		−1,405.065		NC	
710	μ_1	π_1	2.745	1	2.583	0.463	2.810	0.424
	μ_2	π_2			2.886	0.537	5.074	0.318
	μ_3	π_3					0.436	0.259
	AIC		1,781.059		1,761.034		1,764.463	
	BIC		1,790.061		1,783.541		1,800.473	
	Log-likelihood		−888.529		−875.517		−874.231	

NC non-convergence

residence ward known; hospitalizations from hospitals with more than ten discharge episodes and transfers to another hospital were eliminated (to avoid including the inpatient episode twice, as the cause of the transfer was often lack of procedure facilities). The above selection criteria resulted in 1,071 hospitalizations of DRG 714 that took place in 23 hospitals and a total of 637 hospitalizations of DRG 710 in 18 hospitals.

The outcome variable was the logarithm of the number of days between the hospital admission and discharge dates. Patient's demographic characteristics (age, gender, death, and Euclidian distance to the hospital), health relevant factors (urgent admission, number of secondary diagnoses, number of procedures, AIDS as principal diagnosis, and presence of pneumonia) and hospital characteristics (central hospitals, hospital offering more differentiated services) were considered in the analysis.

3.2   Results

Table 1 lists the results from fitting $k = 1, 2$, and 3 components mixture distribution for the logarithm of LOS. Figure 1 shows the empirical distribution of the logarithm of LOS and fitted two-normal mixture distributions for the two DRGs.

Taking into account the covariates, and based on AIC and BIC, the model with two components presents a better fit.

In Tables 2 and 3, we may observe some differences between covariate effects for short-stay and long-stay latent subgroups. Although gender and age are not statistically significant at the 10 % level, in both models they are kept in order to control for possible confounding.

The estimated proportion for the short-stay subgroups is smaller in both DRGs than the estimated proportion of long-stay subgroups.

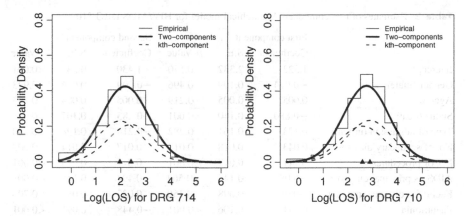

Fig. 1 Empirical distribution of HIV/AIDS LOS and fitted two-component normal mixture model

Table 2 Estimates of two-component hierarchical model for HIV/AIDS DRG 714

	First component			Second component		
	Coefficient	S.E.	p-Value	Coefficient	S.E.	p-Value
Intercept	0.341	0.322	0.290	1.159	0.190	<0.001
Gender (male)	0.036	0.100	0.722	−0.108	0.060	0.071
Age/10	0.057	0.039	0.152	0.025	0.026	0.329
Status (death)	−0.257	0.175	0.143	−0.530	0.094	<0.001
Type of admission (urgent)	0.513	0.134	<0.001	−0.188	0.084	0.026
No. of secondary diagnoses	0.083	0.023	<0.001	0.021	0.010	0.039
No. of procedures	0.131	0.019	<0.001	0.129	0.008	<0.001
AIDS as principal diagnosis	0.290	0.123	0.019	0.137	0.071	0.056
Euclidean distance (median)	−0.089	0.092	0.330	0.153	0.053	0.004
Pneumonia	−0.117	0.111	0.291	0.026	0.062	0.671
Central hospital	−0.156	0.267	0.559	−0.250	0.128	0.051
Random effect (ϕ_{ki}) variance	0.127			0.027		
Mixture proportion (π_k)	0.44			0.56		
AIC	2,482.527					
BIC	2,616.889					

We also plot the hospital random effects in order to observe discrepancies in medical expertise, health care, and other unmeasurable characteristics of the patients hospitalized in different hospitals. Figure 2 shows these effects and their, respectively, 95 % confidence interval (CI) for both components and DRGs.

In 2008, there were 23 hospitals with more than 10 discharges of DRG 714, and there were 18 hospitals (the same as DRG 714) with more than 10 discharges of DRG 710. The hospitals have different effects for the two components. For DRG 714, the random effects of both components of hospital 1 are statistically significant below zero; for DRG 710, hospital 1, 2, and 3 have the random effects of both components statistically significant, and these three hospitals can be considered to

Table 3 Estimates of two-component hierarchical model for HIV/AIDS DRG 710

	First component			Second component		
	Coefficient	S.E.	p-Value	Coefficient	S.E.	p-Value
Intercept	1.322	2.332	0.570	1.430	0.333	<0.001
Gender (male)	−0.074	0.109	0.496	−0.009	0.096	0.923
Age/10	0.005	0.005	0.219	0.005	0.004	0.183
Status (death)	−0.880	0.140	<0.001	−0.285	0.107	0.008
Type of admission (urgent)	0.173	0.161	0.282	−0.214	0.198	0.280
No. of secondary diagnoses	0.043	0.018	0.018	0.017	0.014	0.003
No. of procedures	0.141	0.017	<0.001	0.014	0.013	<0.001
AIDS as principal diagnosis	−0.083	0.144	0.562	0.494	0.146	<0.001
Euclidean distance (median)	−0.089	0.098	0.150	0.032	0.082	0.700
Pneumonia	0.141	0.106	<0.001	−0.448	0.091	<0.001
Central hospital	−0.325	2.328	0.889	−0.573	0.216	0.008
Random effect (ϕ_{ki}) variance	0.225			0.103		
Mixture proportion (π_k)	0.29			0.71		
AIC	1,586.613					
BIC	1,706.945					

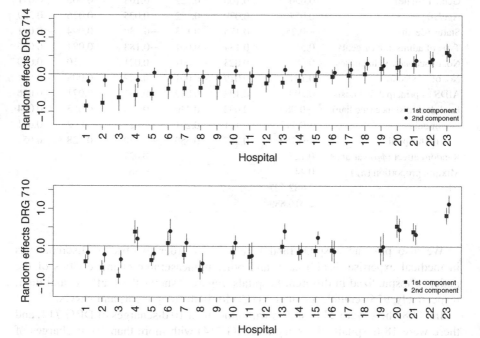

Fig. 2 Random effects and their 95 % CI for each hospital

be the most efficient. In contrast, three hospitals for DRG 714 (21, 22, and 23) and two hospitals for DRG 710 (21 and 23) exhibited large positive effects prolonging hospitalizations.

4   Discussion

The application of finite mixture distribution analysis emphasizes the heterogeneity of LOS DRGs by locating two latent subgroups. The information of the relevant covariates is important for analysing the LOS similarities and dissimilarities between the short-stay and long-stay latent subgroups.

Patient's gender and age are not statistically significant variables at the 10 % level in explaining HIV/AIDS LOS. For both DRGs, deaths are associated with the early days of the hospitalization, meaning that there is high mortality among those patients who arrive at hospital in more severe and advanced states of AIDS-related illness. On the other hand, the number of secondary diagnoses and the number of procedures have a positive value, indicating a long-duration hospitalization, as presumed. A great number of diagnoses or procedures usually indicate a more severe condition of the patient and consequently, a delayed discharge [16, 19]. When AIDS is identified as the principal diagnosis, the hospitalizations tend to be prolonged. Living distant from the hospital was only important to explain DRG 714. Patients from distant areas may have delayed their admission to hospital due to transportation difficulties, which may prolong the hospitalization, mainly for those in long-stay subgroup [4, 18]. Urgent admission was also only relevant to explain DRG 714 and pneumonia was only important to explain DRG 710. The present study also found that differentiated hospitals contributed to a curtailment of prolonged hospitalization in the long-stay subgroup. Central hospitals are the only ones that have communicable diseases departments, which can contribute to reducing the LOS, especially in the case of the long-stay subgroup.

The hierarchical normal mixture model presented here suggests differences amongst hospitals that emphasize the need for further research. DRG data violate the independence assumptions of classical regression analysis, because patients are nested within hospitals on the basis of their own choices (place of residence, trust in a particular doctor, the hospital's reputation, etc.). As a result, hierarchical modelling is strongly advocated as a more appropriate statistical method for dealing with multilevel structured data, such as patients clustered within hospitals [7]. Moreover, the hospital random effects, which acknowledge unmeasured factors that are nonetheless important, should be interpreted as differences in hospital quality/performance.

Hospitals 1, 2, and 3 are the most efficient in terms of risk-adjusted LOS. They are large hospitals in metropolitan areas that offer more differentiated services. Notwithstanding, three other hospitals exhibit large positive effects extending hospitalizations. Two of these (21 and 22) are small hospitals that do not offer differentiated services, located in small areas, as opposed to hospital 23, which is a large hospital in a metropolitan area, offering differentiated services. These hospitals can be classified as less efficient. This study identifies factors associated with short-stay and long-stay latent subgroups considering that the patients are nested within hospitals.

5 Conclusion

The hierarchical normal mixture methodology has demonstrated to be useful for analysing HIV/AIDS LOS, suggesting the presence of a finite mixture of two subpopulations. This stands in contrast to linear regression (one-component model), which imposes an average effect of LOS predictors for short-stay and long-stay latent subgroups, thereby leading to misdirected interpretations. Accounting for clustered observations, the hierarchical structure provides correct inferences about the regression coefficients (β).

Our findings are of great interest for clinical practice, discharge planning, and the efficient management of LOS. For healthcare policy purposes, our identification of "atypical" hospitals should caution policymakers not to regard all hospitals equally. By targeting relevant factors influencing HIV/AIDS LOS, appropriate policies can be developed to manage the hospital care and resources, as well as promote the early prediction of HIV/AIDS patients requiring a longer period of hospitalization, and the higher costs thus incurred. The drawback of this study resides in the use of the LOS logarithm transformation, which in future research will be analysed in the original scale, considering a generalized linear mixture approach.

Acknowledgements This research is partially supported by grant number SFRH/BD/41007/2007, by Project PTDC/MAT/118335/2010 and by Pest-OE/MAT/UI0006/2011, all financed by FCT Portugal.

References

1. Atienza, N., Garcia-Heras, J., Munoz-Pichardo, J.M., Villa, R.: An application of mixture distributions in modelization of length of hospital stay. Stat. Med. **27**(9), 1403–1420 (2008)
2. Dempster, A.P., Laird, N.M., Rubin, D.B.: Maximum likelihood from incomplete data via EM algorithm. J. R. Stat. Soc. Ser. B Methodol. **39**(1), 1–38 (1977)
3. Dias, S.S., Andreozzi, V., Martins, M.O., Torgal, J.: Predictors of mortality in HIV-associated hospitalizations in Portugal: a hierarchical survival model. BMC Health Serv. Res. **9** (2009)
4. Eastaugh, S.R.: Organizational determinants of surgical lengths of stay. Inquiry **17**(1), 85–96 (1980)
5. Everitt, B.S., Hand, D.J.: Finite Mixture Distribution. Monographs on Applied Probability and Statistics. Chapman & Hall, London (1981)
6. Fetter, R.B., Youngsoo, S., Freeman, J.L., Averill, R.F., Thomson, J.D.: Case-mix: definition by diagnosis related groups. Med. Care **18**, 1–53 (1980)
7. Greenland, S.: Principles of multilevel modelling. Int. J. Epidemiol. **29**, 158–167 (2000)
8. Grun, B., Leisch, F.: Fitting finite mixtures of generalized linear regressions in R. Comput. Stat. Data Anal. **51**(11), 5247–5252 (2007)
9. Lee, A.H., Gracey, M., Wang, K., Kelvin, K.W.: A robustified modeling approach to analyze pediatric length of stay. Ann. Epidemiol. **15**(9), 637–677 (2005)

10. Leisch, F.: FlexMix: A general framework for finite mixture models and latent class regression in R. J. Stat. Softw. **11**(8), (2004)
11. Leung, K.M., Elashoff, R.M., Rees, K.S., Hasan, M.M., Legorreta, A.P.: Hospital- and patient-related characteristics determining maternity length of stay: a hierarchical linear model approach. Am. J. Public Health **88**(3), 377–381 (1998)
12. Leyland, A.H., Boddy, F.A.: Measuring performance in hospital care: length of stay in gynaecology. Eur. J. Public Health **7**(2), 136–143 (1997)
13. McLachlan, G.J., Basford, K.E.: Mixture Models: Inference and applications to clustering. Marcel Dekker, New York (1988)
14. Pèrez-Hoyos, S., Ballester, F., Tenias, J.M., Merelles, A., Rivera, M.L.: Length of stay in a hospital emergency room due to asthma and chronic obstructive pulmonary disease: implications for air pollution studies. Eur. J. Epidemiol. **16**, 455–463 (2000)
15. Saez-Castillo, A.J., Olmo-Jimenez, M.J., Sanchez, J.M.P., Hernandez, M.A.N., Arcos-Navarro, A., Diaz-Oller, J.: Bayesian analysis of nosocomial infection risk and length of stay in a department of general and digestive surgery. Value Health **13**(4), 431–439 (2010)
16. Singh, C.H., Ladusingh, L.: Inpatient length of stay: a finite mixture modeling analysis. Eur. J. Health Econ. **11**(2), 119–126 (2010)
17. Team, R.D.C.: R: A Language and Environment for Statistical Computing. The R Foundation for Statistical Computer, Vienna (2008)
18. Xiao, J., Lee, A.H., Vemurri, S.R.: Mixture distribution analysis of length of stay for efficient funding. Socio Econ. Plann. Sci. **33**, 39–59 (1999)
19. Wang, K., Yau, K.K.W., Lee, A.H.: A hierarchical Poisson mixture regression model to analyse maternity length of stay. Stat. Med. **21**, 3639–3654 (2002)

Volatility and Returns of the Main Stock Indices

Thelma Sáfadi and Airlane P. Alencar

Abstract

In this work we studied the association between cumulative returns and estimated volatilities for the main world stock market indices. The analyzed series were daily values from January 4th, 2008 to April 11th, 2011 for the indices: S&P500 (US), Shanghai Comp Index (China), FTSE100 (UK), CAC40 (France), DAX (Germany), S&P/TSX (Canada), Bovespa (Brazil), Merval (Argentina), and Nikkei 225 (Japan). The volatilities were estimated using APARCH models and a cluster analysis was developed based on the correlation of estimated volatilities and on the distance among estimates of APARCH parameters. Based on the volatility, we identified three groups: Canada, Brazil, and Japan; the USA, UK, France, Germany, and Argentina; and only China. Considering the cluster analysis based on the APARCH parameter estimates, four groups were identified. One of these groups is more distant from the others, composed by Brazil and Argentina, and this group presents higher estimates of the ARCH parameter and lower leverage effect. Based on the cumulative changes of the indices in relation to 2008, we identified four groups: Brazil and Argentina, who already have profits in relation to early 2008, France and Japan, recovering slowly; China that fell dizzying in 2008, with cumulative loss of -39% in the period and the other countries which were recovering in 2011.

T. Sáfadi (✉)
DEX, Federal University of Lavras, Lavras, Brazil
e-mail: safadi@dex.ufla.br

A.P. Alencar
IME, University of São Paulo, São Paulo, Brazil
e-mail: lane@ime.usp.br

A. Pacheco et al. (eds.), *New Advances in Statistical Modeling and Applications*,
Studies in Theoretical and Applied Statistics, DOI 10.1007/978-3-319-05323-3_22,
© Springer International Publishing Switzerland 2014

1   Introduction

The association between the indices of main stock indices has been studied by several researchers with different objectives. A possible approach is focused on risk and may be measured by an estimate of volatility for each country based on daily returns of stock market indices, since larger volatilities may be associated with larger probabilities of loss. In this study, we considered the series S&P500 (U.S.), Shanghai Comp Index (China), FTSE100 (UK), CAC40 (France), DAX (Germany), S&P/TSX (Canada), Bovespa (Brazil), Merval (Argentina), and Nikkei 225 (Japan) during the period from January 4th, 2008 to April 11th, 2011.

In [9] the same daily returns from January, 2008 to May 10th, 2010 were analyzed using a dynamic factor model with three factors. They noted that the first factor indicated that the financial crisis mainly associated with the USA was felt by all other markets in the world. The second factor is more associated with Asian countries, China and Japan, and the third factor associated with the European countries.

Considering S&P500 returns, Ding et al. [4] concluded that there is no obvious reason why to assume that the process of the conditional variance is a linear function of quadratic returns, as in GARCH model, or that the process of the conditional standard deviation is a linear function of the absolute returns as in the Taylor model [10]. That is, other powers of the conditional residuals exhibited significant temporal dependencies. The Asymmetric Power ARCH, APARCH, model was proposed by Ding et al. [4] to generalize the GARCH model [2]. The idea of this model is based on empirically finding out that absolute returns present higher autocorrelation than returns and squared returns. Then it was proposed a model that includes a parameter δ that corresponds to the power of the standard deviation. Therefore, in the APARCH model, the conditional standard deviation is modeled as a Box-Cox power transformation of the conditional standard deviation process. Also, it includes the leverage component that considers different effects for positive and negative shocks.

The conditional standard deviation may be estimated using the APARCH model, and these series may be used to cluster the stock indices based on the correlation between each pair of volatilities series. Alencar and Sáfadi [1] estimated the volatilities using APARCH models and used correlation coefficients to cluster these indices. This analysis allows identifying that estimated conditional volatilities present similar behavior in Brazil, Canada, and Argentina, and in another group the USA, UK, Germany, and France. Also the estimated conditional volatility is completely different in China, where it is lower and more stable.

In this study, we present an alternative cluster analysis based on the APARCH estimates of the leverage, GARCH and ARCH components for stock indices in each country. A completely different approach is based on the profitability perspective is a cluster analysis based on the percentage changes of the stock indices in relation to the indices registered in the beginning of 2008.

In this paper, the main goal is to measure the association among these nine stock indices based on the volatilities and parameter estimates for a comparable APARCH model to estimate the conditional standard deviations ($\delta = 1$) and also based on the percentage changes of the stock indices.

2   Methods

As returns we calculated $r_t = \ln Y_t - \ln Y_{t-1}$, where Y_t is the stock market index at day t. In order to accommodate possible autocorrelations of the returns, the following equation is included in the model:

$$r_t = \sum_{i=1}^{p} \phi_i r_{t-i} + \epsilon_t, \tag{1}$$

where $\phi_i, i = 1, \ldots, p$, satisfy the stationarity conditions with the roots B of $1 - \sum_{i=1}^{p} \phi_i B^i$ outside the unit circle [6].

Considering the asymmetric t-Student distribution for the errors, the APARCH model is defined as

$$\epsilon_t = z_t \sigma_t,$$

$$z_t \sim t_{\xi,\nu},$$

$$\sigma_t^\delta = \omega + \sum_{i=1}^{p} \alpha_i \left(|\epsilon_{t-i}| - \gamma_i \epsilon_{t-i} \right)^\delta + \sum_{j=1}^{q} \beta_j \sigma_{t-j}^\delta, \tag{2}$$

with $\omega \geq 0$, $\alpha_i \geq 0$, $\beta_i \geq 0$, $\delta > 0$, and $|\gamma_i| \leq 1$, where σ_t is the conditional standard deviation of the response variable and is known as volatility and $t_{\xi,\nu}$ corresponds to the asymmetric t-Student distribution with asymmetric coefficient equal to ξ and ν degrees of freedom. The asymmetric Student t distribution was parameterized as in [5]. Some restrictions in the parameter space must be imposed to guarantee the positivity of the transformed conditional volatility σ_t^δ, as discussed in [4, 8], but in this paper we include the restriction $\delta = 1$ because all parameters δ are not statistically different from 1 in [1] and under $\delta = 1$ the estimates of all parameters are comparable. If the δ parameter is not constant, the parameters ω, α, β, and γ are not explaining always the same quantity like the variance or the standard deviation. The residual analysis indicated that the assumptions of the model are not violated.

The proposed models were fitted using the maximum likelihood method, implemented in the library fGarch [12] in the free statistical software R [7]. The maximization was performed using the Nelder–Mead algorithm after obtaining the BFGS start values (option algorithm="lbfgsb+nm" using garchFit command, details and references in [12]).

Based on the estimates of ω, α, γ, and β for these nine indices, a hierarchical cluster analysis with the Euclidean distance of the parameter estimates and the complete linkage method was performed to identify groups of countries based on the volatility model [11]. Caiado and Crato [3] proposed a similar cluster analysis using the estimates, but their model is the broadly used GARCH(1,1) model. We used only the estimates since our objective is a cluster analysis of only 9 indices, instead of 27 indices analyzed in [3]. We decided not to include the estimates of the density parameters (ξ and v) to follow the proposal of [3], which included only the α and β estimates of the GARCH(1,1) model, and also because they are not directly related to the estimated volatilities.

An alternative approach is to identify groups of countries based on the direct effects of the crisis and on their recovery. For this purpose another cluster analysis is proposed to cluster the percentage cumulative changes of each daily stock market index in relation to the index in the first day of 2008. These cumulative changes measure the losses and gains of an investor who applied money in the stock in the beginning of 2008. The cluster analysis used the complete method and the Spearman correlation index [11].

3   Results

In [1], an APARCH model was fitted for the same returns and there was no evidence to reject the hypothesis $\delta = 1$ for all countries. This is in agreement with [4], where the autocorrelation of the absolute returns of SP&500 is larger than the autocorrelation of $|r_t|^d$, for $d = 0.125, 0.25, 0.50, 0.75, 1.25, 1.5, 1.75, 2, 3$. The APARCH(1,1) models with $\delta = 1$ were fitted to the series of returns and the estimates are presented in Table 1. Only for the Japanese returns, the first autoregressive parameter was significant ($\hat{\phi}_1 = -0.0879$, standard error $= 0.0387$, $p = 0.0230$). The estimation results are presented in Table 1. Figure 1 presents the log-returns and the volatilities defined as the estimated conditional standard deviations of the log-returns. The autocorrelation of the standardized residuals and their squared values indicate that all assumptions are met.

Almost all the estimates of the leverage parameter γ are close to 1, indicating that positive errors ϵ_t will cause almost no effect in the volatility, unlike the negative shocks that may impact the volatility in the same instant and in the future instants depending on β. Less leverage effects are estimated in Argentina ($\hat{\gamma} = 0.45$). The parameters β are responsible for the inertia in the volatility, and values close to 1 indicate that the σ_t^δ will be close to σ_{t-1}^δ. Most estimates of β are close to 0.9, the exceptions are the smaller value 0.83 in Argentina and 0.99 in China, indicating, respectively, less and more inertia (stability) in the volatilities, which agree with the estimated volatilities presented in Fig. 1.

The estimates of the skewness parameter ξ are all smaller than 1 and they are all very similar, yielding a nearly symmetric error distribution, with the smallest value for Canada (the most asymmetric distribution). Argentina is the country with the larger kurtosis (smallest degree of freedom), followed by China.

Table 1 Estimates, standard errors, and p-values corresponding to the parameters of the APARCH(1,1) with $\delta = 1$ model for each country

		ω	α	γ	β	ξ	ν
US	Estimate	0.0005	0.0984	1.0000	0.8980	0.8386	6.0722
	Std. error	0.0001	0.0180	0.0179	0.0182	0.0410	1.5356
	p	0.0003	<0.0001	<0.0001	<0.0001	<0.0001	0.0001
UK	Estimate	0.0005	0.0931	1.0000	0.8956	0.9718	8.9284
	Std. error	0.0002	0.0169	0.0141	0.0194	0.0503	3.2013
	p	0.0009	<0.0001	<0.0001	<0.0001	<0.0001	0.0053
France	Estimate	0.0007	0.1018	1.0000	0.8798	0.9209	12.2600
	Std. error	0.0002	0.0173	0.0152	0.0218	0.0507	5.8500
	p	0.0008	<0.0001	<0.0001	<0.0001	<0.0001	0.0362
Germany	Estimate	0.0005	0.0911	1.0000	0.9013	0.9231	7.9197
	Std. error	0.0002	0.0175	0.0211	0.0195	0.0464	2.4520
	p	0.0019	<0.0001	<0.0001	<0.0001	<0.0001	0.0012
Canada	Estimate	0.0003	0.0978	0.7973	0.9039	0.7755	9.3423
	Std. error	0.0001	0.0219	0.2037	0.0199	0.0474	2.9316
	p	0.0011	<0.0001	0.0001	<0.0001	<0.0001	0.0014
Brazil	Estimate	0.0006	0.1108	0.6417	0.8901	0.9151	6.9164
	Std. error	0.0002	0.0255	0.1623	0.0255	0.0484	1.7763
	p	0.0218	<0.0001	0.0001	<0.0001	<0.0001	0.0001
Argentina	Estimate	0.0017	0.1622	0.4472	0.8265	0.8993	3.4722
	Std. error	0.0014	0.0832	0.1632	0.1075	0.0376	0.5599
	p	0.2456	0.0512	0.0061	<0.0001	<0.0001	<0.0001
Japan	Estimate	0.0005	0.0769	0.8730	0.9109	0.8457	10.4258
	Std. error	0.0002	0.0220	0.2892	0.0221	0.0455	3.6167
	p	0.0024	0.0005	0.0025	<0.0001	<0.0001	0.0039
China	Estimate	0.0000	0.0144	1.0000	0.9873	0.8879	4.8710
	Std. error	0.0000	0.0030	0.0307	0.0023	0.0396	0.9690
	p	0.4890	<0.0001	<0.0001	<0.0001	<0.0001	<0.0001

The smallest estimate of ω is 3×10^{-5} in China resulting in the lowest level of unconditional volatility, completely different from other countries, as may be visualized in the graphs of Fig. 1. The estimates for the APARCH models reveal that China presents a completely different behavior with lower volatility level and more stability and some countries present similar behavior as the USA, UK, France, and Germany.

The cluster analysis of the estimated volatilities used the complete method and the Spearman correlation (as suggested in [1]) and the corresponding dendrogram is shown in Fig. 2a. In this graph there are three groups: the first consists of Canada, Brazil, and Argentina, the second by the USA, UK, France, Germany, and Japan and the third only with China.

Based on the estimates of ω, α, γ, and β for these nine indices (Table 1), the dendrogram of the hierarchical cluster analysis with the Euclidean distance of the

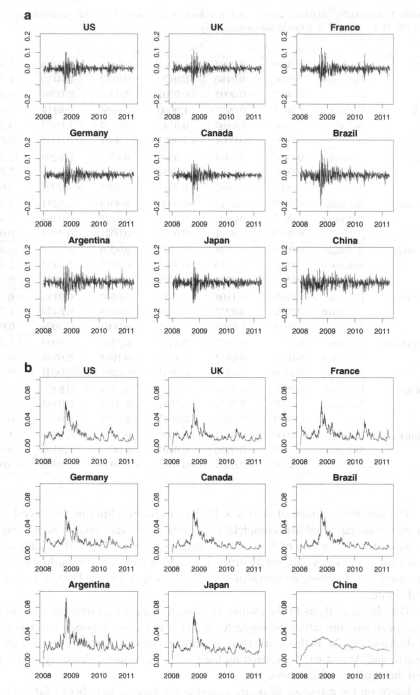

Fig. 1 Log returns and estimated volatility from APARCH $(1,1)$ with $\delta = 1$. (**a**) Log returns, (**b**) volatility estimated

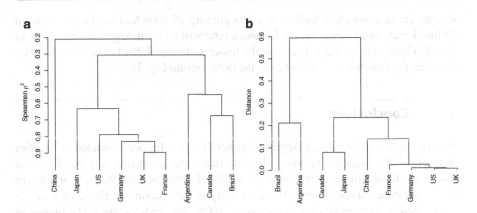

Fig. 2 Dendrogram of (**a**) estimated volatilities and of (**b**) estimates of the parameters

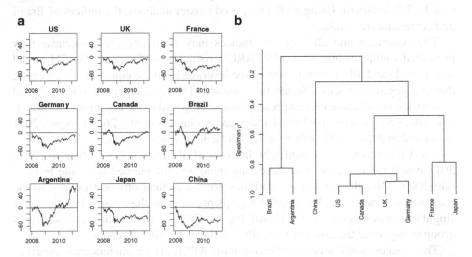

Fig. 3 Percentage cumulative change in relation to 2008 and its corresponding dendrogram. (**a**) Volatility, (**b**) dendrogram

parameter estimates and complete linkage method is presented in Fig. 2b. The most distant group is composed by Argentina and Brazil and both of them present higher estimates of the ARCH parameter and lower leverage estimates, indicating that the volatility depends on the signal of the shock and is more susceptible to shocks. China forms another group with the smallest level of volatility (lowest $\hat{\omega}$) and higher inertia (highest $\hat{\beta}$ and lowest $\hat{\alpha}$). All other countries present similar estimates.

The crisis known as subprime crisis reached the stock prices in October 2008. Instead of presenting the indices, which varies in different ranges, Fig. 3a presents the percentage change of each index in relation to the index observed in the first day of 2008. The cumulative percentage changes in April, 11th are, respectively, -6.2, -4.6, -25.9, -7.7, 1.6, 11.7, 62.0, -33.8, and -39.2%. An investor who started an investment in these stock indices in the beginning of 2008 had the larger losses in China, Japan, and France. He could have a return of 62 % in Argentina and 12 % in Brazil. Other indices did not recover the losses until April 2011. These similarities between the countries are observed in the dendrogram Fig. 3b.

4   Conclusions

We conclude that Brazil and Argentina generally have higher estimated volatilities throughout the analyzed period, indicating that these countries may be at greater risk. Also these countries present higher ARCH estimated coefficients and lower leverage estimated coefficients. However, only in these countries the stock markets have already recovered after the crisis of 2008. The indices are 62 % higher in Argentina compared to its value in the first day of 2008, and this cumulative change was 11.7 % in Brazil. Using all the proposed cluster analysis, the indices of Brazil and Argentina are similar.

The American and all European indices may form one group because they presented similar estimates of all APARCH parameters (for the APARCH model with $\delta = 1$) and all of them presented the leverage estimate equals to 1, indicating that the negative shocks affected the variance. They presented similar estimated volatilities and all these indices present losses compared to the beginning of 2008. It is worth to mention that the French index accumulated loss of -26 % in April, 2011 compared to 2008 and the others indices presented smaller losses.

The Chinese index is completely different of all other indices, since in April, 2011 it still did not exhibit an increasing trend and maintained the same level of mid-2009. Also, the estimated volatilities are lower than others. Contrary to the usual stylized facts, the Chinese returns graph presents no cluster of volatilities. This singularity was responsible for identifying the Chinese stock index as a separated group using either the estimated volatility or the accumulated return.

The Japanese and Canadian estimates of the APARCH parameters are similar, but Canada index reached in April 2011 the same value it registered before the crisis. Meanwhile, the Japanese index showed a devaluation of -33 % compared to 2008.

Acknowledgments The authors acknowledge the financial support of CAPES/ PROCAD grant 177/2007, FAPESP grant 2008/51097-6 and FAPEMIG.

The authors also thank an anonymous referee for very useful suggestions that improved the paper and the estimation of the model.

References

1. Alencar, A.P., Sáfadi, T.: Volatility of main stock indexes: similarities and differences. Int. J. Stat. Econ. **9**, 1–12 (2012)
2. Bollerslev, T.: Generalized autoregressive conditional heteroskedasticity. J. Econ. **31**, 307–327 (1986)

3. Caiado, J., Crato, N.: A GARCH-based method for clustering of financial time series: international stock markets evidence. In: Proceedings of the XIIth Applied Stochastic Models and Data Analysis International Conference (2007)
4. Ding, Z., Engle, R.F. , Granger, C.W.J.: A long memory property of stock market returns and a new model. J. Empir. Finance Amsterdam **1**, 83–106 (1993)
5. Fernandez, C., Steel, M.F.: On Bayesian modeling of fat tails and skewness. J. Am. Stat. Assoc. **93**, 359–371 (1998)
6. Morettin, P.A., Toloi, C.M.: Análise de Séries Temporais. Editora Edgard Blücher, São Paulo (2004)
7. R Development Core Team: R: A Language and Environment for Statistical Computing. R Foundation for Statistical Computing, Vienna. http://www.R-project.org (2010). ISBN 3-900051-07-0
8. Rodriguez, M.J., Ruiz, E.: GARCH models with leverage effect: differences and similarities. Working Paper 09-03. Universidad Carlos III de Madrid (2009)
9. Sáfadi, T., Alencar, A.P., Morettin, P.A.: The dynamic factor model: an application to stock market indexes. Int. J. Stat. Econ. **7**, 127–141 (2011)
10. Taylor, S.: Modelling Financial Time Series. Wiley, New York (1986)
11. Wichern, D.W., Johnson, R.A.: Applied Multivariate Statistical Analysis, 6th edn. Prentice-Hall, Upper Saddle River (2007)
12. Wurtz, D., Chalabi, Y., Luksan, L.: Parameter estimation of ARMA models with GARCH/APARCH errors: An R and SPlus software implementation. Technical report (2009)

Using INLA to Estimate a Highly Dimensional Spatial Model for Forest Fires in Portugal

Isabel Natário, M. Manuela Oliveira, and Susete Marques

Abstract

Within the context of accessing the risk of forest fires, Amaral-Turkman et al. (Environ. Ecol. Stat. 18:601–617, 2011) have proposed a spatio-temporal hierarchical approach which jointly models the fire ignition probability and the fire's size, in a Bayesian framework. This is recovered and applied to Portuguese forest fires data, with some necessary modifications in what concerns the format of the data (not available in a regular lattice over the territory) and also because of the estimation complications that arise due to the high dimensionality of the neighbouring structure involved. To address the latter, as it compromises the estimation via Markov Chain Monte Carlo (MCMC) methods, and having the model be recognized as a latent Gaussian model, it was chosen to do the Bayesian estimation also using an Integrated Nested Laplace Approximation approach, with real computational advantages. Corresponding methodologies and results are described and compared.

I. Natário (✉)
Faculdade de Ciências e Tecnologia (UNL) and CEAUL, Quinta da Torre,
2825-114 Caparica, Portugal
e-mail: icn@fct.unl.pt

M.M. Oliveira
Universidade de Évora and CIMA, Rua Romão Ramalho 59, 7002 Évora, Portugal
e-mail: mmo@uevora.pt

S. Marques
Instituto Superior de Agronomia (UTL) and CEF, Tapada da Ajuda, 1349-017 Lisboa, Portugal
e-mail: smarques@isa.utl.pt

A. Pacheco et al. (eds.), *New Advances in Statistical Modeling and Applications*,
Studies in Theoretical and Applied Statistics, DOI 10.1007/978-3-319-05323-3_23,
© Springer International Publishing Switzerland 2014

1   Introduction

Forest fires are a major disturbance of the Mediterranean landscape [9], that has considerably increased in the last decades, especially in Portugal where the burned area over the last three decades corresponds to nearly 40 % of the total area of mainland Portugal [6]. Several studies have shown that forest fires in Portugal present high temporal as well as high spatial variability [7]. In order to choose and apply effective measures to combat this problem it is crucial to understand which factors contribute to it. The use of geographic information systems as well as adequate statistical models can make an important contribution.

In the context of accessing the risk of fire, Amaral-Turkman et al. [1] have proposed a spatio-temporal hierarchical model which jointly models the probability of fire ignition and the size of the fire, in a Bayesian setting, for Australian data.

In the present work this method is recovered to model the risk of fire in mainland Portugal. In Amaral-Turkman and colleagues' approach [1] the region under consideration is divided into small areas by a regular lattice, and the data are collected in each one of these areas. However, the data in this study could only be available within administrative units that divide the country, *freguesias*, forming an irregular lattice. Nevertheless it was still possible to use the model.

The data available for analysis refer to the period from 2002 to 2007, although we have chosen to work only, in a first approach, with data from 1 year, 2005. Data comprehend the annual percentage of burned area in each *freguesia*, as well as annual values of several covariates, from atmospheric conditions, topographic information, road proximity and population density.

Because the number of *freguesias* that divide mainland Portugal is very high (3,424 *freguesias*), the estimation via Markov Chain Monte Carlo (MCMC) techniques [3], usual in the Bayesian models applied to spatio-temporal data, becomes computationally very heavy or even unfeasible. In order to overcome this problem it has been opted to do the estimation via an Integrated Nested Laplace Approximation (INLA) approach [8], which presents very important computational advantages. This approximated approach has been gaining relevance for the Bayesian estimation of models that fit within the sub-class of regression models with additive structure, the latent Gaussian models.

We describe in detail the genesis of the INLA inferential procedure, implement it for the estimation of the above model in the application of forest fires in Portugal, and we point out which concrete advantages/disadvantages we had, in relation to the MCMC methods.

This paper organizes as follows: in Sect. 2 we present the model for forest fires proposed by Amaral-Turkman et al. [1], in Sect. 3 we briefly describe the MCMC and INLA methods for Bayesian inference, Sect. 4 encloses the application of this model to the forest fires in mainland Portugal and finally we conclude the paper in Sect. 5.

2 A Model for Forest Fires

In this section the spatio-temporal hierarchical model proposed by Amaral-Turkman et al. [1], for simultaneously modeling the proportion of burned area within small divisions (areal units) of the region under study and the probability of ignition, is described in its spatial-only version. In the original work the areal units were defined through a regular lattice, unlike what happens here, where Portugal mainland is taken to be divided by an irregular lattice into *freguesias*. Other references for the distribution of fires are available, as e.g., [10] or the ones in [1].

Let $Y(i)$ represent the proportion of burned area in *freguesia* i and let $R(i)$ be the dichotomous variable indicating whether there has been a fire in *freguesia* i, $i = 1, \ldots, 3{,}424$, in a given year of interest (2005, in the application). A transformation of the non-null Ys towards gaussianity is further considered. The model follows:

$$R(i) \mid p(i) \sim \text{Bernoulli}(p(i)) \qquad \text{logit}(p(i)) = \beta + \mathbf{X}(i)^T \boldsymbol{\eta} + V_2(i)$$

$$Z(i) = \begin{cases} \log\left(\frac{Y(i)}{1-Y(i)}\right), & 0 < Y(i) < 1 \\ 0, & Y(i) = 0 \end{cases}$$

$$Z(i) \mid R(i) = 1 \sim N\left(\mu(i), \sigma^2\right) \qquad \mu(i) = \alpha + \mathbf{X}(i)^T \boldsymbol{\delta} + V_1(i)$$

$$V_1(i) = v_1 W_0(i) + v_2 W_1(i) \qquad V_2(i) = W_1(i)$$

$$W_0 \sim \text{ICAR}(\tau_0) \qquad W_1 \sim \text{ICAR}(\tau_1)$$

where $p(i)$ is probability of ignition in *freguesia* i, $\mathbf{X}$ are covariates of interest, not necessarily the same for $p(i)$ and $\mu(i)$, $V_2(i)$ are unobserved explanatory spatial variables influencing fire ignition, $V_1(i)$ are unobserved explanatory spatial variables influencing fire size—$V_1(i)$ and $V_2(i)$ are dependent spatial random effects. This dependence is induced by the shared spatial latent process W_1, but might proof not to be relevant in case the v_2 parameter is not significantly different from zero. Intrinsic autoregressive priors are considered for the independent spatial latent processes W_0 and W_1. Gaussian prior distributions are taken for the covariate effects and the v parameters. For precision parameters the usual gamma prior choice is made. Prior independence between these parameters is considered. More information on priors and other model details can be found in [1]. Let $\phi(x, \mu, \sigma^2)$, $\phi_2(\mathbf{x}, \boldsymbol{\mu}, \mathbf{C})$ and $\Gamma(x, a, b)$ represent, respectively, the density function of a Gaussian random variable, the density function of a bivariate Gaussian random vector and the density function of a gamma distribution; let $\bar{w}_0(i)$ and $\bar{w}_1(i)$ represent, respectively, the local mean of the W_0 and of the W_1 values in the neighbourhood of area i, with n_i neighbours. The corresponding posterior distribution of the parameters, given below in simplified notation, is proportional to the model likelihood times the parameters prior distributions:

$$\pi(\Theta|\mathbf{Z}, \mathbf{R}) \propto f(\mathbf{Z}, \mathbf{R}|\Theta) \times \pi(\Theta) = f(\mathbf{Z}|\mathbf{R}, \Theta) \times f(\mathbf{R}|\Theta) \times \pi(\Theta) =$$

$$= \prod_i \left[\phi\big(z(i), \mu(i), \sigma^2\big)\right]^{r(i)} \times p(i)^{r(i)}(1 - p(i))^{1-r(i)} \times$$

$$\times \phi(\beta, 0, 10^5) \cdot \prod_j \phi(\eta_j, 0, 10^5) \cdot \phi(\alpha, 0, 10^5) \cdot \prod_k \phi(\delta_k, 0, 10^5) \cdot$$

$$\cdot \phi_2 \left(\begin{pmatrix} v_1(i) \\ v_2(i) \end{pmatrix}, \begin{pmatrix} v_1\bar{w}_0(i) + v_2\bar{w}_1(i) \\ \bar{w}_1(i) \end{pmatrix}, \begin{pmatrix} v_1^2\frac{\tau_0}{n_i} + v_2^2\frac{\tau_1}{n_i} & v_2\frac{\tau_1}{n_i} \\ v_2\frac{\tau_1}{n_i} & \frac{\tau_1}{n_i} \end{pmatrix}\right) \cdot$$

$$\cdot \prod_{m=1}^{2} \phi(v_m, 1, 10) \cdot \prod_{l=0}^{1} \Gamma(\tau_l, 0.01, 0.01) \cdot \Gamma(\frac{1}{\sigma^2}, 0.01, 0.01) \qquad (1)$$

3　Bayesian Estimation

For the kind of complex, multidimensional hierarchical models as the one just described, usually, the Bayesian estimation of the marginal posterior distributions of the parameters involved is not possible from (1), not in a closed form. For the last 30 years this problem has been addressed through approximate methods or, more frequently, through iterative simulation methods, only made possible for the huge development of computer power.

3.1　Markov Chain Monte Carlo

MCMC methods are based on the idea of simulating a (ergodic) Markov chain that has as its limiting distribution the target distribution $\pi(\theta|\mathbf{y})$—the posterior distribution of the model's parameters θ, in this case—known up to a constant term (one cannot sample directly from it). From an initial distribution a $\theta^{(0)}$ value is sampled. Next, a $\theta^{(1)}$ value is sampled from the distribution given by the Markov chain kernel calculated for $\theta^{(0)}$ and so forth, iteratively. As the number of iterations increases the sampled θ values become closer and closer to sampled values from the limiting distribution π and can be considered as so, as well as all the subsequent sampled values, due to stationarity.

The Gibbs sampler [4, 5] is the most commonly used algorithm to produce the upper cited Markov chains with chain transition kernel defined by the complete conditional distributions of θ_i.

The MCMC sample methods for this type of hierarchical models, with many dependencies defined between the high number of parameters of the model, tend to have a very weak performance precisely because of that dependencies. However, the MCMC approach can obtain precise marginal posterior distributions, once the MCMC errors can be arbitrarily small for an arbitrarily large computational time [8]. More efficient MCMC algorithms have been proposed [3], many based on Gaussian approximations, but for some complex models convergence might be very difficult to attain within a reasonably amount of time [11].

3.2 Integrated Nested Laplace Approximation

A recent approach based on some old approximate methods has recently become available for doing Bayesian estimation, the INLA, [8]. The core of this approach is that it provides deterministic approximations of the posterior marginal distributions in a computational efficient way. Its caveat is that being an approximation, the estimation is never bias free.

This estimation approach is to be used whenever the model to be estimated is a member of the special sub-class of structured additive regression models, the latent Gaussian models, characterized for having the response variable Y_i in the exponential family with expected value μ_i, related to a structured additive linear predictor η_i through a "well-behaved" link function $g(\mu_i) = \eta_i = \alpha + \sum_j f^{(j)}(u_{ij}) + \sum_k \beta_k z_{ki} + \varepsilon_i$, where $f^{(j)}$ are unknown functions of the covariates $\mathbf{u}$, β_k are the linear effects of the covariates $\mathbf{z}$ and ε are the non-structured terms (errors). For all the parameters in the predictor, α, $f^{(j)}$, β_k and ε_i, prior Gaussian distributions can be assumed.

Let $\mathbf{x} = (\alpha, f^{(j)}, \beta_k, \varepsilon_i)$ be the vector of all latent Gaussian variables, $\pi(\mathbf{x}|\theta_1) \equiv \mathcal{N}(\mathbf{0}, \mathbf{Q}(\theta_1))$, where $\mathbf{Q}(\theta_1)$ is a precision matrix. Let $\mathbf{y}$ denote the vector of the n_d observations with density $\pi(\mathbf{y}|\mathbf{x}, \theta_2)$, whose elements are assumed to be independent, conditional on $\mathbf{x}$ and θ_2; finally, let $\theta = (\theta_1, \theta_2)$ be the vector of the hyperparameters (not necessarily Gaussian). In this setting, the posterior distribution $\pi(\mathbf{x}, \theta|\mathbf{y}) \propto \pi(\theta)\pi(\mathbf{x}|\theta) \prod_i \pi(y_i|x_i, \theta)$ becomes:

$$\pi(\mathbf{x}, \theta|\mathbf{y}) \propto \pi(\theta)|\mathbf{Q}(\theta)|^{\frac{1}{2}} \exp\left\{ -\frac{1}{2}\mathbf{x}^T\mathbf{Q}(\theta)\mathbf{x} + \sum_i \log(\pi(y_i|x_i, \theta)) \right\}.$$

This INLA approach makes use of some (realistic) assumptions that enables the use of numeric methods, faster and more efficient, namely that $\mathbf{x}$ (frequently high dimensional, 100–100,000 elements) has properties of conditional independence (i.e. is a Gaussian Markov random field, GMRF) and that the number of hyperparameters is small (lets say ≤ 6). The objective is to approximate the marginal posterior distributions $\pi(x_i|\mathbf{y})$, $\pi(\theta|\mathbf{y})$ and $\pi(\theta_j|\mathbf{y})$ through nested approximations:

$$\tilde{\pi}(x_i|\mathbf{y}) = \int \tilde{\pi}(x_i|\theta, \mathbf{y})\tilde{\pi}(\theta|\mathbf{y})d\theta \qquad \tilde{\pi}(\theta_j|\mathbf{y}) = \int \tilde{\pi}(\theta|\mathbf{y})d\theta_{-j},$$

where $\tilde{\pi}$ is an approximate (conditional) density.

The approximations of $\pi(x_i|\mathbf{y})$ are obtained approximating $\pi(x_i|\theta, \mathbf{y})$ and $\pi(\theta|\mathbf{y})$ and then using numerical integration (finite sum) to integrate out θ—which is possible of course given the low dimensionality of θ. The same for $\pi(\theta_j|\mathbf{y})$.

The Laplace approximation proposed for $\pi(\theta|\mathbf{y})$ is $\tilde{\pi}(\theta|\mathbf{y}) \propto \left.\frac{\pi(\mathbf{x},\theta,\mathbf{y})}{\tilde{\pi}_G(\mathbf{x}|\theta,\mathbf{y})}\right|_{\mathbf{x}=\mathbf{x}^*(\theta)}$, where $\tilde{\pi}_G$ is the Gaussian approximation to the complete conditional distribution of $\mathbf{x}$ and $\mathbf{x}^*(\theta)$ its mode, for a given θ.

4   Application: Forest Fires in Mainland Portugal

Mainland Portugal has a total area of $89,000\,\mathrm{km^2}$, divided into 3,424 *freguesias*, with an altitude that varies from the average level of the sea to 2,000 m above it, an average annual temperature that varies from 7 to 18°C, increasing NW-SE, an average annual precipitation varying between 400 and 2,800 mm, decreasing NW-SE. Forests and woods cover about $\frac{1}{3}$ of the country (80 % of the forest area is occupied by Maritime pine, eucalypt, cork oak and holm oak), 25 % is covered by shrublands and about 30 % are agriculture fields [2].

The analysed data consist on the *2005 geo-referenced forest fires* (obtained by the remote detection lab of the Instituto Superior de Agronomia, by semi-automatic classification of remote detection of high resolution data, Landsat Multi-Spectral Scanner, Landsat Thematic Mapper and Landsat Enhanced). Additionally there is the following covariate information: *land cover type maps* (produced using Landsat Thematic Mapper Images from 2000, produced by the Remote Sensing Group from Instituto Geográfico Português, at a scale of 1:100,000), *altitude, slope, proximity to roads, population density* and *precipitation*. Variable selection and their segmentation in ordinal levels were based on a detailed preliminary data analysis [6].

The estimation via MCMC (Gibbs sampling) was done with WinBUGS software, which was only capable of handling the north of mainland Portugal (2,654 *freguesias*). After a burn in of 1,000 iterations, the estimation was based on 5,000 iterations as the process in WinBUGS often stopped due to program traps. By this process the estimation time is about 2 h.

The estimation via INLA was done with R-INLA http://www.r-inla.org/—11 min of computational time for the north mainland data. However, because this is still a software in progress, the model as it was proposed could not be estimated via this method yet because of the linear combination of ICARs that constitute the V_1 component—so we have set $v_1 = v_2 = 1$. Although a solution for this question has been advanced by R-INLA authors, by noting that v_1 is only re-scaling W_0, until this point it was not possible to implement it.

Naturally, in terms of the application, the choice of $v_1 = v_2 = 1$ is not an interesting one as it forces dependence between the spatial random effects for the proportion of burned area and ignition risk. From the MCMC estimation we have obtained $\hat{v}_1 = 0.8$ with standard error (SE) of 0.22 and $\hat{v}_2 = 0.2$ (SE 0.04), revealing some degree of dependence between the spatial random effects but not so much as we are assuming now. Nevertheless, for comparison purposes, we proceed like this, further running the MCMC estimation for this sub-model.

Table 1 Covariate posterior effects estimated by MCMC and INLA and corresponding standard errors, associated with each response; non-significant effects were replaced by (**)

Response	Covariate: level	MCMC estimate (SE)	INLA estimate (SE)
Prop. of burned area	Land Cover: Hard Wood	(**)	0.69 (0.11)
Prop. of burned area	Land Cover: Hard and Soft Wood/Eucalyptus	0.60 (0.10)	0.68 (0.10)
Prop. of burned area	Land Cover: Shrubs	0.84 (0.12)	−0.35 (0.13)
Prop. of burned area	Land Cover: Soft Wood	0.31 (0.10)	0.54 (0.10)
Prop. of burned area	Altitude: Over 700 m	(**)	(**)
Prop. of burned area	Altitude: 200–400 m	0.32 (0.11)	0.29 (0.11)
Prop. of burned area	Altitude: 400–700 m	0.41 (0.10)	(**)
Prop. of burned area	Slope: 0–5 %	1.47 (0.12)	−0.94 (0.12)
Prop. of burned area	Slope: 5–10 %	1.61 (0.14)	−0.70 (0.14)
Prop. of burned area	Pop: Over 100 hab/km^2	−0.95 (0.11)	−2.35 (0.12)
Prop. of burned area	Pop: 25–100 hab/km^2	−0.86 (0.10)	−1.78 (0.11)
Prop. of burned area	Road Proximity: > 1 km	0.86 (0.11)	(**)
Prop. of burned area	Precipitation: 0–10 days of precipitation > 1 mm/year	−0.47 (0.17)	−1.50 (0.15)
Ignition	Land Cover: Hard Wood	−1.14 (0.14)	−0.85 (0.11)
Ignition	Land Cover: Hard and Soft Wood/Eucalyptus	−0.76 (0.14)	−0.25 (0.11)
Ignition	Land Cover: Shrubs	0.48 (0.20)	0.53 (0.15)
Ignition	Land Cover: Soft Wood	−0.79 (0.13)	−0.47 (0.11)
Ignition	Altitude: Over 700 m	−0.88 (0.15)	−0.74 (0.12)
Ignition	Altitude: 200–400 m	−0.36 (0.16)	(**)
Ignition	Altitude: 400–700 m	−0.38 (0.17)	−0.33 (0.12)
Ignition	Slope: 0–5 %	1.22 (0.20)	1.05 (0.15)
Ignition	Slope: 5–10 %	1.87 (0.22)	1.48 (0.16)
Ignition	Pop: Over 100 hab/km^2	(**)	−1.27 (0.12)
Ignition	Pop: 25–100 hab/km^2	(**)	−1.11 (0.11)
Ignition	Road Proximity: > 1 km	(**)	(**)
Ignition	Precipitation: 0–10 days of precipitation > 1 mm/year	−1.28 (0.19)	−1.31 (0.14)

The estimated effects of the covariates of the considered sub-model, produced by the two estimation methods, are depicted in Table 1, as well as their standard errors. Most of the MCMC and INLA estimates agree in direction and in significance, although not so much in value. The main conclusions from here are that the larger proportions of estimated burned areas seem to be related to the land cover (worst for hard and soft wood trees, eucalypts and shrubs), medium altitudes, larger slopes, smaller population sizes, larger distance to roads and smaller precipitation levels. The estimated ignition probability seems to be related to smaller slopes, precipitation levels and altitudes and with land cover (worse for shrubs).

The spatial random effects estimated by MCMC and by INLA are depicted, respectively, in Figs. 1 and 2, suggesting that there are still some spatial effects to be accounted for. The ones estimated by MCMC are greater in magnitude, which might have to do with the fact that the two estimation methods have not chosen as

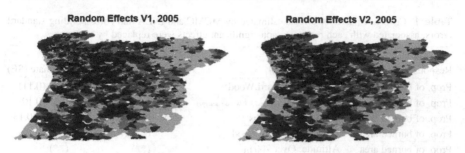

Fig. 1 Random effects estimated via MCMC. Greyscale for cut point values $(-\infty, -1, -0.5, 0, 0.5, 1, +\infty)$

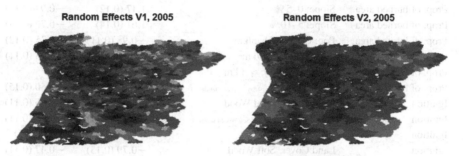

Fig. 2 Random effects estimated via INLA. Greyscale for cut point values $(-\infty, -1, -0.5, 0, 0.5, 1, +\infty)$

significant the same set of covariates. Also note that within each estimation method the V_1 and V_2 estimates are quite similar and, being V_2 part of V_1, this diminishes the importance of the W_0 spatial effect, suggesting that a single spatial random effect might be enough to model both the proportion of burned area as well as the ignition risk.

For both estimation methods the Deviance Information Criterion (DIC) was obtained not only for the sub-model under consideration but also for other models as, for example, the one with no spatial effects (only covariates), which always proved to be worse. However the DIC values were quite different between methods, not being useful for that comparison: for the MCMC estimation of the sub-model the DIC was $-8,070.02$ and of the only covariates model the DIC was $-6,399.11$; for the INLA estimation the corresponding DIC values were $12,797.25$ and $13,140.94$, respectively.

5   Concluding Remarks

In this paper we have described computational advantages of using INLA over MCMC approach in Bayesian estimation, for complicated hierarchical models, within the context of forest fires—from enabling the use of models with many parameters (common in areal data applications where each area corresponds to at least one parameter) to the huge savings in computational time. However there are still some issues to be addressed as, for example, a better understanding of whether the estimates differences obtained between the two methods are due to Winbugs limitations to implement MCMC or due to the INLA approximations, which kept us from estimating the model for each year of the 2002–2007 period for which data were available. After figuring this up, we hope to be able to proceed for the next step of the task of modeling the fire data in Portugal, that is a spatio-temporal analysis.

Acknowledgements The authors acknowledge Professor José Borges for providing access to data collected by the Remote Sensing Laboratory of Instituto Superior de Agronomia, Professor M. Antónia Amaral-Turkman for sharing the WinBugs code of the model in paper [1] and to the anonymous referee for very useful and pertinent comments on a previous paper version. Research partially sponsored by national funds through the Fundação Nacional para a Ciência e Tecnologia, Portugal—FCT, under the projects PEst-OE/SAU/UI0447/2011 and PEst-OE/MAT/UI0006/2011.

References

1. Amaral-Turkman M.A., Turkman K.F., Le Page Y., Pereira J.M.C.: Hierarchical space-time models for fire ignition and percentage of land burned by wildfires. Environ. Ecol. Stat. **18**, 601–617 (2011)
2. Direcção Geral dos Recursos Florestais. Resultados IFN 2005/2006. Lisboa (2006)
3. Gamerman D., Lopes H.F.: Markov Chain Monte Carlo: Stochastic Simulation for Bayesian Inference, 2nd edn. Chapman & Hall/CRC, Boca Raton (2006)
4. Gelfand A.E., Smith F.M.: Sampling-Based Approaches to Calculating Marginal Densities. J. Am. Stat. Assoc. **85**, 398–409 (1990)
5. Geman S., Geman D.: Stochastic Relaxation, Gibbs Distributions, and the Bayesian Restoration of Images. IEEE Trans. Pattern Anal. Mach. Intell. **PAMI-6**, 721–741 (1984)
6. Marques S., Borges J.G., Garcia-Gonzalo J., Moreira F., Carreiras J.M.B., Oliveira M.M., Cantarinha A., Botequim B., Pereira J.M.C.: Characterization of wildfires in Portugal. Eur. J. Forest Res. **130**, 775–784 (2010)
7. Pereira M.G., Malamud B.D., Trigo R.M., Alves P.I.: The history and characteristics of the Portuguese rural fire database. Nat. Hazards Earth Syst. Sci. **11**, 3343–3358 (2011)
8. Rue H., Martino S., Chopin N.: Approximate Bayesian inference for lateny Gaussian models by using integrated nested Laplace approximations. J. R. Stat. Soc. B. **71**, 319–392 (2009)
9. Rundel P.W.: Landscape disturbance in Mediterranean ecosystems: an overview. In: Rundel P.W., Monenegro G., Jaksic F.M. (eds.) Landscape disturbance and biodiversity in Mediterrean-Type ecosystems, pp. 3–22. Springer, Berlin (1998)
10. Schoenberg F.P., Peng R., Woods J.: On the distribution of wildfire sizes. Environmetrics **14**, 583–592 (2003)
11. Simpson, D., Lindgren, F., Rue, H.: Fast approximate inference with INLA: the past, the present and the future. arXiv:1105.2982v1 [stat.CO]

Forecast Intervals with Boot.EXPOS

Clara Cordeiro and M. Manuela Neves

Abstract

Boot.EXPOS is an automatic computational procedure for forecasting time series developed in the ℝ environment joining two very popular methodologies: the exponential smoothing and the bootstrap. Results achieved in previous studies showed that this "mix scheme" seems to be a good approach to obtain point forecasts. This paper investigates the use of Boot.EXPOS in forecast intervals through the application to some well-known data sets. Results obtained show a very good performance of Boot.EXPOS in comparison with its "direct competitors," the exponential smoothing methods.

1   Introduction

A time series is a sequence of observations indexed by time, usually ordered in equally spaced intervals. The most interesting and ambitious task in empirical time series analysis is to forecast future values on basis of its recorded past, and also to calculate prediction intervals. Adequate models need to be fitted to the series. The search for the best model that describes the stochastic behavior of a time series is done through statistical procedures. That model should be able to capture the dynamics of the time series in order to be used in the analysis of the structure of the process or for obtaining predictions.

Exponential smoothing refers to a set of methods that can be used to model and to obtain forecasts. This is a versatile approach that continually updates a forecast,

C. Cordeiro (✉)
FCT of University of Algarve and CEAUL, Campus de Gambelas, 8005-139 Faro, Portugal
e-mail: ccordei@ualg.pt

M.M. Neves
ISA, Technical University of Lisbon and CEAUL, Tapada da Ajuda, 1349-017 Lisboa, Portugal
e-mail: manela@isa.utl.pt

A. Pacheco et al. (eds.), *New Advances in Statistical Modeling and Applications*,   249
Studies in Theoretical and Applied Statistics, DOI 10.1007/978-3-319-05323-3_24,
© Springer International Publishing Switzerland 2014

Table 1 EXPOS classification

Trend component	Seasonal component		
	N (None)	A (Additive)	M (Multiplicative)
N (None)	N,N	N,A	N,M
A (Additive)	A,N	A,A	A,M
Ad (Additive damped)	Ad,N	Ad,A	Ad,M
M (Multiplicative)	M,N	M,A	M,M
Md (Multiplicative damped)	Md,N	Md,A	Md,M

emphasizing the most recent experience, i.e., to recent observations is given more weight than to the older observations. Exponential smoothing methods (EXPOS) stand out among other methods due to their versatility in the wide choice of models that they include. Their widespread dissemination made them the most widely used methods for modeling and forecasting time series. A first classification of EXPOS, due to Pegel's [22], considers the trend and seasonal patterns that a series reveals as none, additive (linear) or multiplicative (nonlinear). Since then many researchers, see [10, 16], have investigated and developed EXPOS methods in a total of fifteen methods, see Table 1.

Analogous to many other areas in statistics, the analysis of time series has been benefiting from the use of computer-intensive procedures to help in modeling and predicting in the most complex analytic situations. Among those procedures the bootstrap methodology is one of the most well known.

In time series, bootstrap is most frequently applied to residual resampling. Sieve bootstrap, introduced in [1], is a model-based approach that considers an autoregressive process fitted to a stationary time series, and then resamples the approximately i.i.d. residuals, see [1, 17]. In previous works, Cordeiro and Neves [4, 5] compared the use of EXPOS methods and bootstrap methodology. [6,7] present a first sketch of an algorithm joining EXPOS methods and bootstrap. Boot.EXPOS was constructed combining the use of EXPOS with the bootstrap methodology for modeling and obtaining forecasts. It was applied in time series forecasting competitions, see [7, 8, 19], standing among the best forecasting procedures [18]. Up to now only point forecasts were considered. In [20] an extension of Boot.EXPOS for obtaining forecast intervals was briefly presented but not developed, which is the proposal of this paper.

Here, part of an extensive study [3] with several different examples of time series is presented. Due to an improvement of the previous algorithm, more accurate point forecasts and forecast intervals were obtained. Section 2 presents the connection between the EXPOS and the bootstrap procedure leading to the Boot.EXPOS. The case studies appear in Sect. 3 and finally, some concluding remarks are pointed out in Sect. 4.

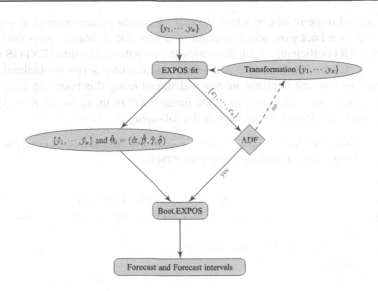

Fig. 1 The computational scheme

2 Bootstrap and EXPOS Together: Boot.EXPOS, a Team Work

Consider a time series $\{y_1, y_2, \cdots, y_n\}$ to which one of the EXPOS models in Table 1 is fitted. Model selection is based on the Akaike's criterion (AIC). This model selection criterion is preferable when compared to other selection criteria because of the parsimonious model penalty, see [15] for more details. The estimates of the exponential smoothing parameters, $\theta = (\alpha, \beta, \gamma, \phi)$, are obtained by minimizing the mean squared error (MSE) of the one-step-ahead forecast errors $(e_t = y_t - \hat{y}_t, t = 1, \cdots, n)$ over the fitted period. As it is known, a good model must have residuals that do not present significant patterns in its correlation structure. Usually, the structure of white noise in the EXPOS context is not investigated but it is common to observe some pattern left in the error term. So, autoregressive models (AR) are suitable to model this error sequence. Given the, approximately, i.i.d nature of the AR residuals, the IID bootstrap [9] can easily be extended to the dependent case.

A scheme of the automatic algorithm that includes Boot.EXPOS is given in Fig. 1. This computational scheme starts by selecting the most appropriate EXPOS model (Table 1) and then investigates the stationarity of the EXPOS residuals $\{e_1, \cdots, e_n\}$, using the ADF (Augmented Dickey–Fuller) test [21]. If the null hypothesis of nonstationarity is not rejected, an adequate transformation (Box-Cox and differencing) is performed on $\{y_1, y_2, \cdots, y_n\}$ and the process starts again.

Boot.EXPOS is ready to start when the component "EXPOS residuals" can be assumed as stationary (rejection of the ADF null hypothesis). Boot.EXPOS starts by performing an autoregressive (AR) adjustment on the EXPOS residuals.

These second stage residuals, which we call *AR residuals*, are resampled and used afterwards in a backward reconstruction of a new *AR residuals* series (using the estimated AR coefficients). With this series and with the initial fitted EXPOS values $\{\hat{y}_1, \cdots, \hat{y}_n\}$, a bootstrap sample path of the original time series is obtained. Point forecasts and forecast intervals are now calculated using this bootstrap time series and also the parameters estimated in the initial EXPOS fit, $\hat{\theta}_0 = (\hat{\alpha}, \hat{\beta}, \hat{\gamma}, \hat{\phi})$. The description of the Boot.EXPOS steps is the following:

Step 1:   Adjust an AR model to $\{e_1, \cdots, e_n\}$, by the AIC selection criterion;
Step 2:   Obtain the *AR residuals* and center them;

    For $b = 1, \cdots, B$
Step 3:   Draw a random sample from the centered residuals;
Step 4:   Use AR model recursively for obtaining a bootstrap series of the residuals;
Step 5:   Construct a "bootstrapped" time series $\{\hat{y}_1^*, \cdots, \hat{y}_n^*\}$ using $\{\hat{y}_1, \cdots, \hat{y}_n\}$ and the bootstrap series of the residuals;
Step 6:   From $\{\hat{y}_1^*, \cdots, \hat{y}_n^*\}$ and using $\hat{\theta}_0 = (\hat{\alpha}, \hat{\beta}, \hat{\gamma}, \hat{\phi})$, obtain:
    Point forecasts: h step-ahead forecasts $\tilde{y}_{bh}$;
    Forecast intervals: draw a random sample e_h^* of size h from $\{e_1, \cdots, e_n\}$
    and obtain $\tilde{y}_{bh}^* = \tilde{y}_{bh} + e_h^*$ or $\tilde{y}_{bh}^* = \tilde{y}_{bh}(1 + e_h^*)$ according to the error type;

Step 7:   For each h forecast horizon, calculate

    Point forecast: $\tilde{y}_h$ as the average of $\{\tilde{y}_{bh}, b = 1, \cdots, B\}$;
    Forecast intervals: bootstrap percentile intervals for $\{\tilde{y}_{bh}^*, b = 1, \cdots, B\}$;

As we referred to above the **initial step** before applying the Boot.EXPOS procedure is to select the "best" EXPOS method using the AIC criterion and to test the stationarity of EXPOS residuals. Time series fitted values $\{\hat{y}_1, \cdots, \hat{y}_n\}$ and the optimized smoothing parameters $\hat{\theta}_0 = (\hat{\alpha}, \hat{\beta}, \hat{\gamma}, \hat{\phi})$ are kept for later use at Boot.EXPOS backstage, while the EXPOS residuals series $\{e_1, \cdots, e_n\}$ (Fig. 1) has a leading role, since it is the input time series in Boot.EXPOS algorithm.

2.1   Forecast Intervals in EXPOS

The EXPOS method is a procedure that provides only a point forecast. So, in order to obtain prediction intervals, a stochastic data generating process must be provided by a statistical model. In this context, all the methods in Table 1 have an underlying state space model [16]. For each method in Table 1, there are two possible state space models: a model with additive errors and another one with multiplicative errors. Thus, there are thirty EXPOS models associated with the methods in Table 1. Additive and multiplicative models give identical point forecast (provided that the same parameters are used), but their prediction intervals will differ [16].

It is useful to compute the associated prediction distribution for each model, i.e., the distribution of a future value of the series given the model. One simple way of obtaining the prediction distributions is through a simulation approach (not developed in this paper). Another way is to derive the distributions analytically (already done in [16]). For linear models with homoscedastic errors the prediction distribution is clearly Gaussian. In case of linear models with heteroscedastic errors and models with multiplicative errors and multiplicative seasonality but additive trend, the prediction distributions are non-Gaussian because of the nonlinearity of the state space equations. However, prediction intervals based on the Gaussian formula will give reasonably accurate results, see [16] for more details.

2.2   Forecast Intervals in Boot.EXPOS

In Cordeiro and Neves past studies, attention was dedicated to obtain point forecasts, but a forecast interval gives a clearer indication of future uncertainty [2]. So to calculate prediction intervals is necessary to incorporate some uncertainty into the point forecast through forecasting errors obtained after fitting a model to past data. Attention must be given now to Step 6 in the Boot.EXPOS algorithm where forecast intervals are not calculated, despite the designation in the step. What is done there is simply to "add" error uncertainty to the point forecast and only after performing the B replications these intervals are calculated, as described in Step 8. Let F_h be the empirical distribution function of the $\{\tilde{y}_{bh}^*, b = 1, \cdots, B\}$. Considering the percentile bootstrap methodology, [9], the $(1 - \alpha) \times 100\,\%$ confidence intervals are given by $[F_h^{-1}(\alpha/2), F_h^{-1}(1-\alpha/2)]$. For a 95 % confidence interval and $B = 1,000$ replications, the percentiles for determining the interval are $F_h^{-1}(0.025) = \tilde{y}_{bh}^{*(25)}$ and $F_h^{-1}(0.925) = \tilde{y}_{bh}^{*(975)}$.

3   Case Study

The case study presented here is limited to a group of six time series plotted in Fig. 2. Each time series is splitted into two parts: the fitting set and the validation set:

$$\underbrace{y_1, y_2, \cdots, y_{n-k}}_{\textit{fitting set}}, \overbrace{y_{n-k+1}, \cdots, y_n}^{\textit{validation set}}.$$

To the fitting set $\{y_1, \cdots, y_{n-k}\}$ is applied the Boot.EXPOS procedure and the EXPOS too. For these family of methods, Hyndman and Khandakar [14] have developed the **ets**() function in ®R [13] that chooses the model (among those in Table 1) that better fits the data, to obtain forecasts and forecast intervals. The exponential smoothing procedure led to the following classification: time series

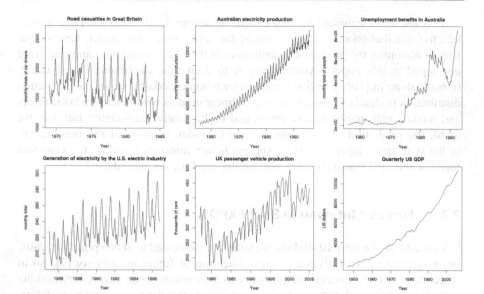

Fig. 2 ℝ data sets: *UKDriverDeaths* (datasets [23]); *elec, dole,* and *uselec* (fma [12]); *ukcars, usgdp* (expsmooth [11])

dole—ETS[1](A,Ad,A); *ukcars*—ETS(A,N,A) and *usgdp*—ETS(A,Ad,N), so all of them are linear models with homoscedastic errors; *UKDriverDeaths*—ETS(M,N,A) also a linear model but with heteroscedastic errors; *elec*—ETS(M,Ad,M) and *uselec*—ETS(M,N,M) are models with multiplicative errors and seasonality, but additive trend. For the series under study, several accuracy measures have been calculated, and Table 2 presents RMSE (Root Mean Squared Error), MAE (Mean Absolute Error), and MAPE (Mean Absolute Percentage Error).

Figure 3 shows the exponential smoothing and the Boot.EXPOS forecast intervals (all these are 95 % confidence intervals). As it can be seen forecast intervals based on the proposed procedure are narrower than those obtained with **ets()**.

4   Conclusions

This article concentrates on forecast intervals achieved by the Boot.EXPOS procedure. This promising procedure has been developed for point forecasting as a first purpose, but after a safe basis is being established, investigation went on in terms of forecast intervals. These intervals are based on a nonparametric approach, while those produced by EXPOS are based on a parametric approach. As showed in the

---

[1]ETS stands for Error, Trend, Seasonality.

Table 2 Accuracy measures for time series in Fig. 1

Time series	n-h	s	h	ets	function	Accuracy measures		
						RMSE	MAE	MAPE
UKDriverDeaths	180	12	12	(M,N,A)	ets	205.63	198.49	14.68
					Boot.EXPOS	84.93	67.79	4.88
elec	464	12	12	(M,Ad,M)	ets	348.87	305.88	2.19
					Boot.EXPOS	333.90	300.85	2.17
dole	427	12	12	(A,Ad,A)	ets	15,271.15	10,927.08	1.45
					Boot.EXPOS	11,419.08	8,223.98	1.07
uselec	130	12	12	(M,N,M)	ets	5.68	4.35	1.72
					Boot.EXPOS	4.03	3.04	1.20
ukcars	105	4	8	(A,N,A)	ets	19.46	16.05	3.95
					Boot.EXPOS	15.58	11.56	2.88
usgdp	229	4	8	(A,Ad,N)	ets	59.08	43.12	0.38
					Boot.EXPOS	38.70	24.98	0.22

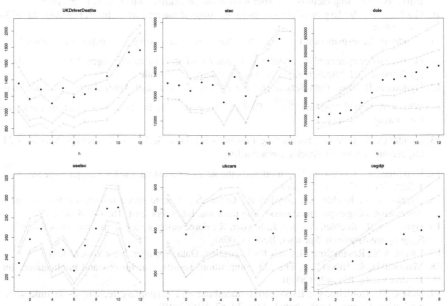

Fig. 3 Forecast intervals comparison: ets (*solid line with triangle*), Boot.EXPOS (*plus sign with solid line*) and validation set (*filled circle*)

case study, the Boot.EXPOS approach produces narrower intervals when compared to the other method.

Acknowledgments Research partially funded by FCT, Portugal, through the project Pest-OE/MAT/UI0006/2011.

References

1. Bühlmann, P.: Sieve bootstrap for time series. Bernoulli **3**, 123–148 (1997)
2. Chatfield, C.: The Analysis of Time Series. An Introduction. Chapman & Hall/CRC, Boca Raton (2004)
3. Cordeiro, C.: Métodos de reamostragem em modelos de previsão. Ph.D. thesis (2011)
4. Cordeiro, C., Neves, M.: The bootstrap methodology in time series forecasting. In: Rizzi, A., Vichi, M. (eds.) Proceedings of CompStat2006, pp. 1067–1073. Springer, Heidelberg (2006)
5. Cordeiro, C., Neves, M.: Resampling techniques in time series prediction: a look at accuracy measures. In: Gomes, M.I, Pestana D., Silva P. (eds.) ISI-2007 - Book of Abstracts, p. 353. CEUL, INE and ISI Editions, Lisbon (2007)
6. Cordeiro, C., Neves, M.: Bootstrap and exponential smoothing working together in forecasting time series. In: Brito, P. (ed.) Proceedings in Computational Statistics (COMPSTAT 2008), Porto, pp. 891–899. Springer, Berlin (2008)
7. Cordeiro, C., Neves, M.: Forecasting time series with Boot.EXPOS procedure. Revstat Stat. J. **7**(2), 135–149 (2009)
8. Cordeiro, C., Neves, M.: Boot.EXPOS in NNGC competition. In: Proceedings of the International Joint Conference on Neural Networks (IJCNN2010), World Congress on Computational Intelligence, pp. 1135–1141. IEEE, Barcelona (2010)
9. Efron, B.: Bootstrap methods: another look at the jackknife. Ann. Stat. **7**, 1–26 (1979)
10. Gardner, E.S., Jr.: Exponential smoothing: the state of the art-part II. Int. J. Forecast. **22**, 637–666 (2006)
11. Hyndman, R.: expsmooth: Data sets from the book "Forecasting with exponential smoothing: the state space approach" by Hyndman, Koehler, Ord and Snyder (Springer, 2008). ℝ Package Version 2.01. http://cran.r-project.org/package=expsmooth
12. Hyndman, R.: fma: Data Sets from Forecasting: Methods and Applications by Makridakis, Wheelwright and Hydman. ℝ Package Version 1.21. http://cran.r-project.org/package=forecasting
13. Hyndman, R.: Forecast: Forecasting Functions for Time Series. ℝ Package Version 1.21. http://cran.r-project.org/package=forecasting
14. Hyndman, R., Khandakar, Y.: Automatic time series forecasting: the forecast package for ℝ. J. Stat. Softw. **27**(3), 1–22 (2008)
15. Hyndman, R., Koehler, A., Snyder, R., Grose, S.: A state space framework for automatic forecasting using exponential smoothing methods. Int. J. Forecast. **18**, 439–454 (2002)
16. Hyndman, R., Koehler, A., Ord, J., Snyder, R.: Forecasting with Exponential Smoothing: The State Space Approach. Springer, Berlin (2008)
17. Lahiri, S.N.: Resampling Methods for Dependent Data. Springer, New York (2003)
18. Makridakis, S., Hibon, M.: The M3 competition: results, conclusions and implications. Int. J. Forecast. **16**, 451–476 (2000)
19. Neves, M., Cordeiro, C.: Exponential smoothing and resampling techniques in time series prediction. Discuss. Math. Probab. Stat. **30**, 87–101 (2010)
20. Neves, M., Cordeiro, C.: Computational intensive methods for predtction and imputation in time series analysis. Discuss. Math. Probab. Stat. **31**, 121–139 (2011)
21. Patterson, K.: Unit Root Tests in Time Series. Palgrave Macmillan, New York (2011)
22. Pegels, C.C.: Exponential smoothing: some new variations. Manag. Sci. **12**, 311–315 (1969)
23. ℝ Development Core Team: A Language and Environment for Statistical Computing. Software available at http://www.r-project.org

Table-Graph: A New Approach to Visualize Multivariate Data. Analysis of Chronic Diseases in Portugal

Alexandra Pinto

Abstract

Chronic diseases are the world's major cause of death and disability (WHO: Chronic Diseases and Health Promotion, http://www.who.int/chp/about/integrated-cd/en/index.html, 2012). The prevalences of chronic diseases in Portugal were calculated from the fourth National Health Survey (NHS) and then presented in different graphs, trying to find one that best represents this data. There are several two-dimensional representations used to visualize multivariate data. In this paper I propose and implement the Table-graph, a graphical representation suitable for multivariate data. Table-graph can be used as an alternative to multi-line charts and radar plots, when there are several variables to present, especially if their values are similar, leading to confusing charts. In this study, Table-graph proved to be an important technique in data visualization and it was used to present data of chronic diseases.

According to NHS, high blood pressure, rheumatic pain and chronic pain are the most prevalent chronic diseases in Portugal. In general, prevalences of chronic disease obtained from NHS are underestimated because they are not provided by medical diagnosis.

1 Introduction

Chronic diseases are long duration diseases and are the cause of 63 % of all deaths [15]. Among chronic diseases, the main cause of mortality are: cardiovascular diseases, cancer, chronic respiratory diseases, diabetes and obesity [15]. In developing countries about 79 % of deaths are attributed to chronic diseases [16].

A. Pinto (✉)
Faculty of Medicine of Lisbon, Laboratory of Biomathematics, Av. Professor Egas Moniz,
1649-028 Lisboa, Portugal
e-mail: apinto@fm.ul.pt

A. Pacheco et al. (eds.), *New Advances in Statistical Modeling and Applications*,
Studies in Theoretical and Applied Statistics, DOI 10.1007/978-3-319-05323-3_25,
© Springer International Publishing Switzerland 2014

257

In the fourth Portuguese National Health Survey (NHS) individuals were inquired about 15 chronic diseases. All data obtained is based on individual answers about chronic diseases. About 90 % of the diagnoses of chronic diseases were made by doctors and nurses, except for obesity that occurred in 64.2 % of individuals [1]. When we tried to visualize this data, some issues arose because of its high dimensionality.

In recent years, data visualization has become a widely used tool to extract relevant information from data [12]. Data visualization is the process of converting data into images [8] and is a powerful analysis tool that helps us discover patterns and trends hidden in the data [11]. The main purpose of data visualization is to extract as much information as possible from a data set, in a quick, clear and precise way [14]. The major difficulties experienced in the visualization field are related to the organization of data and the choice of the best graphical representation to present the data [1, 2], which is becoming more complex due to the large variety of available plots. Once these difficulties are overcome, researchers face the challenge to produce a clear and appealing plot, in order to present an interesting insight on their work [3].

To deal with the difficulty of analysing graphs with a high number of variables or/and groups [7], I developed and implemented the Table-graph method, first proposed by Tufte [13]. Table-graph is a tool to visualize multivariate data and it looks like a mix between a table and a graph, providing a visual image of each element of the chart, without loss of information. Tufte's version (which he called Table-graphic), not implemented, is more focused on a table. My version is an improved version more focused in the graph. I propose three new approaches to present data interactively, allowing the visualization of a progressive transition between a common multi-line chart and a Table-graph. I also built a GUI (Graphical User Interface) with the aim of producing a user-friendly tool that will soon be available for researchers.

In order to compare graphical representations I also present data in a common radar plot and multi-line charts.

2 Objectives, Material and Methods

The aim of this study is to propose and implement a new and interactive graphical representation that solves some clutter caused by data. The new graph is presented with data from the fourth NHS (2005/2006). The NHS is an instrument that deals with health conditions of the population (individuals living in their usual residence during the period of data collection). The fourth NHS is the first survey to cover the autonomous regions of Açores and Madeira. In this survey, 41,193 individuals were inquired and more than 400 variables were collected.

In this study I applied radar plot, multi-line charts and Table-graphs, to prevalences of chronic diseases obtained from the fourth NHS.

These prevalences were calculated using the appropriate weight factors, provided by INE,[1] in order to get values related to all resident population [5]. The data set was ordered by prevalences' mean of the seven NUTS II[2] regions of Portugal: Norte, Centro, Lisboa e Vale do Tejo (LVT), Alentejo, Algarve, Açores and Madeira. Then prevalences were presented graphically.

Table-graph was implemented in Matlab 7. Radar plot and multi-line charts were produced in MS-Excel. The prevalences were calculated using IBM SPSS 18.

3   Table-Graph

Table-graph is a graphical representation of multivariate data, similar to a line chart, that increments a leap value for each variable to prevent line intersection. The leap is a value artificially added to the data, only for representation purposes; it is a new gap between series[3] that helps transforming a multiple line graph into a Table-graph. With this leap lines do not intersect and the variations between them become more clear. For each series, the value added grows in multiples of the leap (*Values of series n = Values of series n + n * leap*). To calculate the leap value, a simple algorithm was created. It finds the mean range of the series' values which is then multiplied by a factor. This factor was found by simple trial and error, and by comparing the outputs. I found that value 5 provides the right gap between series, for several examples tested. The formula to calculate the leap value is:

$$Leap\ value = ((mean(Series\ max[n]) - mean(Series\ min[n])) * 5$$

$$*Leap\ factor\ slider.$$

Where:
- $n = 1, \ldots,$ *number of groups (regions)*;
- *Series max[n]* is the vector of maximum of each series;
- *Series min[n]* is the vector of minimum of each series;
- *Leap factor slider* is an extra factor (varying between 0 and 1) that places each line closer or farther away from the others. This factor can be controlled in GUI.

In the software created, the *Leap factor slider* controls the Table-graph's leap, from no leap (resulting in a multi-line chart) to the maximum leap (Table-graph with the highest gap between series). This factor is controlled in GUI, with a slider, providing an interactive graph, which is a major advantage of this graph.

The proposal software requires Excel data files as input.

---

[1] National Institute of Statistics of Portugal.

[2] Standard Nomenclature of Territorial Units for Statistics purposes. NUTS II refers to regions larger than districts.

[3] In this paper "series" are the prevalences of each pathology along the seven regions.

I propose three variants of Table-graph that enable different and optimized re-presentations to analyse multivariate data. Researchers can choose the most suitable representation for the data:

1. *Regular*: In this case, the series name and value labels are placed on a virtual line defined by the mean range of each series. This representation is more suitable when data series have small ranges. It deals well with some intersection.
2. *On the series*: In this case, the series names are positioned near the first value and the value labels are placed on the series line. This is the most similar approach to Tufte's version. This variant is more suitable for large ranges.
3. *On the reference line*: In this case, the series name and value labels are placed at a reference line (an extra line of reference values of the database; there is one reference line for each series and it has only one position for each series). The reference values are optional and this variant is not displayed if there are no reference values as input (from Excel data file). This variant is suitable for those case studies where reference values are important to make comparisons.

4   Results and Discussion

High blood pressure is the most prevalent chronic disease in Portugal, both in general and in each NUTS II region. Rheumatic pain and chronic pain have also high prevalence in all the country (Figs. 1, 2, 3, 4, and 5).

A radar plot (radial, spider or star plot) is a two-dimensional chart of multivariate data that provides an easy way to analyse patterns across groups, showing differences or disparities within and between them [6,10]. Figure 1 shows a radar plot with too many groups and similar values, leading to a cluttered and illegible graph. In this case study, the solution would be creating one plot for each region. Nevertheless this procedure highly increases eyes movement to perform comparisons between charts, which are referred by Tufte as a negative issue that should be carefully considered and avoided in data visualization [13].

The multi-line chart is an appropriate display when the number of variables is not so high and values are not too similar [9]. The big disadvantage arises when lines intersect each other, or even overlap (Fig. 2).

If we add value labels to Fig. 2, instead of providing more information we will present an unreadable chart (Fig. 3).

The radar plot and the multi-line chart became even more cluttered after adding a marker to distinguish each line, due to the need to make displays in a gray scale.

In Figs. 4 and 5 are shown two variants of Table-graph for chronic diseases data. We can analyse prevalences of all chronic diseases mentioned in the NHS, for each region, in both variants. The *Regular* variant does not fit this data properly. The reference line (Fig. 5) presents the overall prevalences in Portugal and allows comparisons between these values and prevalences of each region.

Table-graph tool deals well with data up to 15 variables but for more variables the chart will show very flat series. This fact can be improved increasing chart size.

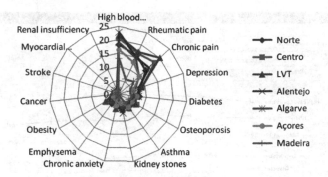

Fig. 1 Radar plot—prevalences (%) for the 15 chronic diseases in each region

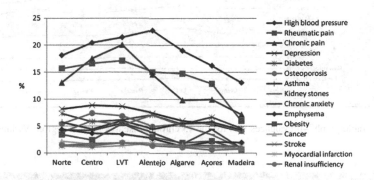

Fig. 2 Multi-line chart with markers—prevalences (%) for the 15 chronic diseases in each region

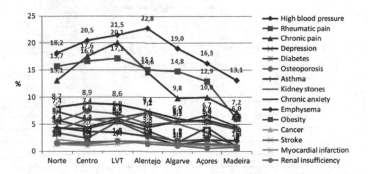

Fig. 3 Multi-line chart with markers and value labels—prevalences (%) for the 15 chronic diseases in each region

The Table-graphic proposed by Tufte is presented in Fig. 6. In this chart the gaps between series are not calculated by an algorithm, they are chosen arbitrary (case to case) to avoid intersection. This method is not interactive, neither is implemented. Tufte's method is design in a way that the limit for number of variables is the length of the sheet.

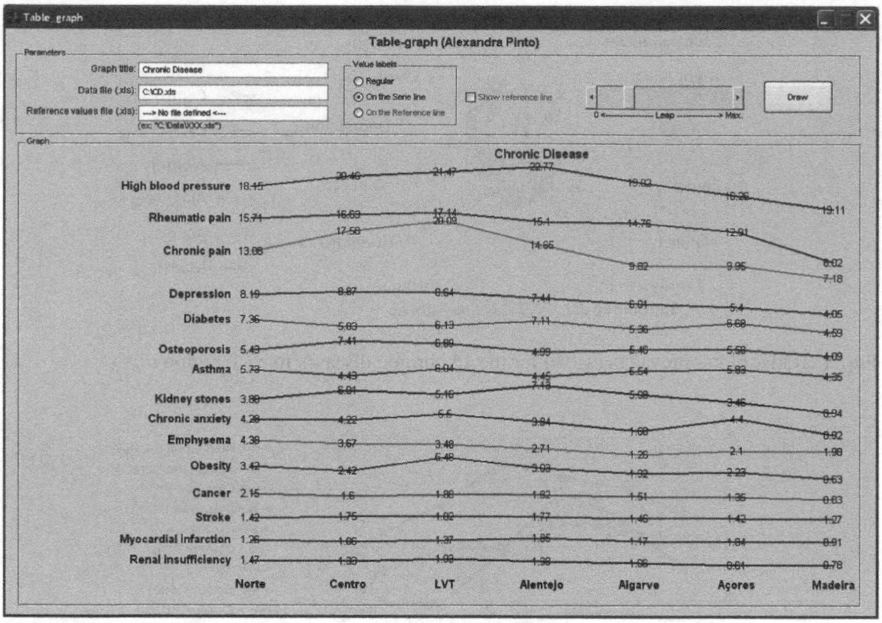

Fig. 4 Table-graph—prevalences (%) for the 15 chronic diseases in each region. Variant: *On the Series*

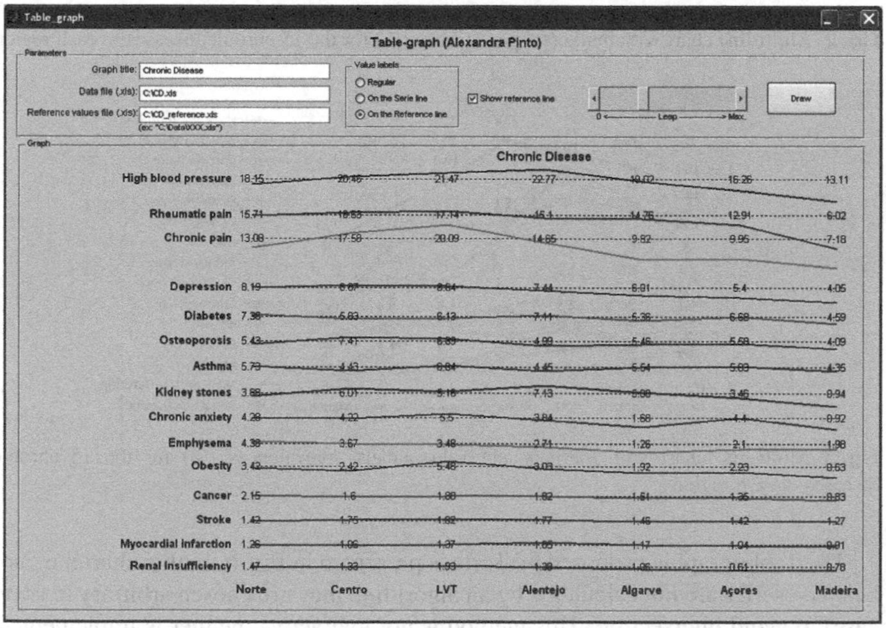

Fig. 5 Table-graph—prevalences (%) for the 15 chronic diseases in each region. Variant: *On the Reference line*

Fig. 6 Table-graph proposed
by Tufte with oncologic data
as example. *Source*: Tufte
[13]

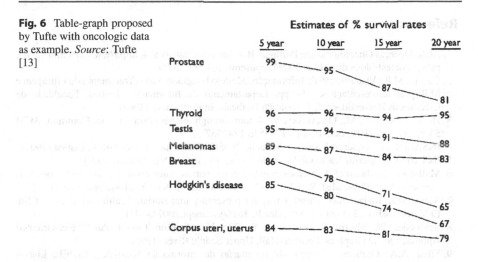

In chronic diseases data obtained from NHS, some prevalences are underestimated, especially for obesity, directly asked to the interviewed individuals, reaching only 3.8 % overall. However, when this prevalence is calculated through BMI (body mass index), almost a fourfold higher prevalence is found [1, 4].

5 Conclusion

High blood pressure, rheumatic pain and chronic pain are the most prevalent chronic diseases in Portugal, according to the NHS of 2005/2006. Some prevalences of chronic diseases are underestimated, especially obesity, because they were not estimated by a doctor diagnosis.

In this study I propose the Table-graph, an interactive graph for multivariate data. Table-graph allows us to make transversal comparisons between and within regions, in the same representation.

When the original position of values is needed to make general comparisons or to identify groups, Table-graph can be presented with no leap, leading to a common multi-line chart. Researchers can switch (gradually or not) between a Table-graph and a common multi-line chart. Researchers can choose the variant of Table-graph that best fits each data set. Table-graph's main advantage is the interactivity and the possibility of analysing series without overlay. On the other hand the main disadvantage of Table-graph is the flat effect when the number of series increases.

Table-graph is an improvement because it provides an alternative tool, presenting data more clearly than common tools, when there are several variables and their values are similar.

Acknowledgements I thank to Victor Lobo for his suggestions on an earlier version of this paper. I also thank to anonymous reviewers for their helpful comments and suggestions.

References

1. ACS: Doenças Oncológicas em Portugal. Boletim Informativo n°4. http://www.acs.min-saude.pt/wp-content/blogs.dir/1/files/2009/05/pnsemfoco4.pdf (2009)
2. Carmo, M.B.: Visualização de Informação. Modelo Integrado para o Tratamento de Filtragem e Múltiplas Representações, 314 pp. Departamento de Informática. Lisboa, Faculdade de Ciências da Universidade de Lisboa, Ph.D. thesis, in Portuguese (2003)
3. Heer, J., Bostock, M., Ogievetsky, V.: A tour through the visualization zoo. Commun. ACM 53(6), 59–67 (2010). doi:10.1145/1743546.1743567
4. INSA/INE: 4° Inquérito Nacional de Saúde 2005/2006: Destaque. INSA/INE, Lisboa (2007)
5. INSA/INE: Inquérito Nacional de Saúde 2005/2006. INSA/INE, Lisboa (2009)
6. Mallinson, T., Hammel, J.: Measurement of participation: intersecting person, task, and environment. Arch. Phys. Med. Rehabil. 91(1), S29–33 (2010). doi:10.1016/j.apmr.2010.04.027
7. Saary, M.J.: Radar plots: a useful way for presenting multivariate health care data. J. Clin. Epidemiol. 61(4), 311–317 (2008). doi:10.1016/j.jclinepi.2007.04.021
8. Schroeder, W., Martin, K., Lorensen, B.: The Visualization Toolkit: An Object-Oriented Approach to 3D Graphics. Prentice-Hall, Upper Saddle River (1998)
9. Silva, A.A.: Gráficos e Mapas. Representação de informação estatística. LIDEL, Lisbon (2006). ISBN: 978-972-757-340-0
10. Stafoggia, M., Lallo, A., Fusco, D., Barone, A.P., D'Ovidio, M., Sorge, C., Perucci, C.A.: Spie charts, target plots, and radar plots for displaying comparative outcomes of health care. J. Clin. Epidemiol. 64(7), 770–778 (2011). ISBN: 0895-4356
11. Talbert, N.: Toward human-centered systems. IEEE Comput. Graph. Appl. 17(4), 21–28 (1997)
12. Tominski, C., Schulze-Wollgast, P., Schumann, H.: Visual methods for analyzing human health data. Information Science Reference (2008). ISBN: 978-1-59904-889-5
13. Tufte, E.R.: Beautiful Evidence. Graphics Press, Cheshire (2006). ISBN: 978-0961392178
14. Ware, C.: Information Visualization: Perception for Design (Interactive Technologies). Ed. Morgan Kaufman, San Francisco (2004). ISBN: 978-1558608191
15. WHO: Chronic Diseases. WHO, Geneva. http://www.who.int/topics/chronic-diseases/en/ (2012)
16. WHO: Chronic Diseases and Health Promotion. WHO, Geneva. http://www.who.int/chp/about/integrated-cd/en/index.html (2012)

Application of Item Response Theory to Mathematics High School Exams in Portugal

Gonçalo Jacinto, Paulo Infante, and Claudia Pereira

Abstract

Item response theory (IRT) provides statistical models that relate an examinee's response to a test item to an underlying latent trait that is measured by the items. Unlike classical test theory, IRT models focus on the responses to individual questions instead on the total score obtained in the test.

In this paper we apply IRT models to the results of the quiz part of the first and second calls of mathematics exams accomplished by high school graduating students in Portugal, in the years 2008–2010. We concluded that the quiz part of the first call exams had a difficulty level lower than the students median ability, whereas the quiz part of the second call exams had a difficulty level higher than the students median ability.

1   Introduction

The IRT provides models that relate the probability of a given question being answered correctly to the unobservable latent traits (usually called ability or proficiency level in the field under evaluation) of the individuals by means of the way that responses are given to the questions (usually called items) in a test.

Unlike the classical test theory where only the final score in the exam is considered as a measure of the proficiency level of the students, in IRT the items of a test are the central elements. One of the advantages of IRT over the classical test theory is that IRT allows for the comparison between two populations when

G. Jacinto (✉) • P. Infante
CIMA-UE and ECT/DMAT of University of Évora, Évora, Portugal
e-mail: gjcj@uevora.pt; pinfante@uevora.pt

C. Pereira
MMEAD/ECT of University of Évora, Évora, Portugal
e-mail: claudia15360@gmail.com

A. Pacheco et al. (eds.), *New Advances in Statistical Modeling and Applications*,
Studies in Theoretical and Applied Statistics, DOI 10.1007/978-3-319-05323-3_26,
© Springer International Publishing Switzerland 2014

the tests have some items in common, or it can compare individuals from the same population even when the tests are completely different.

The different models used in IRT take into account the following characteristics: (1) the nature of the question (if it is dichotomous or not); (2) the number of populations involved (one or more); (3) the number of latent traits under study. The models used in this paper only consider one latent trait or ability, being called unidimensional models. These models are used in several countries, being powerful instruments used by education institutions and in another fields where some kind of ability needs to be measured.

We can see, for example, [1, 3, 7] for some applications of IRT. In [3] IRT was applied to smaller-sized class examinations and [7] made a comparison between the classical test theory and IRT, for the analysis of the results of a statistical reasoning test. In [1] is proposed a methodological approach based on the two-parameter logistic model to obtain a common test metric for future use that assures the comparability of student scores over time.

In Portugal, as far as we know, the only existent studies applying IRT models to the results of national exams [5] were made by GAVE (Gabinete de Avaliação Educacional, an entity belonging to the Portuguese Ministry of Education), for the exam results made by students of the fourth grade in the years 2008 and 2009. In this study, a two-parameter logistic model was fitted to the responses given to 27 questions, being the results to the questions transformed in dichotomous or polychotomous values (there is no information how that is made), and where the goodness-of-fit of the model was not presented.

In this paper we applied IRT models to the results of the quiz part of the first and second calls of mathematics exams accomplished by high-school graduating students in Portugal, in the years 2008–2010. A random sample of 10,000 students from each year was obtained from GAVE, being the sample proportion of students from each call equal to the true proportion. We only consider IRT models for dichotomous responses, and for that reason we have only analysed the first part of the exams which have eight quiz questions. We assumed that only one latent trait is under study: the proficiency level in mathematics. Using IRT models we concluded that the first call exams had a difficulty level lower than the students median ability, whereas the quiz part of the second call exams had a difficulty level higher than the students median ability. Due to space restrictions, the results obtained for 2009 exams are not presented. However, the model fitted for the years of 2008 and 2010 was adjusted for 2009 and the same general conclusions were obtained.

This paper is organized as follows. In Sect. 2 the item response logistic model for dichotomous items is presented. In Sect. 3 the results obtained for each call and year are illustrated. Finally, in Sect. 4, we conclude and make some remarks.

2 Unidimensional IRT Models for Dichotomous Responses

To model the quiz part of the exams we fitted unidimensional IRT models for dichotomous responses, mainly the one-, two-, and three-parameter logistic models. For each call from each year we selected the model that fits the data best.

In each case we checked the two key assumptions: unidimensionality and local independence. We then proceed to analyse the estimated parameters.

The most simple model is called the one-parameter logistic model (ML1) or Rasch model. This model considers the item difficulty as the only parameter to explain the relation between the item itself, the ability of the individual and the response given to the item. The item difficulty represents the location of the item on the ability scale. That is, for dichotomous items, it is defined as the point in the ability range where an individual has a probability equal to 0.50 to give the correct answer.

Let β_i denote the difficulty of item i and θ_j the ability of the individual j, then $P(U_{ij} = 1|\theta_j)$ denote the probability of individual j with ability θ_j to answer correctly to question i, and the logistic model ML1 is given by

$$P(U_{ij} = 1|\theta_j) = \frac{1}{1 + e^{-(\theta_j - \beta_i)}}. \tag{1}$$

In this model the item discrimination, that refers to how sharply an item differentiates among different individuals who have different ability levels, is the same across all items. However, the discrimination varies across the items, that is, if an item has a high discrimination value it is likely to have different responses among individuals that have different abilities, while if an item has a low discrimination value it is likely to have similar responses among individuals that have different abilities. When the discrimination value varies along the items, we have the two-parameter logistic model (ML2). If α_i denotes the discrimination value of item i, the model ML2 is given by

$$P(U_{ij} = 1|\theta_j) = \frac{1}{1 + e^{-\alpha_i(\theta_j - \beta_i)}}. \tag{2}$$

Finally, if we include in the model the guessing parameter γ_i, that denotes the probability that an individual with low ability answer correctly to the item, we have the three-parameter logistic model (ML3), given by

$$P(U_{ij} = 1|\theta_j) = \gamma_i + (1 - \gamma_i)\frac{1}{1 + e^{-\alpha_i(\theta_j - \beta_i)}}. \tag{3}$$

To fit each model we use the software R project with the package ltm [6]. This package uses the Marginal Maximum Likelihood Estimation (MMLE).

Formally, the model parameters are estimated by maximizing the observed data log-likelihood obtained by integrating out the latent variables; the contribution of the mth sample unit is

$$l_m(\theta) = log \int p(x_m|z_m, \theta)p(z_m)dz_m,$$

where $p(\cdot)$ denotes the probability density function, x_m denotes the vector of responses for the mth sample unit, z_m denotes the individuals level and $\theta = (\alpha_i, \beta_i, \gamma_i)$. This integral is approximated using the Gauss–Hermite quadrature rule with 21 points when one latent variable is specified. The maximization of the integrated log-likelihood with respect to θ is achieved using BFGS algorithm [2].

As mentioned above we must verify two model assumptions: unidimensionality and local independence. Unidimensionality means that all items in the test represent the same latent trait. This is achieved by a procedure proposed in [2] for examining the latent dimensionality of dichotomously scored item responses. The statistic used for testing unidimensionality is the second eigenvalue of the tetrachoric correlations matrix of the dichotomous items. The tetrachoric correlations are computed and the largest one is taken as the communality estimate. A Monte Carlo procedure is used to approximate the distribution of this statistic under the null hypothesis.

The assumption of local independence means that for the trait being measured the items are uncorrelated among them. As referred in [4], unidimensionality implies local independence, since the unique cause of response of the student is the assumption of the dominant latent trait. Then if the unidimensionality assumption is verified, local independence also is verified.

To investigate the fit of each model, we examine the two-way and three-way chi-square residuals produced by the margins method. These residuals are calculated by constructing all possible 2×2 contingency tables for the available items and checking the model fit in each cell using the Pearson's chi-square statistic.

3   Critical Analysis of the Obtained Results

All the exams have eight quiz questions. The first three questions are about probability (usually probability theory or random variables), the next three questions are about trigonometry and calculus (usually include questions about logarithms, limits and differential calculus) and the final two questions are about complex numbers. Usually the same topics, despite not in the same order, are evaluated in both calls and years.

3.1   Results for the First and Second Calls of the 2008 Exams

In the phase of model fitting, questions 3, 6 and 8 were withdrawn from both exams (by lack of model fitting) and the best model for both exams (first and second calls) was the ML3 model. The estimated parameters are presented in Table 1, and both models have verified the required assumptions: the chi-square tests for the two-way and three-way margins are not significant at 10 %; the test to evaluate the unidimensionality assumption had a p-value equal to 0.71 for the first call and a p-value equal to 0.60 for the second call, which allowed us to assume unidimensionality.

Regarding the first call exam, the easiest questions were questions number 2, 4 and 7 (about probabilities, logarithms and complex numbers, respectively), with a

Table 1 Parameters of the IRT model for the 2008 Exams (C_1—first call, C_2—second call)

Item	P C_1	C_2	Disc. C_1	C_2	Difficulty C_1	C_2	Guessing C_1	C_2	$P_{X\mid Z}$ C_1	C_2	Biserial Corr. C_1	C_2
1	0.58	0.74	2.14	1.53	0.47	0.00	0.34	0.47	0.52	0.74	0.60	0.52
2	0.85	0.80	2.24	2.13	−1.02	0.54	0.27	0.70	0.93	0.77	0.57	0.45
4	0.86	0.69	2.34	1.35	−0.96	−0.81	0.37	0.00	0.94	0.75	0.55	0.58
5	0.54	0.40	1.94	3.40	0.45	1.17	0.27	0.30	0.48	0.31	0.62	0.55
7	0.72	0.42	1.84	1.14	−0.65	0.46	0.12	0.03	0.79	0.40	0.64	0.60

P denotes the sample proportion of correct answers to the items; *Disc.* denotes the discrimination parameters of each item; $P_{X\mid Z}$ denotes the probability that a median level student answers correctly to the question, $P(X = 1 \mid Z = 0)$

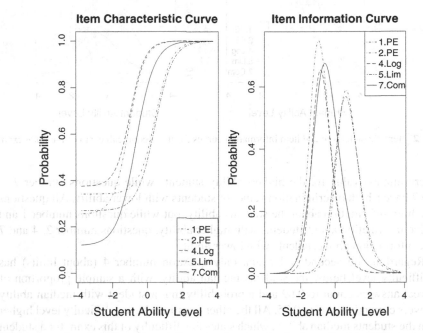

Fig. 1 Item characteristic and item information functions for the first call of the 2008 exam

sample proportion of students with a correct answer equal to 0.85, 0.86 and 0.72, respectively, and a probability that a median level student answers correctly equal to 0.93, 0.94 and 0.79, respectively. Note that the difficulty parameter of each of these questions is negative, showing that their difficulty level is lower than the students median ability.

The most difficult questions were questions number 1 and 5 (about probabilities and limits), with a probability that a median ability student answers correctly equal to 0.52 and 0.48, respectively, and a difficulty parameter of 0.47 and 0.45, respectively.

In Fig. 1 we can observe the item discrimination and the item information functions, where we can conclude that questions number 1 and 5 have a high

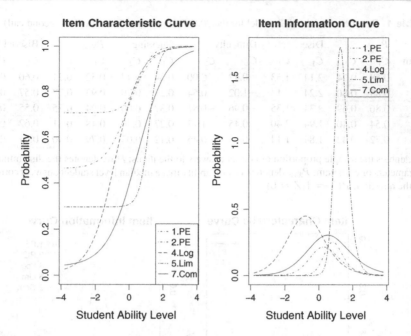

Fig. 2 Item characteristic and item information curves for the second call exam of the 2008 exam

discrimination value for the highest ability students while questions number 2, 4 and 7 have a high discrimination value for students with lower ability. All questions give high information about the students ability, but while questions number 1 and 5 give information about students with higher ability, questions number 2, 4 and 7 give information about students with lower ability.

Regarding the second call exam, only question number 4 (about limits) has a difficulty level below the students median ability, with a sample proportion of correct answers equal to 0.69 and a probability that a student with median ability answers correctly equal to 0.75. All the other questions have a difficulty level higher than the students median ability, which states the difficulty of this exam for a student with median ability. The most difficult questions were questions number 5 and 7 (about limits and trigonometry, respectively), where only about 40 % of the students gave the correct answer and with a probability that a student with median ability answers correctly equal to 0.31 for question number 5 and equal to 0.40 for question number 7.

In Fig. 2 we can observe the item discrimination and item information functions, where we can conclude that questions number 1 and 2 (both about probabilities), and question number 4, have a very low discrimination value, being useless to evaluate the ability level of the students. On the other hand, question number 5 has a very high discrimination value being able to distinguish the top 25 % of students. Only one question (question number 4) gives information about the students with low

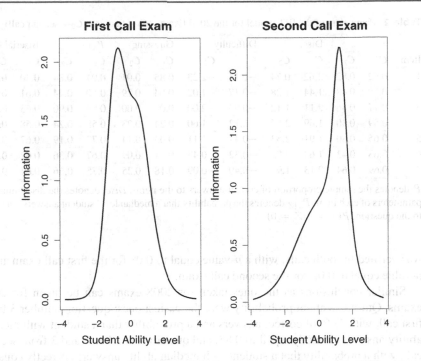

Fig. 3 Item information curve for the first call (*left*) and second call (*right*) of the 2008 exams

ability, being question number 5 the one with highest information value for students with higher ability.

Finally in Fig. 3 we compare the test information functions of both calls. We can observe that the second call exam gives essential information about higher ability students, while the first call exam gives essential information about the low ability students. Therefore, unlike the first call exam, the second call exam is designed for students with ability higher than the median. As a note, the Portuguese Mathematical Society in a report made about both exams in 2008 classified the first call exam a very easy one, with elementary questions while the second call exam was considered much more difficult than the first call, being exactly what we have concluded by modelling the quiz part of the exams by an IRT model.

3.2   Results for the First and Second Calls of the 2010 Exams

In the phase of model fitting question 8 was withdrawn from both models (by lack of model fitting) and the best model for both exams (first and second call) was the ML3 model. The estimated parameters are presented in Table 2, and both models have verified the required assumptions: the chi-square tests for the two-way and three-way margins are not significant at 1 % level, and the unidimensionality assumption

Table 2 Parameters of the IRT model for the 2010 Exams (C_1—first call, C_2—second call)

Item	P C_1	C_2	Disc. C_1	C_2	Difficulty C_1	C_2	Guessing C_1	C_2	$P_{X\mid Z}$ C_1	C_2	Biserial Corr. C_1	C_2
1	0.92	0.82	2.62	0.74	−0.14	−2.23	0.85	0.01	0.94	0.84	0.30	0.41
2	0.58	0.48	1.44	1.28	−0.17	1.02	0.04	0.29	0.60	0.44	0.61	0.50
3	0.77	0.86	2.11	1.12	−0.93	−2.00	0.02	0.00	0.88	0.90	0.63	0.42
4	0.59	0.30	1.69	2.65	0.12	1.60	0.24	0.23	0.58	0.24	0.58	0.46
5	0.65	0.30	1.91	2.53	−0.47	1.11	0.03	0.14	0.72	0.19	0.67	0.54
6	0.63	0.43	1.63	2.05	−0.32	0.43	0.11	0.09	0.67	0.36	0.62	0.61
7	0.69	0.64	2.13	1.94	−0.40	−0.09	0.18	0.25	0.75	0.66	0.64	0.58

P denotes the sample proportion of correct answers to the items; $Disc.$ denotes the discrimination parameters of each item; $P_{X\mid Z}$ denotes the probability that a median level student answers correctly to the question, $P(X = 1 \mid Z = 0)$

was verified for both calls, with a p-value equal to 0.08 for the first call exam and a p-value equal to 0.02 for the second call exam.

Similar conclusions as the ones taken for 2008 exams can be taken for 2010 exams. Questions about probability were the easiest ones: question number 3 from first call, with 77 % of correct answers and a probability that a student with median ability answers correctly equal to 0.88; and questions number 1 and 3 from second call, with a probability that a student with median ability answers correctly equal to 0.84 and 0.90, respectively.

In the first call exam only question number 4 (about differential calculus) has a difficulty parameter higher than the students median ability, and all questions have a probability higher than 0.58 that a student with median ability answers correctly. On the other hand, for the second call exam, only questions number 1 and 3 (about probabilities) and question number 7 (about complex numbers) have a difficulty below the students median ability, being question number 4 (about differential calculus) the most difficult one.

Observing the discrimination and difficulty parameters from Table 2, for the first call exam we can conclude that question number 1 (about probabilities) is completely useless since it presents a very low discrimination value, being question number 3 (about probabilities) the one that gives the higher information about the worst students. It should be noted that none of the questions give information about the high ability students. Contrasting with these results, three questions from the second call exam allow to distinguish the high ability students: question number 4 (about differential calculus), question number 5 (about limits) and question number 6 (about differential calculus).

Finally, in Fig. 4 we present the test information functions for both calls. As in 2008 exams, we can conclude that second call exam gives information essentially about students with ability higher than the median, while the first call exam gives information essentially about students with ability lower than the median. The

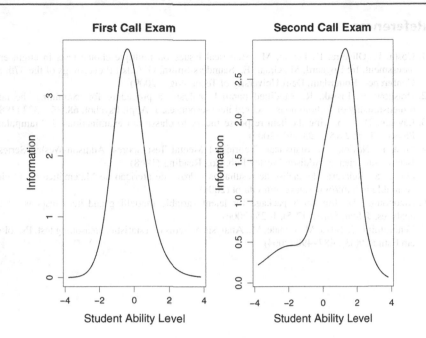

Fig. 4 Item information curve for the first call (*left*) and second call (*right*) of the 2010 exams

Portuguese Mathematical Society in its report about 2010 exams stated the higher difficulty of the second call exam, which is in line with the conclusions we have obtained.

4   Some Remarks

We have used IRT models to analyse the results of the quiz part of the first and second calls of mathematical exams accomplished by high-school graduating students in Portugal, in the years 2008–2010.

We have concluded that, for the 3 years, the quiz part of the first call exams had a difficulty level lower than the students median ability, whereas the quiz part of the second call exams had a difficulty level higher than the students median ability.

We also concluded that in the 3 years there were always useless questions. In all the first call exams there was a very easy question about probabilities, and there was always a question about probabilities that gives information about the students with lower ability. Despite we have only analysed the quiz part of the exams, the conclusions obtained were the same as the ones produced by the Portuguese Mathematical Society. As future work, it is our intention to apply IRT models to analyse the non-dichotomous questions and to compare the tests over the 3 years.

References

1. Costa, P., Oliveira, P., Ferrão, M.: Statistical issues on multiple choice tests in engineering assessment. In: Bogaard, M., Graaf, E., Saunders-Smith, G. (eds.) Proceedings of the 37th Sefi Conference, Rotterdam, Delft University of Technology (2009)
2. Drasgow, F., Lissak, R.: Modified parallel analysis: a procedure for examining the latent dimensionality of dichotomously scored item responses. J. Appl. Psychol. **68**, 363–373 (1983)
3. Lawson, D. : Applying the item response theory to classroom examinations. J. Manipulative Physiol. Ther. **29**(5), 393–397 (2006)
4. Lord, F., Novick, M.: Statistical Theories of Mental Test Scores. Addison-Wesley Series in Behavioral Science. Addison-Wesley Pub. Co., Reading (1968)
5. Olivia, S.: Relatório de análise de resultados - Prova de aferição de Matemática de 1° ciclo. Available in http://www.gave.min-edu.pt (2008)
6. Rizopoulos, D.: ltm: an R package for latent variable modelling and item response theory analyses. J. Stat. Softw. **17**(5), 1–25 (2006)
7. Vendramini, C., Silva, M., Canale, M.: Analysis of items of a statistical reasoning test. Psicologia em Estudo **9**(83), 487–498 (2004)

Évora Residents and Sports Activity

Luísa Carvalho, Paulo Infante, and Anabela Afonso

Abstract

A healthy life is promoted by active practices that decrease sedentary lifestyles, reducing the risks of occurrence of various diseases. The World Health Organization recognizes the importance of physical activity in physical, mental, and social health. This paper presents some of the results obtained from a survey realized in collaboration with the Sports Division of Municipality of Évora to characterize the physical activity of the residents. After an exploratory data analysis, and some comparisons using parametric and nonparametric tests, we adjust a logistic regression model to identify enhancers' factors of physical activity. We conclude that the residents in this municipality have a high physical activity rate and that there are no significant differences in physical activity rate between genders. Physical activity rate decreases drastically with age, varies with the professional situation and increases with qualification. The more likely practitioner is a young male, self-employed without higher education that knows some sports initiatives and that is satisfied with municipality sports offer.

1 Introduction

The World Health Organization (WHO) recognizes the importance of sports activity for our physical, mental and social health [5]. It refers the need of policies that take into account the necessities and people resources, trying to integrate physical

L. Carvalho, (✉)
University of Évora, R. Romão Ramalho 59, 7000-671 Évora, Portugal
e-mail: carvalh.luisa@gmail.com

P. Infante • A. Afonso
Department of Mathematics and Research Center of Mathematics and Applications (CIMA-UE),
University of Évora, Évora, Portugal
e-mail: pinfante@uevora.pt; aafonso@uevora.pt

A. Pacheco et al. (eds.), *New Advances in Statistical Modeling and Applications*,
Studies in Theoretical and Applied Statistics, DOI 10.1007/978-3-319-05323-3_27,
© Springer International Publishing Switzerland 2014

activity in all age groups and social sectors. In this way, WHO recommends a daily minimum of 30 min of moderate-intensity physical activity for adults and at least 60 for young people.

The sedentary lifestyle, coupled with the increasing use of technology in everyday life, is leading people to higher levels of inactivity in all age groups. In this context, different municipalities have developed several initiatives to promote the sport activity, seeking to create infrastructures that attempt to meet this objective and the needs and desires of its residents.

In a comparative study [4] of sports activity in Portugal between 1988 and 1998, there was a slight decline in the overall percentage of practitioners (27–23 %). The sports participation depends on several factors such as age, gender, educational level and occupational status. Young people practice more sport than older generations and the rate of sports participation decreases with age. The physical activity in females is generally less than in men.

The European Union (EU) has made some studies related with this theme. The most recent [2] shows that in Portugal 33 % of citizens practice physical activity at least once a week, a value similar to the EU percentage. Nordic countries, such as Sweden and Finland, present a sports level activity of 72 %.

Several studies about sports activity have been made in some municipalities of Portugal but they are mainly descriptive. The novelty of our work is the estimation of the proportion of practitioners by age and area in the municipality of Évora and also the construction of the practitioner profile. We start by characterizing the physical activity of the residents in this municipality, which consists of 11 rural and 8 urban parishes' councils and has a very high rate of ageing population. In Sect. 2 we described the methodology used in collecting the data. In Sect. 3 we present some results that characterize the physical activity in this municipality. In Sect. 4, we give the profile of the practitioner using a logistic regression model. Finally, we conclude with some remarks.

2 Methodology

Between June 24 and August 11, 2011, a sample survey was applied to the population resident in Évora having a fixed phone and aged 15 years and over, by the sports section of the municipality of Évora.

According to National Institute of Statistics data, in 2001 Census, Évora municipality had 48,097 residents aged 15 or more years, living mostly in urban areas (40,550 residents). Given that the population structure of the municipality of Évora presents differences between rural and urban areas, and knowing from previous studies [2, 4] that sports in Portugal differs between age classes, we decided to break the population into strata, whose differentiating variables are the age group and the area of residence.

We used a stratified random sampling design [1] where the variable of interest was the rate of physical activity in the municipality of Évora. To calculate the global sample size we considered as initial estimates the global indices obtained for Portugal in [2], since there were no available rates for the municipality of Évora, a maximum margin of error equal to 3.5 %, for a confidence level of 95 %. We used proportional allocation to get the sample size by stratum, based on data from Census 2001.

We must point out that, despite of this study design was based on data available to the sample survey application date (2001 Census), data from Census 2011, meantime available, allow us to conclude that the sample taken remains representative. Indeed, despite the aging increase population, the populational differences between 2011 and 2001 on each stratum are small, and the samples sizes would not present relevant changes. To analyze data we used exploratory data analysis and some parametric and nonparametric tests. To obtain a profile for the practitioner it was adjusted a logistic regression model. The significance level for all tests has been fixed at 5 and at 10 % for the significance of the variables of logistic regression model.

3   Sports Practice Characterization

This study had the participation of 653 residents in the municipality of Évora, most of them were females (54 %). Participants age vary from 16 to 92 years, age average is equal to 47 years ($SD = 19.58$) with females slightly older. Also, 80 % lives at urban area, 55 % had education at the high school and superior, 55 % are workers with 39 % employed by someone else.

With 95 % confidence, over half of the residents practice some physical activity (56 %), with an error of the estimate equal to 3.7 %. The practitioner has, on average, about 43 years of age, half of the practitioners have 28–57 years, and the older practitioners have more than 80 years. About one in four residents already practiced physical activity earlier and 18 % never practiced physical activity.

Physical activity rate decreases with the increasing of age: 72 % of young practitioners against 38 % of elderly practitioners and did not significantly differ from urban to rural areas ($z = 0.704$, p-value $= 0.482$). In rural areas the rate who has never be a practitioner is 29 %, higher than in the urban area (18 %).

In urban area the rate of male practitioners tends to decrease with age (Fig. 1) but is not statistically different between the age classes 25–39 and 40–64 (Z test with Bonferroni correction: $z = -0.70$, p-value $= 0.460$), and the physical activity rate is significantly lower in elderly women (Z tests with Bonferroni correction: p-values < 0.05). In rural areas the estimated proportion of practitioners of both genders is much lower for elderly people. In women the estimated rate decreases with age, while for men higher percentages of practitioners were estimated for individuals between 25 and 64 years old. We do not present the confidence intervals for the rural area, since they were very wide due to the small number of respondents.

Individuals with lower levels of education have lower rates of physical activity; retired persons and unemployed have the lowest percentages of practitioners;

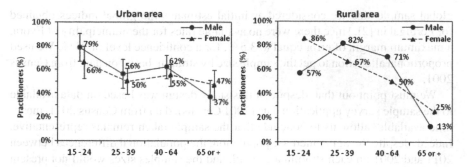

Fig. 1 Estimated physical activity rate by gender, age and area of residence, and correspondent 95 % confidence interval

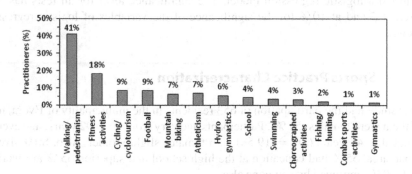

Fig. 2 Physical activities more practiced

workers are those who have higher rates of physical activity (48 %) and the tertiary sector is the one who have more practitioners.

We have identified a set of 25 individual and collective sport activities. Among practitioners, 79 % have only one physical activity. Figure 2 shows the sports most frequently cited by respondents, standing out walking (41 %) and fitness activities (18 %).

As expected, young people, between 15 and 20 years, practice physical activities at school (Fig. 3). The age of combat sports activities practitioners is also very low, on average 22.8 years, probably because of the recent availability of some of these modalities for the residents in the municipality of Évora. The football players are on average about 27 years ($SD = 8.4$), half of the swimmers have a maximum of 32 years, and only 25 % are over 35.5 years. The activities hydro gymnastics, walking/pedestrianism, and gymnastics are associated with elderly people. Note that there are at least two swimmers that stand out from other swimmers because they are a little older, and there is a practitioner of combat sports activities who is much older than the others.

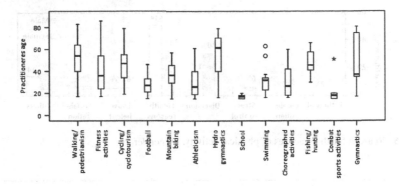

Fig. 3 Practitioners age for the physical activities more practiced (at least 4 %)

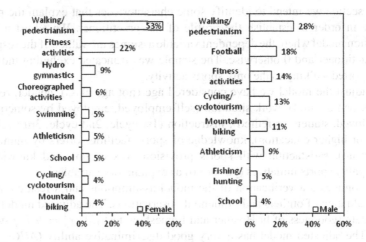

Fig. 4 Physical activities more practiced by gender

There is a greater diversity of activities practiced in males and there are different preferences in each gender: females tend to opt for maintenance activities and males for leisure or competitive activities (Fig. 4).

The main reasons for physical activity identified by practitioners were: 40 % health-related and 24 % entertainment. Figure 5 shows significant difference was found between genders (Homogeneity Chi-square: $\chi_6^2 = 20.14$, p-value $= 0.003$) where health reasons were the more mentioned by females (51 %), while males are divided between health reasons (32 %) and entertainment reasons (30%).

Almost 70 % of the respondents who have practiced physical activity in the past reported two reasons for the abandon of physical activity and only 4 % have not present any justification. The main reason mentioned, by both genders, was the lack of time. In general, the lack of time was due to family obligations (18 %), professional/scholar reasons (24 %) and 58 % have not specify why.

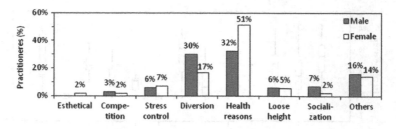

Fig. 5 Reasons for the practices of physical activity, by gender

4   Practitioner Profile

In this section we intend to identify some characteristics that explain the physical activity in order to establish the profile of the practitioner. We adjust a logistic regression model where the dependent variable assumes the value 1 if the respondent is a practitioner and 0 otherwise. The sample was truncated excluding individuals who reported walking as the only sports activity.

To adjust the model we have considered age (not age class), gender, residence zone (urban or rural), work situation (self-employed, employed by someone else, unemployed, student or retired), instruction (1st cycle, 2nd cycle, 3rd cycle, high school or higher education), knowledge of sports facilities offers by municipality (yes or no), satisfaction with sports provision (yes or no), and knowledge of municipality sports initiatives (yes or no) as explanatory variables.

We conducted a verification of the model assumptions and a residual analysis with evaluation of outliers and influential observations. The adjusted model fit well to the data (goodness of fit Hosmer and Lemeshow test [3]: $\chi_8^2 = 9.7$, p-value $= 0.29$). The adjusted model has a very good discriminative ability ($AUC = 0.80$, 95 % CI :]0.76; 0.84[) with a sensitivity equal to 74 % and a specificity equal to 75 % if we use a cutoff point equal to 0.49.

Later, since the sample is relatively small ($n = 522$) to leave out some individuals for further validation of the model, we choose to conduct a cross-validation by bootstrap, where different models were adjusted to 10 random samples consisting of 90 % of individuals from original sample. Values were estimated by each model for 10 % of individuals who were left out in each model and also we have recorded the values of Chi-square statistical test of each model. The average value of the Chi-square for the 10 models was equal to 9.4 with a minimum of 6.2 and a maximum of 13.9. To validate the model we obtain an $AUC = 0.77$ (95 % CI :]0.73; 0.81[) that reveals its internal consistency.

The enhancers' factors of physical activity (and therefore increase the likelihood of an individual be a practitioner) are: be male, be self-employed, do not live in São Mamede or Senhora da Saúde, have lower age, have knowledge of municipality sports initiatives, and be satisfied with sports provision of the municipality of Évora (Table 1).

Table 1 Estimated coefficients ($\hat{\beta}$), standard errors (SE), and p-value in the partitioner logistic regression model

Variable	$\hat{\beta}$	$SE(\hat{\beta})$	p-Value
Age	−0.1035	0.0324	0.0014
Sex[a]	0.6056	0.2124	0.0044
Behalf of others[b]	−0.8885	0.3638	0.0150
Unemployed[b]	−1.4203	0.4977	0.0043
Student or reformed[b]	−0.1441	0.3735	0.6997
Higher education[c]	−1.7373	0.6737	0.0099
Know equipment	−1.1710	1.1505	0.3087
Satisfied with offer	0.7868	0.2533	0.0019
Know initiatives	0.9718	0.3763	0.0098
Area[d]	0.5118	0.2754	0.0631
Age × Higher education	0.0557	0.0162	0.0006
Age × Know equipment	0.0586	0.0328	0.0745[e]
Constant	1.3246	1.2741	0.2985

[a] Female sex as the reference category
[b] The variable working situation has self-employed as the reference
[c] In the final model, instruction variable has two categories: higher education and others, the latter being the reference
[d] In the final model Senhora da Saúde and São Mamede parish are the reference category and area is formed by the remaining parishes
[e] p-Value equal to 0.03 using likelihood ratio test

Assuming that other values are fixed, the following conclusions can be drawn for the residents in the municipality of Évora which physical activity is not only walking:

1. A male person has about twice the possibility of being a practitioner than a female person (odds ratio 95 % CI :]1.2; 2.8[);
2. A self-employed has more than twice the possibility of being a practitioner than employed by someone else (odds ratio 95 % CI :]1.2; 4.9[) and about four times more possibility than an unemployed (odds ratio 95 % CI :]1.6; 11.0[);
3. For someone who knows some sports initiative of the municipality, the possibility of being a practitioner has about twice the possibility than who knows no sports initiative (odds ratio 95 % CI :]1.3; 5.5[);
4. A person satisfied with sports offer has twice the possibility of being a practitioner comparatively to an unmet person (odds ratio 95 % CI :]1.3; 3.6[);
5. A resident in the parishes Senhora da Saúde or São Mamede has 40 % less possibility of being a practitioner comparatively to a person resident in another parish of this municipality (odds ratio 95 % CI :]6 %; 62 %[);
6. For the one who does not have higher education and knows no sports facilities, an increase of 5 years in age is associated with a 40 % reduction in the possibility of being practitioner (odds ratio 95 % CI :]19 %; 57 %[), while who does not have higher education and knows sports facilities an increase of 5 years in age

is associated with a 20 % reduction of being practitioner (odds ratio 95 % CI :]14 %; 25 %[);

7. In persons over 40 years higher education increases the possibility of practicing sport: At the age 40, the possibility of being practitioner is almost twice higher (odds ratio 95 % CI :]1.0; 2.6[); At the age 50, the possibility of being practitioner is about three times higher (odds ratio 95 % CI :]1.6; 5.1[); At the age 60, the possibility of being practitioner is five times more (odds ratio 95 % CI :]2.3; 11.1[); At the age 70, the possibility of being practitioner is about eight and a half times higher (odds ratio 95 % CI :]3.0; 25.5[); At the age 80, the possibility of being practitioner is about fifteen times higher (odds ratio 95 % CI :]3.9; 59.7[).

5   Final Remarks

In this paper we have made a general characterization of the physical activity in the municipality of Évora based on a survey to residents. On the basis of a logistic regression model we intended to identify some enhancers' factors of physical activity in order to establish the profile of the practitioner.

We can highlight the following results: (a) The residents in this municipality have a high physical activity rate (55 %), which is higher than in Portugal [2]; (b) unlike what was found in Portugal [4], in Évora no significant differences in physical activity rate were found between genders. This rate is similar between areas, decreases drastically with age, varies with the professional situation, and increases with qualification, being important to emphasize that more than half of the individuals with first cycle education never had practiced physical activity; (c) more than 3/4 of the students practiced physical activity, while retired persons and unemployed persons are who less practiced physical activity; (d) the reasons for the practice are not the same for both sexes: the main reason given by females was health, while males are divided between health and entertainment reasons; (e) the more likely practitioner is a young male, self-employed, who knows some sports initiatives and who is satisfied with municipality sports offer.

Acknowledgements This is a join work with Sports Division from Municipality of Évora and had the collaboration of the Organization and Informatics Management Division from Municipality of Évora. PI and AA are members of CIMA-UE, a research center funded by the FEDER program, administrated by FCT Pluriannual Funding.

References

1. Cochran, W.G.: Sampling Techniques, 3rd edn. Wiley, New York (1977)
2. Eurobarometer. Sport and Physical Activity. Special Eurobarometer 334/Wave72.3. TNS Opinion & Social. European Commission. Available via DIALOG. http://ec.europa.eu/public_ opinion/archives/ebs/ebs_334_en.pdf (2010). Accessed 15 Mar 2012
3. Hosmer, D., Lemeshow, S.: Applied Logistic Regression, 2nd edn. Wiley, New York (2000)

4. Mariovet, S.: Práticas Desportivas na Sociedade Portuguesa (1988–1998). In: Actas do IV Congresso Português de Sociologia: Passados Recentes, Futuros Próximos. Available via DIA-LOG. http://www.aps.pt/cms/docs_prv/docs/DPR462e088b86481_1.PDF (2000). Accessed 15 Mar 2012
5. World Health Organization: Global recommendations on physical activity for health. World Health Organization. Available via DIALOG. http://whqlibdoc.who.int/publications/2010/9789241599979_eng.pdf (2010). Accessed 15 Mar 2012

Index

A. Pacheco et al. (eds.), *New Advances in Statistical Modeling and Applications*,
Studies in Theoretical and Applied Statistics, DOI 10.1007/978-3-319-05323-3,
© Springer International Publishing Switzerland 2014